Cambridge astrophysics series

Spectroscopy of astrophysical plasmas

SPECTROSCOPY OF ASTROPHYSICAL PLASMAS

A. DALGARNO & D. LAYZER

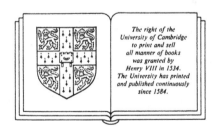

CAMBRIDGE UNIVERSITY PRESS

Cambridge

London   New York   New Rochelle

Melbourne   Sydney

Published by the Press Syndicate of the University of Cambridge
The Pitt Building, Trumpington Street, Cambridge CB2 1RP
32 East 57th Street, New York, NY 10022, USA
10 Stamford Road, Oakleigh, Melbourne 3166, Australia

First published 1987

British Library cataloguing in publication data

Spectroscopy of astrophysical plasmas.

1. Plasma (Ionized gases)   2. Astrophysics
I. Dalgarno, A. II. Layzer, David
523.01   QC718

Library of Congress cataloguing in publication data

Spectroscopy of astrophysical plasmas.

Includes bibliographies.
1. Astronomical spectroscopy. 2. Plasma spectroscopy.
3. Astrophysics. I. Dalgarno, A. II. Layzer, David.
QB465.S64   1987   523.01   86-13003

ISBN 0 521 26315 8 hardcovers
ISBN 0 521 26927 X paperback

Transferred to digital printing 2004

Contents

Contents

List of contributors

Lawrence H. Aller, Astronomy Department, University of California, Los Angeles, CA 90024, USA

E. H. Avrett, Solar and Stellar Physics Division, Center for Astrophysics, 60 Garden Street, Cambridge, MA 02138, USA

J. H. Black, Steward Observatory, University of Arizona, Tucson, AZ 85721, USA

Robert L. Brown, National Radio Astronomy Observatory, Edgemont Road, Charlottesville, VA 22901, USA

A. Dalgarno, Centre for Astrophysics, 60 Garden Street, Cambridge, MA 02138, USA

D. Layzer, Center for Astrophysics, 60 Garden Street, Cambridge, MA 02138, USA

Beverly T. Lynds, Kitt Peak National Observatory, 950 North Cherry Avenue, P.O. Box 26732, Tucson, AZ 85726, USA

Richard A. McCray, Joint Institute for Laboratory Astrophysics, University of Colorado, Boulder, CO 80309, USA

C. F. McKee, Dept. of Physics and Astronomy, University of California, Berkeley, CA 94720, USA

Robert W. Noyes, Solar and Stellar Physics Division, Center for Astrophysics, 60 Garden Street, Cambridge, MA 02138, USA

Donald E. Osterbrock, Lick Observatory, University of California, Santa Cruz, CA 95064, USA

W. H. Parkinson, Center for Astrophysics, 60 Garden Street, Cambridge, MA 02138, USA

Blair Savage, Washburn Observatory, University of Wisconsin, 475 N. Charter Street, Madison, WI 53706, USA

Jack B. Zirker, Sacramento Peak Observatory, Sunspot, NM 88349, USA

Ben Zuckerman, Department of Astronomy, University of California, Los Angeles, CA 90024, USA

Journal titles and abbreviations used in the references

Astrophys. J.   *Astrophysical Journal*
Astrophys. J. Suppl.   *Astrophysical Journal Supplements (Series)*
Astrophys. Lett.   *Astrophysical Letters*
Astron. Astrophys.   *Astronomy and Astrophysics*
Astron. J.   *Astronomical Journal*
Mon. Not. Roy. Astron. Soc.   *Monthly Notices of the Royal Astronomical Society*
Quart. J. Roy. Astron. Soc.   *Quarterly Journal of the Royal Astronomical Society*
Ann. Rev. Astron. Astrophys.   *Annual Review of Astronomy and Astrophysics*
Pub. Astron. Soc. Pacific   *Publications of the Astronomical Society of the Pacific*
J. Hist. Astron.   *Journal for the History of Astronomy*
Bull. Amer. Astron. Soc.   *Bulletin of the American Astronomical Society*
J. Astron. Soc. Egypt   *Journal of the Astronomical Society of Egypt*
Bull. Astron. Soc. India   *Bulletin of the Astronomical Society of India*
Mitt. der Astr. Gessel.   *Mitteilungen der Astronomischen Gesellschaft*
Astr. J. U.S.S.R.   *Soviet Astronomy*
Zeits. f. Ap.   *Zeitschrift für Astrophysik*
Rev. Mod. Phys.   *Reviews of Modern Physics*
Phys. Rev. Lett.   *Physical Review Letters*
Phys. Rev. A.   *Physical Review: A*
Amer. J. Phys.   *American Journal of Physics*
Can. J. Phys.   *Canadian Journal of Physics*
Zeits. f. Phys.   *Zeitschrift für Physik*
Prog. Theo. Phys.   *Progress in Theoretical Physics*
Phys. Scrip.   *Physica Scripta*
App. Opt.   *Applied Optics*

J. Quant. Spec. Rad. Trans. *Journal of Quantitative Spectroscopy and Radiative Transfer*

J. Opt. Soc. Amer. *Journal of the Optical Society of America*

Opt. Comm. *Optics Communications*

Ark. f. Fys. *Arkiv für Fysik*

J. Phys. Chem. Ref. Data *Journal of Physical and Chemical Reference Data*

Chem. Phys. Lett. *Chemical Physics Letters*

Proc. Roy. Soc. London A *Proceedings of the Royal Society of London: A*

Ann. N.Y. Acad. Sci. *Annals of the New York Academy of Science*

Proc. Nat. Acad. Sci. *Proceedings of the National Academy of Sciences of the USA*

Dokl. Akad. Naus. USSR *Doklady Academia Nauka USSR*

J. Geophys. Res. *Journal of Geophysical Research*

Space Sci. Rev. *Space Science Reviews*

Geochim. Cosmochim. Acta *Geochimica et cosmochimica acta*

Preface

The authors and editors of *Spectroscopy of Astrophysical Plasmas* dedicate this book to Leo Goldberg, who, fifty years ago, recognized both the fundamental role of spectroscopy in the observation and interpretation of astrophysical objects and the essential supporting role of basic laboratory and theoretical studies of atomic and molecular spectra. Leo recognized the importance of all regions of the electromagnetic spectrum from radio waves to gamma rays and he made unique original contributions in ultraviolet, visible, infrared, millimeter and radio astronomy. He fostered the careers of many astronomers, several of whom are authors of this book. His understanding of the value of observations at all wavelengths led him to become a persuasive and influential advocate of space astronomy. At Harvard University he created a research group that was at the forefront of ultraviolet observations, particularly of the Sun.

Atoms, Stars and Nebulae, the title of Leo's first book, written with Lawrence Aller, sums up the main themes of Leo Goldberg's remarkable and still flourishing scientific career. In the early 1930s, when Leo embarked on that career, the 'new physics' – quantum mechanics – was still *terra incognita* for most astronomers. A few, however, had recognized its possibilities years earlier. They saw that quantum mechanics could make possible a quantitative understanding of the structure, composition, and physical conditions of stellar atmospheres and interiors, planetary nebulae, and the interstellar medium; and they set out to do something about it. One of these far-sighted and energetic people was Donald H. Menzel, and Leo Goldberg was a member of the first generation of Menzel's students at Harvard.

By the early 1930s it was clear that quantum mechanics could *in principle* describe atomic and molecular structure and spectra, and predict the outcomes of atomic collisions. But in practice very few of the atomic and molecular properties that astrophysicists needed to know could either be calculated or measured in the laboratory. New theories were needed to bridge the gap between 'in principle' and 'in practice', and Donald Menzel and Leo Goldberg were as eager to develop these theories as to apply them. One of Menzel's enduring interests was the prediction of atomic transition probabilities. While Leo was still an undergraduate, Menzel assigned him an 'impossible' problem – to calculate certain transition probabilities not covered by existing rules for making such calculations. In solving the problem (Menzel had neglected to explain that it was impossible) Goldberg invented the important theoretical concept of fractional parentage, later elaborated by Giulio Racah. It was an auspicious beginning to a scientific career in which the spectra of atoms and molecules have always been at the center of the stage.

More than any other participant in the broad scientific endeavor his work has done so much to shape, Leo Goldberg appreciated its diversity as well as its underlying unity. Though himself primarily a theorist, and secondarily a fine observer, his sustained and successful promotion of innovative experimental research – at Michigan, at Harvard, and at the Kitt Peak National Observatory – has earned him the lasting gratitude of spectroscopists, and of the scientific community as a whole. The outcome of these efforts are reflected in the contributions to this volume.

The individual chapters demonstrate the application of spectroscopic methods to the diagnosis of a broad range of astronomical environments. We hope that the presentations will be a useful contribution to the training of the new generation of astronomers, whose education was a matter of vital concern to Leo Goldberg throughout his scientific career.

A. Dalgarno and D. Layzer

1

Optical observations of nebulae

BEVERLY T. LYNDS

1.1 Discovery

In Ptolemy's *Almagest*, six objects are listed as 'stella nebulosa', hazy, luminous spots on the Celestial Sphere. Once viewed telescopically, these six objects were resolved into clusters of stars; however, other nebulous objects were noticed. The first, in Orion, appears to have been discovered by Fabri de Peiresc in 1610. Two years later, Simon Marius noted a nebula in Andromeda that had also been recorded by Al Sufi in the tenth century. The discoveries continued and, in 1781, Charles Messier compiled a list of nebulae and star clusters. The 105 objects he catalogued are still identified by their Messier number; the Orion Nebula is M42 while the Andromeda Nebula is M31.

An all-sky survey carried out by the Herschels at the turn of the nineteenth century resulted in a *General Catalogue* containing over 5000 nebulous objects. In 1888 a *New General Catalogue* was published by J. L. E. Dreyer; later editions included Index Catalogues and tabulated more than 13 000 objects. The Orion Nebula, M42, is also identified as NGC 1976; the Andromeda Nebula, M31, is NGC 224.

About the middle of the nineteenth century, Lord Rosse constructed a six-foot reflector in Ireland with which he made numerous visual sketches of nebulae and applied names to them by which they are still referred. M97 (NGC 3587) is called the 'Owl', while M51 (NGC 5194) is known as the 'Whirlpool', indicative of its spiral form.

During the initial discovery period of nebulae, a debate was ongoing as to whether or not all nebulae could be resolved into stars if a telescope of sufficient light-gathering power and resolution were available. With the advent of astronomical photography, the spiral nebulae were resolved into stars and shown to be 'island universes' at extragalactic distances from our Milky Way system. The term 'Extragalactic Nebulae' was adopted for them. In the course of the twentieth century, this term has been phased out in favor of the more explicit 'galaxies'; the Andromeda Nebula is now more appropriately called the Andromeda Galaxy.

The development of the fine instruments of Yerkes, Lick, Heidelberg, Harvard, and Mt Wilson Observatories at the turn of the century produced a series of magnificent photographs of nebulae still unresolved into stars. Some of these objects appear to be well-defined, circular or elliptical disks resembling planets; Herschel referred to them as Planetary Nebulae. The Owl, shown in Fig. 1.1, is one example. In contrast to the Planetary Nebulae, many nebulae are more diffuse in appearance and are always found close to the plane of the Milky Way. The Orion Nebula, the brightest and most extensively studied, is illustrated in Fig. 1.2.

Photographic surveys of the Milky Way demonstrated the presence of dark regions void of stars. E. E. Barnard published a catalogue of 349 'dark markings'; Fig. 1.3 reproduces a photograph of B33, the 'Horsehead'. The

Fig. 1.1. A CCD Frame of the Owl Nebula. This is a low-excitation planetary nebula of about 3 arc minutes in angular size. The CCD Frame was taken with the KPNO 92-cm telescope and a filter that transmits only the Hydrogen Balmer Beta line at $\lambda4861$ Å.

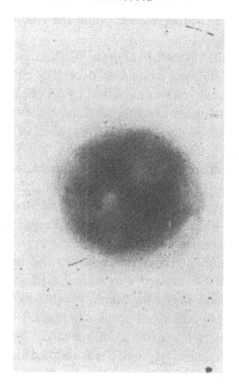

Barnard objects clearly demonstrate that there is obscuration between us and the general background of stars. These 'dark nebulae' have the same galactic distribution as that of the luminous nebulae and are often associated with them, as illustrated by B33. The term galactic nebulae now refers to either luminous or dark diffuse nebulae; their visual appearance depends

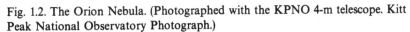

Fig. 1.2. The Orion Nebula. (Photographed with the KPNO 4-m telescope. Kitt Peak National Observatory Photograph.)

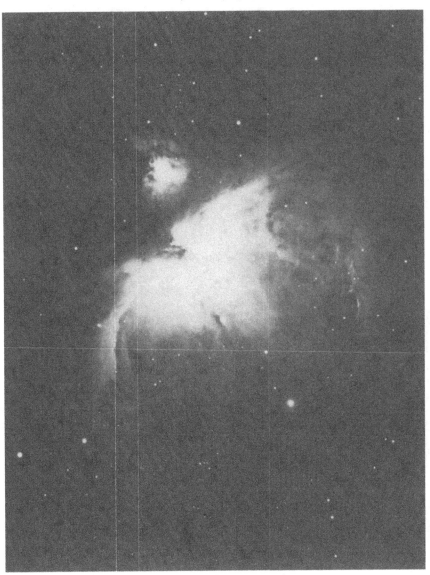

upon the physical conditions both within the nebulae and in the regions of space surrounding them.

In a classical paper published in 1922, Edwin Hubble described the spectra of luminous diffuse nebulae. He found that the spectra defined a series ranging from nebulae showing only bright (emission) lines to objects having a continuous spectrum. Hubble noted that the stars associated with emission

Fig. 1.3. The Horsehead Nebula, B33. (Photographed with the KPNO 4-m telescope. Kitt Peak National Observatory Photograph.)

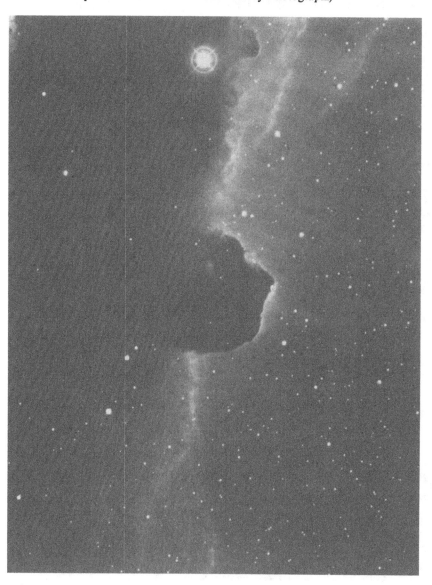

nebulae are those classified as being the hottest of stars (the spectroscopic designation of O to B0, essentially equivalent to black-body radiators of temperatures between 50 000 K and 30 000 K). The spectra of nebulae having only cooler imbedded stars displayed a continuous spectrum. The conclusion is that the stellar radiation determines the nature of the nebular emission; the hotter stars excite the gaseous component of the nebula into bright line emission while the cooler stars, with insufficient ultraviolet energy to affect the gas, have their light reflected by the solid particles in the nebula. Hubble's work was based on slitless spectra; slit spectrograms of the 'Continuum' nebulae confirm that the nebular light has the same spectral features as that of the associated star; the starlight is simply reflected or scattered by particles in the nebula.

1.2 Reflection nebulae and interstellar grains

The material blocking background starlight and producing a dark nebula and the particles reflecting starlight and producing a luminous nebula are believed to be small solid grains. These grains are more efficient in extinguishing blue light than red and produce a 'reddening' of starlight shining through them. The wavelength dependence of the amount of extinction enables us to estimate the size and chemical composition of the grains. If the particles were large compared with the wavelength of light, they would block out all wavelengths equally well. If the particles were small relative to the wavelength of light, they would produce Rayleigh scattering proportional to λ^{-4}, and the background stars would be more reddened than observed. In the optical range, the extinction produced by grains in the interstellar medium is proportional to λ^{-1}, indicating that the grains effective in blocking visible light have sizes of the order of the wavelength of light.

It is necessary to carry out a formal solution to Maxwell's equations in this scattering problem. The Mie Theory provides a rigorous solution for spheres of arbitrary size and is usually adopted, although irregularly-shaped inhomogeneous grains more nearly resemble interstellar grains. Mie's solution yields an efficiency factor $Q_{ext}(2\pi a/\lambda, m)$, expressing how much more efficient the grain is in intercepting light than is a simple geometrical cross section. If I_0 is the intensity of the incident light, then a particle of radius a will intercept $Q_{ext}\pi a^2 I_0$ of the incident beam; Q_{ext} is a function of the size of the particle relative to the wavelength of light and of its index of refraction, m.

Solutions to Maxwell's equations involve the conductivity, σ, dielectric constant ε, and diamagnetic constant (set equal to 1) of the homogeneous sphere. The conductivity and dielectric constant are usually expressed in terms of the complex index of refraction:

$$m = \sqrt{(\varepsilon - 2\lambda i \sigma/c)},$$

where λ is the wavelength of the electromagnetic wave.

The index of refraction expresses the relative magnitudes of the scattering (real term) component and the absorption (imaginary term) component of the extinction. Small metal spheres have indices of the order of

$$m = 37 - 41i,$$

and their absorption exceeds the total scattering, whereas dielectrics have a very small imaginary coefficient of the order of

$$m = 1.33 - 0.05i,$$

and simply reflect (scatter) most of the incident light.

The *albedo*, ω, of the particle is defined by the ratio of the scattered radiation to the total intercepted radiation, or

$$\omega = \frac{Q_{sca}}{Q_{ext}}.$$

If the nebular grains have high albedos, most of the starlight is scattered and a bright reflection nebula is observed; if the albedo is low, most of the starlight is absorbed by the grains. The energy absorbed is thermalized in the grain and reemitted at wavelengths characteristic of the grain temperature.

Small particles do not generally scatter light isotropically, and in scattering problems it is necessary to allow for the non-isotropy of the scattered radiation. This is done by introducing a phase function, $\Phi(\theta)$, normalized to unity, which specifies the fraction of scattered light deflected into the θ direction, where θ is the angle between the incident light and the scattered ray (Fig. 1.4).

The traditional phase function for interstellar particles was introduced by

Fig. 1.4. Light is scattered by a particle into the θ-direction.

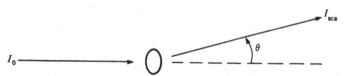

Henyey and Greenstein and is of the form:

$$\Phi(\theta) = \frac{\omega}{4\pi}(1-g^2)\frac{1}{(1+g^2-2g\cos\theta)^{3/2}}.$$

The parameter g measures the asymmetry of the phase function and is essentially the average of the cosine of the angle θ. Isotropic scattering is characterized by a g of 0; a particle that has a strongly forward-throwing phase function may have a g value close to unity.

The conductivity and dielectric constant of a given chemical vary with wavelength; therefore, the scattering properties of interstellar grains will also change with wavelength. It has been found that a combination of different types of grains is needed to account for interstellar extinction as measured from the ultraviolet to the infrared. For example, the observations are reproduced by a mixture of graphite particles having radii of about $0.02\,\mu m$, silicates (SiO_2) of radii $\approx 0.04\,\mu m$, and SiC of radii $\approx 0.07\,\mu m$. Fig. 1.5 reproduces theoretical curves for albedo and g for this mixture of graphite and silicates. The drop in albedo in the ultraviolet ($\lambda^{-1}\approx 8.4\,\mu m^{-1}$ or 2175 Å) is produced by absorbing grains of graphite.

The Pleiades (Fig. 1.6) is a familiar example of a reflection nebula. The illuminating stars are clearly identifiable; measurements of their brightnesses and that of the nebula can be compared with the theoretical predictions based on the physical parameters of grains defined by Fig. 1.5, the number of grains per unit volume, and an assumed geometrical configuration of star and nebula.

For a realistic model of a reflection nebula, it is necessary to allow for

Fig. 1.5. Theoretical curves for albedo and g (from Gilra, D. P. (1971). *Nature*, **220**, 237.)

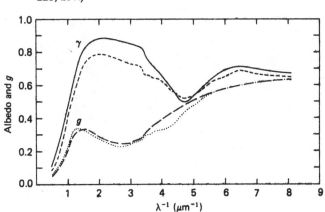

multiple scatterings and to integrate all of the light scattered into the line of sight; that is, to solve an equation of transfer.

We define the specific intensity of radiation, I_v, as the amount of energy passing through a unit area per unit time per unit frequency contained in a unit solid angle. Let there be n_g grains per unit volume in a medium through

Fig. 1.6. Pleiades Open Star Cluster in Taurus showing nebulosity. (Photographed with the KNPO 4-m telescope. Kitt Peak National Observatory Photograph.)

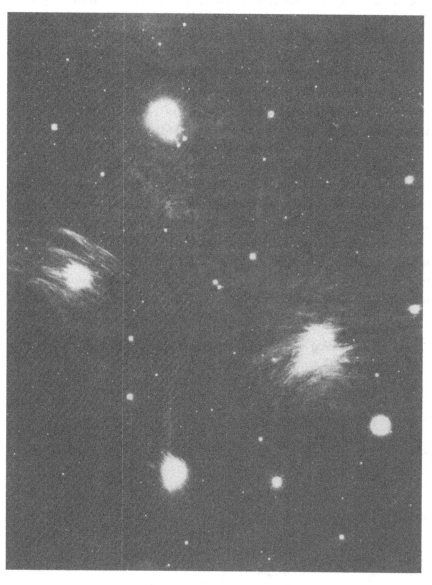

which the radiation is flowing. If I_ν is the incident specific intensity, and dI_ν the change in the intensity as the light passes through a cylinder of unit area and length dl of the medium (Fig. 1.7),

$$dI_\nu = -n_g Q_{\text{ext}} \pi a^2 I_\nu \, dl,$$

or

$$I_\nu = I_0 \, e^{-\tau},$$

where I_0 is the original incident intensity and τ, the *optical depth*, is defined by

$$\tau = \int_0^L n_g Q_{\text{ext}} \pi a^2 \, dl,$$

and L is the total path length of the scattering medium through which the radiation passes.

In a scattering medium, other rays may be scattered into the line of sight, and a *source function*, S, the additional intensity scattered into the line of sight, must be included at each point in the integration. Therefore, the equation of transfer integrated along a ray is of the form

$$I(A) = I(B) \, e^{-\tau_{AB}} + \int_0^{\tau_{AB}} S(t') \, e^{-t'} \, dt',$$

where B is an arbitrary starting point within the nebula and A is a point at which the net intensity is calculated; τ_{AB} is the optical depth between the points AB and S is the source function (Fig. 1.8).

For a theoretical model of a reflection nebula, this equation is solved in a complicated iterative way by assuming a source function, solving the integral equation, predicting the resulting source function, and repeating until agreement is reached between the assumed and calculated source functions.

Fig. 1.7. Light of incident intensity I_0 passes through a volume of unit area and length dl.

$I_\nu \longrightarrow \bigcirc \cdots \bigcirc \longrightarrow I_\nu + dI_\nu$

αl

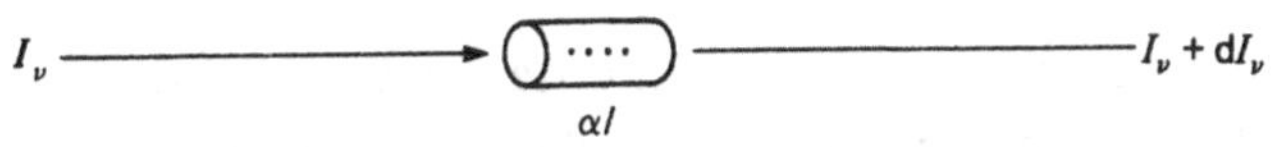

Fig. 1.8. Two points within a reflection nebula scattering light along path AB.

Each model so calculated begins with a specified phase function and albedo and predicts the emergent intensity or surface brightness of the nebula. By comparing the measured brightness of a reflection nebula with the models, the best fit should define the albedo, phase function, and optical depth of the nebular grains. The greatest uncertainty in this analysis arises in the problem of establishing the true three-dimensional geometry of the location of the nebula and star from a two-dimensional photograph of the object.

The Pleiades represent a reflection nebula in which the star is assumed to be in front and the nebula shines by the backscattered light. A relatively high albedo and nearly isotropic phase function are suggested from ultraviolet measurements; but no known distribution of particle sizes can provide these values.

The surface brightnesses of dark nebulae have been measured at a wavelength of 4300 Å. These nebulae reflect the starlight of the Milky Way and models of their brightness predict a composition of grains with albedos of 0.7 and $g = 0.7$, numbers that are not very consistent with the values of Fig. 1.5 but are generally adopted as typical values for grains in the interstellar medium, with the caveat that the issue is quite controversial.

If the coefficient of the imaginary term of the index of refraction is not zero, a grain will absorb some of the incident radiation and convert it into internal energy. Small grains do not radiate efficiently at long wavelengths, and, *if* an energy balance is achieved (energy absorbed equals energy radiated), the grain's temperature will increase until its black-body radiation balances the absorbed radiant energy. A mixture of grains of different sizes and refractive indices probably coexist in a nebula and very likely each type of grain will establish a different temperature, values of which have been calculated to range from about 8 to 200 K, depending on the chemical composition and size of the grain and upon the nearness of the illuminating star. Black bodies at these temperatures radiate primarily in the infrared region of the spectrum; evidence for this type of emission has been found in a number of emission nebulae in which it is assumed that grains exist and are heated by the hot central star.

It is now believed that the very hot, young stars associated with emission nebulae (the O stars) were formed from condensations within a giant dust cloud. During the very early stages in the creation of a star, the young 'protostar' may be completely surrounded by a dense dust shell; as the star begins to shine, its radiation is absorbed by the grains of the dust shell, which begins to glow in the infrared as a consequence of the heating of the grains.

Some infrared objects are coincident with optical nebulae. One interesting object is called the Red Rectangle because of the way it appears on a photograph taken in red light (Fig. 1.9). Photographed in blue light, the object is nebulous, having embedded in it an hourglass-type or biconal red nebula with apex at a central star. The spectrum of the blue nebulosity is identical to that of the central star, indicating that it is a reflection nebula. The infrared emission is a composite of black-body radiation and emission features attributed to the heated grains. The 'hourglass', 'biconal', or

Fig. 1.9. The Red Rectangle. From Cohen, M. *et al.* (1975), *Astrophys. J.* **196**, 179.

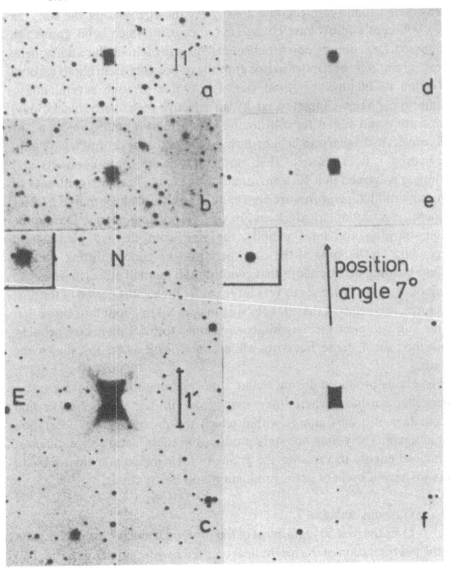

'cometary' appearance of the nebula may be interpreted as a consequence of the angle through which the nebula and star are viewed and/or as a product of constraints imposed by magnetic fields (if present).

The Red Rectangle is only one example of an infrared object that is detectable in the infrared because of grain emission. Other optical objects having similar infrared spectroscopic emission include planetary nebulae, novae, supernovae, and certain very luminous, cool stars called red giants. Contemporary models of stellar evolution predict certain stages in which the internal energy balance of a star will become unstable resulting in a burst or a steady outflow of the outer layers of the star. As the hot gases of the stellar atmosphere expand into the low density region surrounding the star, the gases will cool and a phase change may occur in which solid grains are condensed. Depending upon the chemical mixture of the stellar atmosphere, silicate grains may be condensed, or graphite or possibly amorphous carbon.

During its lifetime, a typical star will evolve through several quasi-equilibrium phases characterized by an effective (essentially black-body) temperature and size. After contracting and forming a young object, a star will convert hydrogen into helium (burn hydrogen) in its central region until the hydrogen is exhausted. Then, the star's helium core contracts and hydrogen is burned in a shell surrounding the core. The core continues to contract until temperatures are high enough to convert helium into heavier elements, e.g. carbon, nitrogen and oxygen, producing energy. During the helium-burning and later periods, the star is in the red giant stage. Depending on the mass of the star, there may be periods during which a helium-burning shell develops that goes through thermal pulses in which the luminosity of the shell rises to very large values for a short period of time. At this stage, it is believed that stars having masses up to about four times that of the Sun will eject their hydrogen envelopes to form planetary nebulae, while their small, dense, hot cores will become visible as the so-called white dwarfs.

The grains produced during stellar atmospheric outflows move into the interstellar medium where they may grow in size by accretion and accumulate into dust clouds within which the process of star formation begins again. The young hot stars produced in these clouds have sufficient ultraviolet energy to vaporize the grains in their immediate environment; thus we have a cycle of grain production and destruction.

1.3   Gaseous nebulae

Over the past 50 years, most of the work on nebulae has concentrated on the interpretation of the bright lines emitted by the nebular gas. Fig. 1.10

Fig. 1.10. Spectrogram of a typical emission nebula. The forbidden lines are identified by [ ] around the ion producing them. The lines identified as NS are produced by OH in the Earth's atmosphere. The [O I] line at $\lambda 6300$ is present both in the night sky emission and in the nebula. The dark band through the center of the spectrogram is produced by scattered starlight in the nebula.

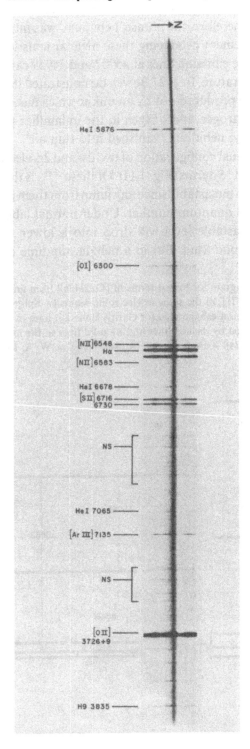

reproduces a spectrogram of a typical emission nebula. The recombination lines produced by the hydrogen atoms are easily identified. The Balmer alpha line ($\lambda 6563$) is the strongest emission in the red spectral region, a fact not surprising because of the overwhelming abundance of hydrogen (see Chapter 4). Other bright emission lines were at first not identifiable with any known terrestrial element. Therefore, the phrase 'nebulium' was introduced for the as-yet unidentified element producing these nebular emissions; the designation N_1 and N_2 for the emission lines at $\lambda 5007$ and $\lambda 4959$ can still be found in the astronomical literature. In 1927 Bowen demonstrated that most of the unidentified lines were produced not by an unknown element but by singly and doubly ionized nitrogen and oxygen in the unfamiliar physical condition of very low density; nebulium 'vanished into thin air'.

For N II and O III, the normal configuration of two 2s- and 2p-electrons is characterized by ^3P-, ^1D-, and ^1S-terms (Fig. 1.11). Of these, $^3P_0$ is the stable ground state while the rest are metastable since any jump from them involves zero change in the azimuthal quantum number. Under normal laboratory conditions, an ion in a metastable level will drop into a lower state by inelastic collisions of the second kind. But in a nebula, the time between

Fig. 1.11. Energy-level diagram for lowest terms of [O III], all from ground $2p^2$ configuration, and for [N II], of the same isoelectronic sequence. Splitting of the ground ^3P term has been exaggerated for clarity. Emission lines in the optical region are indicated by dashed lines and by solid lines in the infrared. From Osterbrock, D. (1974). *Astrophysics of Gaseous Nebulae*. W. A. Freeman and Company.

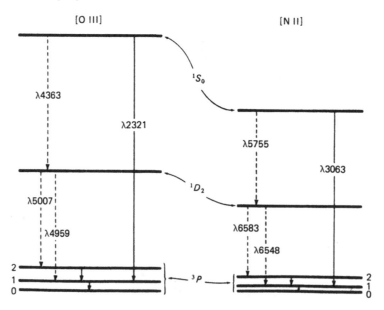

collisions may be from 10^4 to 10^7 seconds and there is time for the ion to return spontaneously to the ground level with the emission of the appropriate forbidden line. The lines are strong in nebular spectra because the levels are populated not by a recombination process but by collisional excitation (inelastic collisions of the first kind) by electrons in the ionized nebular gas.

While the key was being forged to unlock the secrets of the forbidden lines, Zanstra was demonstrating that the hydrogen recombination lines in gaseous nebula could be used to measure the ultraviolet ($\lambda < 912$ Å) emission of the exciting star, one of the most fundamental measurements and one that cannot even now be made directly because of the large optical depth of interstellar hydrogen. It is assumed that hydrogen in the nebula is photoionized by the exciting star

$$\mathrm{H}(ls) + h\nu_{uv} \to \mathrm{H}^+ + \mathrm{e},$$

and that the Balmer lines are formed by recombination:

$$\mathrm{H}^+ + \mathrm{e} \to \mathrm{H}(nl) + h\nu.$$

A hydrogen atom in an n, l excited state will quickly cascade down to the ground state with emissions of the appropriate recombination lines (Fig. 1.12). Zanstra argued that once a nebular hydrogen atom has recombined, any emission produced by a jump directly to the ground level will produce a quantum that has a very high probability of being absorbed by another H atom (the nebula is optically thick to the Lyman lines), but this is not the case for transitions between upper levels. The net result is that one can qualitatively demonstrate that for every ultraviolet quantum absorbed, a quantum corresponding to a Balmer transition is emitted. The strongest of the Balmer lines are in the measurable red and blue regions of the spectrum; therefore, from direct measurement of the number of Balmer quanta emitted per second by the nebula, a lower limit to the ultraviolet flux of the star could be estimated.

Since Zanstra's pioneering studies, more sophisticated analyses of the recombination-line spectrum have been developed. *Statistical equilibrium* (steady state) is assumed, requiring that for any level nl, the population of the level be constant in time. That is, the number of transitions entering the state must be equal to the number of transitions depopulating the state per unit time, or

$$N_e N_+ \alpha_{nl}(T_e) + \sum_{n' > nl'}^{\infty} \sum N_{n'l'} A_{n'l', nl} = N_{nl} \sum_{n''=1l''}^{n-1} \sum A_{nl, n''l''},$$

if we can neglect collisional processes.

The left side of the equation accounts for the manner by which the level is populated; the first term represents the number captured into the state nl, and the second term sums all transitions into the state from other n', l' states. The right side of the equation represents the number leaving by radiative transitions (we have neglected any collisional transitions). Thus, to predict the intensities of the recombination lines, it is necessary to have accurate values for the atomic recombination coefficients (α_{nl}) and transition probabilities ($A_{nl,n'l'}$) and to solve the statistical equilibrium equations for the population, $N_{n,l}$, of all levels of the ion or atom. The parameter N refers to number per cubic centimeter, with the subscript identifying the particular species and T_e is the electron (kinetic) temperature. If collisions are significant (and they usually are), equivalent terms involving collisional cross sections must be added.

If the populations of levels are known, the emissivity (J_{ij}, the energy flux corresponding to the transition $i \to j$ radiated per unit electron density per unit ion density per unit solid angle) can be found by

$$J_{ij} = \frac{h\nu_{ij}}{4\pi} A_{ij} \frac{N_{i,\text{ion}}}{N_{\text{ion}} N_e},$$

Fig. 1.12. Schematic diagram of energy levels of hydrogen with various series identified. From Merrill, P. W. (1956). *Lines of the Chemical Elements in Astronomical Spectra.* Carnegie Institute of Washington Publication 610.

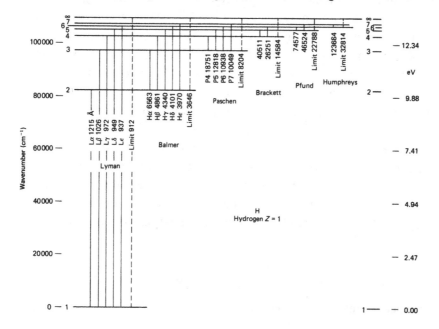

and predicted line ratios can then be compared with those observed (after the observations are corrected for the effects of interstellar reddening).

For hydrogenic ions the atomic coefficients may be calculated exactly but become increasingly difficult with increasing n. Goldberg was the first to apply the theory to neutral helium in nebulae, and contemporary sophisticated quantum-mechanical calculations have produced accurate predictions of the emissivities of hydrogen and helium for ranges of temperatures appropriate to astronomical plasmas.

For the forbidden lines emitted as a consequence of collisional excitation, the equations of statistical equilibrium are much simpler because only a few levels – two to five – need be considered. However, the atomic parameters are less accurately known and theoretically predicted line intensities are compared with those actually observed in a nebula as a check on the accuracy of the ionic parameters. Thus theory and observation proceed in an iterative manner with increasing accuracy of quantum-mechanical calculations providing more accurate diagnostics for the physical conditions and element abundances in emission nebulae, while accurate measurement of line ratios in nebulae provide checks on the theory.

In addition to the equation defining the steady-state population of atomic or ionic energy levels, we may also define an ionization equilibrium equation specifying that the number of ionizations per unit time must equal the number of recombinations per unit time. Or if there were a pure hydrogen gas surrounding a source of uv quanta:

$$N_{\mathrm{H}} \int_{v_0}^{\infty} \frac{4\pi J_v}{hv}\, \alpha_v\, \mathrm{d}v = N_p N_e \alpha_A(T_e),$$

where N_{H} represents the number density of neutral hydrogen, α_v is the ionization cross section, and α_A is the total recombination coefficient. The mean intensity, $4\pi J_v$, will clearly depend on the uv character of the exciting star and the manner in which the radiation is transported to the point at which the ionization equation is being applied. In 1939, Stromgren demonstrated that hydrogen is almost completely ionized up to a specific distance from the ionizing star. Beyond this, there is a very thin shell of partial ionization, followed by the 'normal' region of neutral hydrogen thus defining the H II and H I regions of the interstellar medium. The ionized (H II) zone surrounding a hot star is referred to as the Stromgren sphere. Fig. 1.13 illustrates the resulting balance in the ionization of a nebula of density 10 hydrogen atoms per cubic centimeter with a mixture of 15 percent helium. The radii of the Stromgren spheres for two typical hot stars are normalized to unity. Column 5 of Table 1.1 lists the nebular sizes in parsecs

(1 pc = 3.09 dex 18 cm) calculated for pure hydrogen nebulae on the assumption that the density in the nebula is $\approx 100 \, \text{cm}^{-3}$. The fourth column lists the parameter derivable from theory; r_1 is the radius of the H II region, N_p is the number density of protons; N_e is the electron density. The free electrons come from hydrogen and $N_e = N_p$ if the hydrogen is completely ionized. In a real nebula, each helium ion accounts for one free electron in the zone in which helium is singly ionized (Fig. 1.13). Thus, from a determination of the electron density and the size of a nebula, the temperature of the exciting star may be found from Table 1.1.

Fig. 1.13. Ionization structure of two homogeneous H + He model H II regions. From Osterbrock (1974).

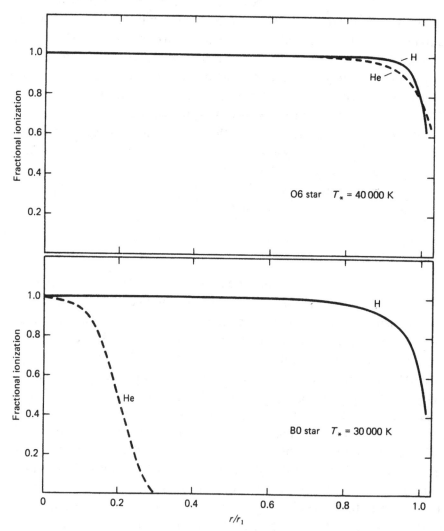

When dust is present in a nebula it competes with the gas in absorbing the ionizing stellar radiation. The grains of the interstellar medium are quite efficient in absorbing uv quanta and 'soften' the ionizing radiation and consequently decreasing the size of the H II region.

The angular size of an emission region can be determined from photographs taken in the light of a strong emission line (usually Hα) or from radio contours if the nebula is heavily obscured. The size in parsecs can then be deduced if the distance to the nebula is known. Distances to nebulae are usually determined by estimating the distance to the exciting star, if identified. If the star's absolute brightness is known and the apparent brightness measured, the distance may be determined based on the inverse-square law of how radiant flux varies with distance. The electron density is determined by measurement of the intensities of certain nebular emission lines.

In the density/temperature range of nebulae, there are certain line ratios that change measurably as the electron density varies, and other ratios that are more sensitive to changes in the electron temperature. By measuring such ratios it is thus possible to deduce the plasma conditions. In particular, the $^4S_{3/2}-^2D_{5/2}$ and $^4S_{3/2}-^2D_{3/2}$ transitions (Fig. 1.14) of O II and S II ($\lambda 3728.8$, $\lambda 3726.1$ and $\lambda 6717.4$, $\lambda 6730.8$, respectively) are sensitive to electron density and from these ratios the most accurate estimates of the electron densities *in the regions giving rise to the forbidden lines* have been obtained.

Fig. 1.15 illustrates typical values measured for these two ratios for several nebulae. Sample electron densities are indicated at the tick marks on the two curves.

Electron temperatures have been determined optically from the higher-energy level transitions that require more energetic collisions for population of the level. Specifically, the $^1D_2-^1S_0$ transition compared with the $^3P_2-^1D_2$ of N II ($\lambda 5754.6/\lambda 6583.4$) and O III ($\lambda 4363.2/\lambda 5006.9$) (see Fig. 1.11) are

Table 1.1

Spectral Type	T_* (K)	Log $Q(H^0)$ (photons/s)	Log $N_e N_p r_1^3$ (cm^{-6} pc)	r_1 (pc)
O5	48 000	49.67	6.07	4.9
O6	40 000	49.23	5.63	3.5
O7	35 000	48.84	5.24	2.6
O8	33 500	48.60	5.00	2.2
O9	32 000	48.24	4.64	1.6
B0	30 000	47.67	4.07	1.1

useful thermometers. These line ratios are quite large (Fig. 1.16) and are difficult to measure. More precise temperature determinations can be made from radio data (see Chapter 2). The temperatures deduced from radio data represent the region emitting the recombination lines, while the optical line ratios yield values of the temperature in the environment producing the forbidden lines. The temperatures for emission nebulae have been measured to be between 5000 and 15 000 K, and are in general agreement with the temperatures predicted for equilibrium model H II regions.

Models defining the detailed structure of H II regions have been constructed by using the equations of ionization, thermal and radiative equilibrium. The energy gained by the plasma is dominantly that of photoionization (see Chapters 9 and 10 for exceptions to this assumption), while the energy losses arise from many terms, including recombinations, free–free emission, and collisional excitation of line radiation. The electron temperature is determined by equating the heating and cooling rates.

Each model specifies the nature of the exciting star and the electron

Fig. 1.14. Energy-level diagram of ground configurations of O II and S II. From Osterbrock (1974).

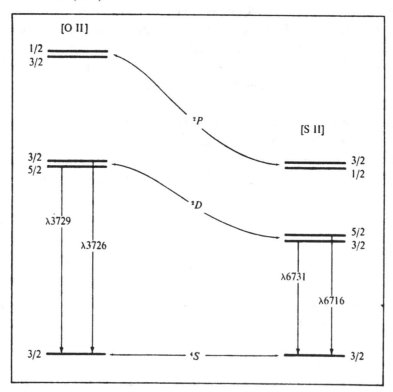

density. The results can be graphically represented in terms of the fractional ionization of each element as a function of the distance from the exciting star (expressed in units of the radius of the H II zone). Fig. 1.17 represents the change in the ionization of nitrogen within a nebula for a model ionized by a star with an effective temperature of 35 000 K. The dotted, dashed, and solid lines refer to nebulae having electron densities of 1, 100, and 10 000 cm^{-3}, respectively. The dominant ion of nitrogen is N III; N II appears primarily in the outer shell of the nebula. Oxygen, having ionization potentials similar to N, behaves in the same fashion. When dust is added to the model, the relative sizes of the zones dominated by higher stages of ionization decrease and zones of lower excitation increase.

Fig. 1.18 sketches the variation in electron temperature for a number of nebular models. The curves labeled *A* represent dust-free models; curves *B–E* are models with increasing amounts of dust. The temperatures in the upper left portions of each diagram are the effective temperatures of the exciting stars; 'IMF' refers to a model using a cluster of young stars. Electron temperatures average about 7000 K and increase slightly in the outer zones of the nebula.

Fig. 1.15. Theoretical (curves) and observational (points) relation between doublet ratios for S II and O II. The tick marks represent values of electron density in cm^{-3}. Two different values of electron temperature are shown.

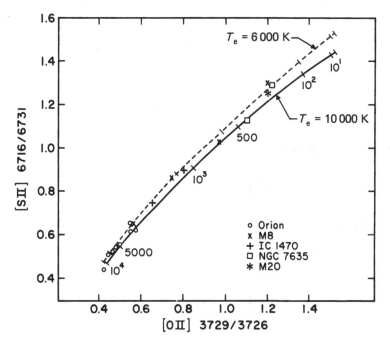

If the relative populations of all energy levels of an ion or atom are known, and if the ionization factor, f, is fixed, then by measuring the relative intensities of the nebular emission lines we can establish the electron temperature, density, and finally the abundance of the element.

Fig. 1.16. Theoretical prediction of ratio of O III lines as a function of electron temperature. From Osterbrock (1974).

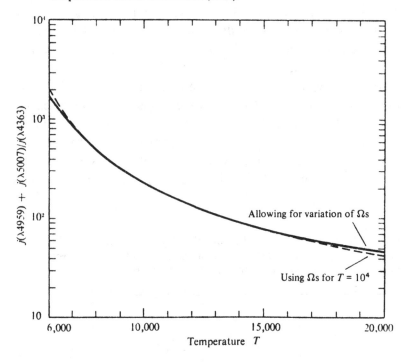

Fig. 1.17. Ionization structure of nitrogen. From Stasinska, G. (1978), *Astron. Astrophys. Suppl.* **32**, 429.

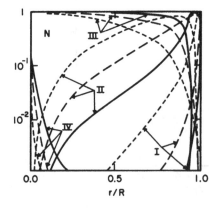

The Trifid Nebula, Fig. 1.19, is a traditional example of a galactic nebula and one we will use to illustrate the application of the physical models.

The Trifid has a hot O star at its center; from the measured apparent brightness of the star, the distance to the object has been established to be 2200 parsecs. The Trifid is characterized by dust lanes dividing it into three segments. Bright emission rims bordering the dark nebulae attest to the fact that the dark clouds are very close to the emission region.

The spectrogram of Fig. 1.10 is of the Trifid, the position of the spectrographic slit is indicated by a black line on Fig. 1.19. With data such as these, intensities of the nebular emission lines can be measured. From the ratios of the [O II] and [S II] doublets, it is found that the electron density in M20 is fairly uniform and has a value of $250 \pm 100\,\mathrm{cm}^{-3}$ (see Fig. 1.15).

Fig. 1.18. Electron temperatures for several models. From Sarazin, C. L. (1977), *Astrophys. J.* **211**, 772.

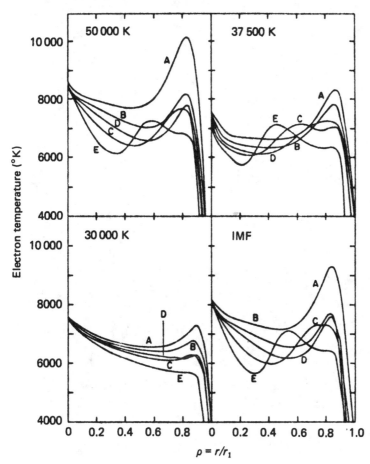
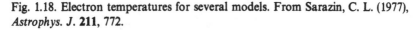

Fig. 1.20 exhibits the change in the ratio [O II] $\lambda3726 + \lambda3729$/[N II] $\lambda6548 + \lambda6583$ across the region sampled by the spectrogram. The ratio of these two ions should be fairly constant in a nebula because their ionization potentials are very similar; therefore, the marked changes in the observed ratio are caused by reddening as the nebular light passes through the dust lanes. The changes in the ratio are coincident with the dust features of Fig. 1.19. If the wavelength variation of the dust extinction is known, it is

Fig. 1.19. The Trifid Nebula. The dark line represents the position of the slit for the spectrogram of Fig. 10. (4-m KPNO photograph.)

possible to measure the extinction in the dust lanes from the data of Fig. 1.20. In this manner, it has been found that the dark clouds associated with the Trifid have about 50 to 100 solar masses and are probably 'wisps' of a massive dark cloud (≈ 1300 solar masses) that lies southwest of the Trifid.

Corrections for interstellar extinction must be made before emission measurements of nebular lines can be matched with models. Usually, predicted line ratios that are relatively insensitive to nebular conditions are used to establish the reddening – the ratio of the higher Balmer lines to Hα, for example, or the ratio of the blue to the 1-micron lines of [S II].

For optical nebulae, the ratio of Hα/[N II] $\lambda6548 + \lambda6583$ is essentially unaffected by reddening; this ratio provides an excellent 'diagnostic' for the nebula. Fig. 1.21 illustrates the ratio for the sampled region of the Trifid; the Hα/[N II] values are found to vary from 1.0 to 3.5; the smaller ratios coincide with the bright emission rims.

The intensity of the hydrogen Balmer line is given by:

$$I_{H\alpha} = \frac{1}{4\pi} \int N_p N_e h\nu_{H\alpha} \alpha_{H\alpha}^{eff} \, ds,$$

Fig. 1.20. Observations of the intensity ratio of [O II]/[N II] for the Trifid Nebula. The position refers to points along the spectrogram and its position in the nebula is shown in Fig. 1.19. The zero of position coincides with the position of the exciting star.

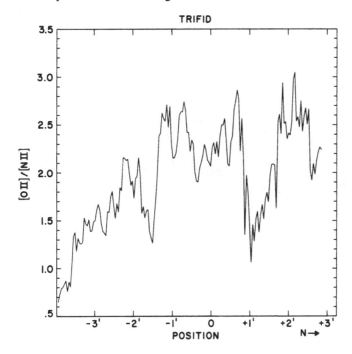

where the integral is over the path length through the nebula. The emissivities of the [N II] lines can be calculated by solving the equations of statistical equilibrium for a five-level ion and using the collisional cross sections and radiative transition probabilities established for N II.

The atomic parameters and emissivities are tabulated in Table 1.2 for a range of electron temperatures and densities.

Table 1.2 can be used to predict the intensity of the [N II] lines if the number of ions per cubic centimeter is known. N_{ion} may be expressed in

Table 1.2

T_e (K)	N_e (cm^{-3})	$\alpha_{H\beta}$ (10^{14})	Hα/Hβ	[N II]J_{4-2} (10^{23})	[N II]J_{4-3} (10^{23})
5000	10^2	5.391	3.032	3.028	9.036
5000	10^4	5.443	3.003	2.583	7.707
10 000	10^2	3.023	2.859	19.40	57.90
10 000	10^4	3.036	2.847	17.17	51.23

Fig. 1.21. Observed ratio of Hα/[N II] as measured on the spectrogram of Fig. 1.10.

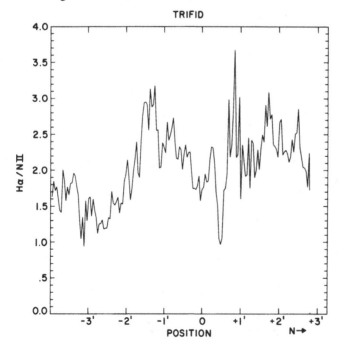

terms of the total abundance of nitrogen multiplied by f_{NII}, the fraction of nitrogen in the first ionized state. Thus, the observed Hα/[N II] ratio and the data in Table 1.2 can be used to deduce either f_{NII} or the abundance of nitrogen once T_e and n_e are established.

Table 1.3 illustrates the manner in which the H/[N II] ratio varies with T_e and f_{NII} if a normal solar abundance is assumed for nitrogen. As demonstrated in Table 1.2, the ratio is much less sensitive to changes in n_e for normal nebular conditions.

The electron temperature of the Trifid has been measured to be approximately 8000 K. The Trifid is a dusty nebula; measurements of the scattered light in the nebula indicate that the dust optical depth is greater than 1. Therefore, the appropriate models are ones having an appreciable optical depth in the dust. The f_{NII} predicted for a dusty model is ≈ 0.4, and the predicted ratio Hα/[N II] of 2.0 (Table 1.3) is reasonably consistent with the observed values of Fig. 1.21. Better agreement is obtained if the abundance of N is decreased to 82 percent of the solar value, thus increasing the ratio to a value of 2.4, the average value of Fig. 1.21.

The simplest interpretation in the change in the Hα/[N II] ratio across the Trifid is that it is a consequence of the changing f_{NII} as the stellar radiation penetrates the denser dust clouds. In the vicinity of the bright ionization rims, the electron density increases by about a factor of 5 and the level of ionization of N drops (see Fig. 1.21), f_{NII} increases, and Hα/[N II] decreases to about 1.0 (see Table 1.3). An alternative explanation, but not as satisfactory quantitatively, is that more of the N is frozen onto grains in the dark cloud and, as the grains are sputtered by the stellar flux, the N atoms evaporate, thus changing the abundance of the gaseous phase of the element.

The radius of the Trifid ionized region is 5.8 arc minutes, corresponding to 3.7 parsecs. Using an electron density of 250 cm^{-3}, we find that log $N_e N_p r_i^3$ is

Table 1.3. Hα/[N II]$\lambda 6548 + \lambda 6583$

T_e (K)	f_{NII}	0.2	0.4	0.6	0.8
5000		23	12	7.7	5.8
6000		10.5	5.2	3.5	2.6
7000		6.6	3.2	2.2	1.6
8000		3.9	2.0	1.3	0.9
9000		2.6	1.3	0.8	0.7
10 000		1.9	1.0	0.6	0.5

6.50, implying that the nebula is too large to be ionized by a single O star, according to Table 1.1.

A Zanastra-type analysis using measurements of radio continuum emission has been used to set a lower limit for the star's uv energy:

$$L_c \geqslant 6.75 \times 10^{48} \quad \text{photons/s,}$$

which is just about the amount radiated by a model O star (Table 1.1). However, the dust in the Trifid absorbs a large amount of the ionizing quanta; nearly six times as much if the optical depth of the dust is near unity. Therefore, more than 4×10^{49} photons/s beyond the Lyman limit must be available to produce the Trifid nebula.

The energy absorbed by the dust is reemitted in the infrared by the grains. Using the measured dust optical depth, we predict that the nebula grains should radiate 1.7×10^{-9} watts/m^2 s in the infrared. An infrared flux of $\geqslant 1.5 \times 10^{-9}$ has been measured, in excellent agreement with the predicted flux.

We conclude that the Trifid nebula requires more ionizing quanta than a single O star is believed to produce. The nebula fits a model having an ionizing source with $T_{\text{eff}} \approx 50\,000\,\text{K}$ embedded in a gaseous medium of electron density $\approx 250\,\text{cm}^{-3}$ with a mixture of dust grains having albedos ≈ 0.6 and an optical depth near unity. The need for additional uv quanta suggests that there is another early-type star hidden behind the dust lanes if stellar radiation is the only energy input.

As the example of the Trifid nebula demonstrates, optical data when combined with radio and infrared observations provide tests for nebular models. The astrophysical plasma associated with young massive stars must be considered to be a dusty one, and the influence of dust absorption must be included in models before the nature of the exciting star and the abundances of the elements in the nebulae can be determined.

1.4   Kinematics

Optical nebulae have been identified in galaxies beyond the Milky Way system. Fig. 1.22 reproduces a photograph of the spiral 'Whirlpool' galaxy, M51, taken through a filter that transmitted the Balmer alpha line of hydrogen. Most of the bright objects visible are H II regions and the striking characteristic of these nebulae is that they are strung out along the spiral arms of the galaxy, like 'beads on a string'. The spiral arms are best defined by the dark dust lanes, with which the H II regions are generally associated.

From velocity measurements, it has been established that a spiral galaxy rotates about its center in such a way as to produce 'trailing' spiral arms. The

Fig. 1.22. Two photographs of M51. The upper one was taken in the light of Hα; the lower one in the red continuum avoiding emission regions. Bright H II regions are evident on the upper photograph and lie along the dust lanes of the galaxy.

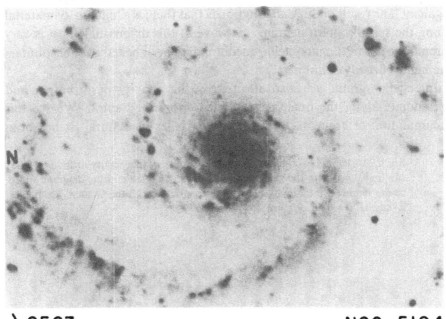

λ6563   NGC 5194

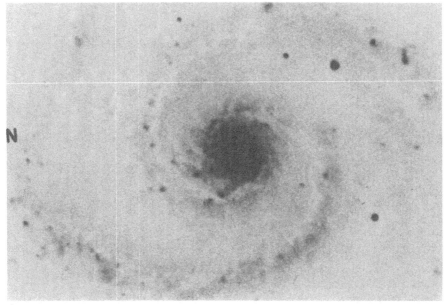

λ6650   NGC 5194

stability of an arm is believed to be maintained by a fluctuation in the general density field produced by the billions of stars making up the galaxy. This 'density wave' exerts a gravitational perturbation on the particles of the interstellar medium as they move in nearly circular orbits about the center of a galaxy. The result of this perturbation is that there is a 'pile-up' of material along the spiral pattern defining the wave. If this differential effect is very strong, the gas of the interstellar medium may be shocked as it encounters the higher density spiral arm.

Interstellar grains are also affected by this gravity perturbation and accumulate along the inner edges of the two-armed spiral. We see this accumulation of grains as the strong dust lanes of a spiral. It has been

Fig. 1.23. Diagram of the distribution of O stars in the vicinity of the sun. The sun lies 9 kpcs from the galactic center, which is off-scale in the diagram. Filled circles represent the bright H II regions associated with the O stars. From Lynds, B. T. (1980), *Astrophys. J.* **85**, 1046.

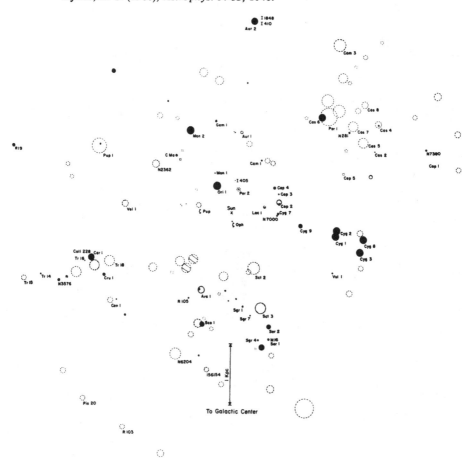

established that heavy molecules (such as CO and H_2CO) exist in abundance in the dust clouds.

The optical nebulae of our own Galaxy appear to display a spatial distribution like that of a typical spiral galaxy. Fig. 1.23 illustrates the relative distribution of nearby O stars and their optical emission nebulae as projected onto the galactic plane (the plane defined by the great circle of the Milky Way). The Sun is believed to lie near the inner edge of a spiral arm about 9000 parsecs from the center of our Galaxy. We can measure O-star distances accurately only to two or three kiloparsecs; therefore Fig. 1.23 represents only a small sample of the whole Milky Way spiral; the center of the Galaxy is off-scale below the figure.

Segments of three arms are identified in the figure. The Orion Nebula is the nearest of the bright emission regions; the spiral arm in which the Sun is located is called the Orion Arm. Closer toward the center of the galaxy are found the bright nebulae most easily seen in the Southern Hemisphere – the fascinating star-producing complex of η Carina, and the beautiful objects in Sagittarius – M17 (the Omega), M8 (the Lagoon), M20 (the Trifid). This

Fig. 1.24. The Rotation Curve of our Galaxy. The estimated curve for our Galaxy is between curves A and B. The dotted lines are typical velocities measured for other galaxies. From Caldwell, J. A. R. and Ostriker, J. P. (1981), *Astrophys J.* **251**, 61.

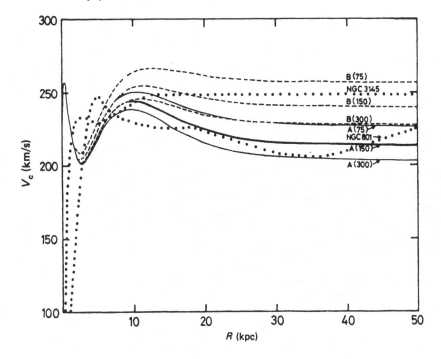

inner arm is often referred to as the Sagittarius arm. Similarly, at distances beyond the Sun are a number of H II regions in the constellation Perseus; these are in the so-called Perseus arm.

Most of the young stellar objects and the material of the interstellar medium move in nearly circular orbits about the center of the galaxy in a manner analogous to that of the planets and with orbital velocities similar to those measured in other galaxies. An object's radial distance from the galactic center will determine its orbital velocity; this dependence is usually presented in the form of a curve relating circular velocity to galactic distance – the so-called Rotation Curve (Fig. 1.24). If the orbital velocity of an object can be established, its distance can be determined from the rotation curve. In this manner, kinematic distances to many distant nebulae have been determined and the 'spiral pattern' of our Galaxy extended to greater distances from the Sun.

The Orion Nebula, being the brightest (by about an order of magnitude) emission nebula and also one of the nearest, has been studied extensively. It

Fig. 1.25. Schematic representation of the 'blister' model for IC1318 and the dark nebula L 889. The numbers are velocities along the line of sight. From Fountain, W. F., Gary, G. A. and O'Dell, C. R. (1983), *Astrophys. J.* **269**, 164.

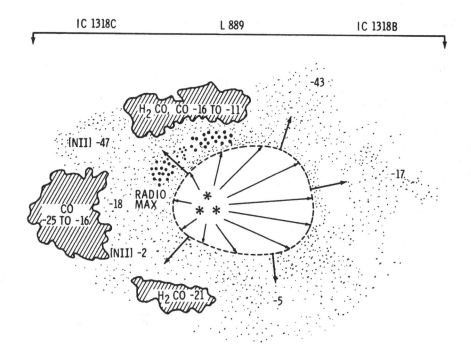

is a very complex region having a cluster of hot, young stars at its center (the Trapezium). Of particular interest are several infrared objects that may represent regions in which stars are presently being formed. The O stars have ionized a portion of a giant molecular cloud complex, forming a 'blister' on its edge.

Many other galactic nebulae have the appearance of blisters on the edges of molecular clouds. When a very hot star is produced, the immediate surroundings are heated; and if the formation process occurs near the edge of a molecular cloud the ionized region expands into the lower density interstellar medium. Fig. 1.25 illustrates a model of IC1318; fragments of a dense molecular cloud are shown by the shaded regions. The three hot stars have produced an ionized region that is expanding into the lower-density region identified by the stippled areas. The velocities in the line of sight are given in the figure; negative velocities represent velocities of approach. For models like this, the molecular cloud is assumed to have the standard orbital velocity; streaming motions are deduced from the differential velocities of cloud and emission region. Dynamic models of the turbulent interstellar medium are ones now being proposed (see Chapters 9 and 10).

References

Section 1

An example of the early photographic surveys of bright and dark nebulae is

Barnard, E. E. (1927). *Photographic Atlas of Selected Regions of the Milky Way*, ed. E. B. Frost & M. R. Calvert. Carnegie Institution of Washington.

The classical papers on the spectra of nebulae are

Hubble, E. (1922). *Astrophys. J.* **56**, 162, 400.

Section 2

A basic monograph on scattering is

van de Hulst, H. C. (1957, 1981). *Light Scattering by Small Particles*. Dover Publications.

The classical paper on interstellar scattering is

Henyey, L. G. & Greenstein, J. L. (1941). *Astrophys. J.* **93**, 70.

More recent discussions are

Greenberg, J. M. (1968). *Interstellar Grains, Stars and Stellar Systems, VII*, ed. B. M. Middlehurst & L. H. Aller, p. 221. University of Chicago Press.

Savage, B. D. & Mathis, J. S. (1979). Observed properties of interstellar dust. *Ann. Rev. Astron. Astrophys.* **17**, 73.

An interesting monograph on theories of star formation is

Roger, R. S. & Dewdney (eds.) (1982). *Regions of Recent Star Formation*. Reidel Publishing Co.

Section 3

The material in this section is more thoroughly discussed in

Osterbrock, D. E. (1974). *Astrophysics of Gaseous Nebulae.* W. A. Freeman & Co.

The classical papers on ionization are

Bowen, I. S. (1927). *Nature,* **120,** 473.

Bowen, I. S. (1972). *Astrophys. J.* **97,** 1.

Zanstra, H. (1927). *Astrophys. J.* **65,** 50.

Stromgren, B. (1939). *Astrophys. J.* **89,** 526.

Recent models of dusty plasmas are

Sarazin, C. L. (1977). *Astrophys. J.* **211,** 772.

Petrosian, V. & Dana, R. A. (1980). *Astrophys. J.* **241,** 1094.

Mathis, J. S. (1985). *Astrophys. J.* **291,** 247.

Section 4

A summary of theories of spiral structure is

Toomre, A. (1977). *Ann. Rev. Astron. Astrophys.* **15,** 437.

2

Radio observations of H II regions

ROBERT L. BROWN

2.1 Thermal equilibrium

Our understanding, and subsequent interpretation, of the continuum emission from H II regions at all frequencies is greatly facilitated by the recognition that H II regions are very nearly in thermal equilibrium. Simply, the velocity distributions of all the species comprising an H II region – atoms, ions and electrons – are Maxwellian and all can be characterized locally by a single kinetic temperature. This situation obtains because the mean energy of an electron in a 10^4 K nebula, $\frac{1}{2}mv^2 = \frac{3}{2}kT \approx 1.5$ eV is far less than that needed to excite common nebular atoms and ions. Consequently most collisions between electrons and ions or atoms are purely elastic Coulomb collisions. To see this we can compare the time t_c for an electron to lose its initial translational kinetic energy through the cumulative effect of electrostatic attractions and repulsions (Coulomb collisions) with the time t_{ex} between inelastic collisions in which the electron loses its energy via collisional excitation of an atom or ion. Here

$$t_c = \frac{1.24 \times 10^{-18}\, v^3}{n\pi_s}\ \text{s} \tag{2.1}$$

where v is the electron velocity (cm s^{-1}), n is the density (cm^{-3}) of positive charges (mostly protons) and π_s is a logarithmic factor, $\pi_s \approx 17$ under typical nebular conditions. On the other hand, the time for inelastic collisions to produce an excited atom or ion is

$$t_{ex} = (n_i \sigma_{ex} v)^{-1} \tag{2.2}$$

where σ_{ex} is the cross section for such a collision (commonly $\approx 10^{-16}$ cm^{-2}) and n_i is the number density of target ions.

If now we perturb the initial Maxwellian velocity distribution of the nebular gas, that perturbation will be shared by all constituents via Coulomb collisions, and thermal equilibrium reestablished, in a time t_c. On the other hand, inelastic collisions will tend to preserve the perturbation since the excess energy is radiated away not redistributed within the nebula. Thus, one measure of the departure from thermal equilibrium in an H II region, Δ_{eq}, is given by the ratio of these two time scales,

$$\Delta_{eq} \approx t_c/t_{ex} \approx 7 \times 10^{-36} \, v^4(n_i/n). \tag{2.3}$$

As $v = 10^6$–10^7 cm s^{-1} while the ion to proton ratio $n_i/n < 10^{-3}$, thermal equilibrium is the rule in H II regions.

The thermal radiation emitted by an object in equilibrium is black-body radiation when that object is optically thick. Thus, the specific intensity I_v of the radiation from an H II region of large optical depth is accurately given by the Planck function $B_v(T)$

$$I_v \equiv B_v(T) = \frac{2hv^3/c^2}{\exp(hv/kT) - 1}, \tag{2.4}$$

which, at any frequency, is uniquely characterized by the temperature. Black-body radiation is isotropic and unpolarized.

At radio frequencies $hv/kT \approx v(\text{GHz})/20\,T(\text{K}) \ll 1$ for typical H II region temperatures. In this case we may expand the exponential in equation (2.4) and obtain

$$B_v(T) = 2v^2kT/c^2 \tag{2.5}$$

which is the Rayleigh–Jeans approximation to the Planck function. In radio astronomy it is convenient (as described below) to use (2.5) to express the specific intensity as a temperature, the *brightness temperature* T_B, defined by $I_v = B_v(T_B)$. The equation for transfer of thermal emission with these definitions in the Rayleigh–Jeans regime takes the simple form

$$\frac{dT_B}{d\tau_v} = -T_B + T_e \tag{2.6}$$

where T_e is the kinetic temperature, or *electron temperature*, of the H II region and τ_v is the optical depth through the nebula along the line of sight. For an isothermal nebula, the emergent specific intensity, expressed as a brightness temperature as defined above, is

$$T_B(v) = T_e(1 - e^{-\tau_v}) + T_B(0). \tag{2.7}$$

Here $T_B(0)$ is the temperature of the background sky at the frequency of interest. Note that as the optical depth becomes very large the brightness temperature approaches the kinetic temperature of the nebula as, of course, it must for an object in thermal equilibrium.

2.2 Thermal bremsstrahlung radiation

The brightness, or specific intensity, of an H II region can be evaluated at radio frequencies from (2.7) once we know the source optical depth. The opacity is provided by free–free absorption – absorption of radiation by free (unbound) electrons in the electrostatic field of a positive charge. Here we compute this absorption by analysis of the inverse process, free–free emission, and application of Kirchhoff's law.

At radio frequencies the emission from an H II region is principally thermal bremsstrahlung or free–free radiation. Bremsstrahlung arises from random encounters of the fast-moving electrons with the much slower nebular protons and positive ions. In these encounters the electron, electrostatically deflected in its path, experiences a net acceleration and subsequently radiates away the energy gained. Classically, the positive charge is regarded as fixed during the encounter and the acceleration experienced by the electron is regarded as occurring instantaneously. Looking at the radiation resulting from this classical collision in the frequency domain – i.e., by considering the Fourier transform – such an instantaneous event transforms to a broadband spectrum of frequencies limited at the high frequency end only by the condition that the energy of the photon cannot exceed the total energy of the incident electron. Thus we expect the bremsstrahlung emission coefficient j_v (energy radiated per unit time per unit volume, per unit frequency per steradian) to be proportional to the number of electron–ion collisions that occur per unit time, that is, proportional to the number density of electrons and ions, but independent of frequency as long as $hv \ll kT$ where we are far from the exponential cut-off of the Maxwellian electron velocity distribution.

The bremsstrahlung emission coefficient for electrons of mass m interacting with ions of charge $Z_i e$ is

$$j_v = \frac{8}{3}\left(\frac{2\pi}{3}\right)^{1/2} \frac{Z_i^2 e^6}{m_e^{3/2} c^3 (kT)^{1/2}} g_{ff} n_e n_i \, e^{-hv/kT} \tag{2.8}$$

$$= 5.44 \times 10^{-39} Z_i^2 n_e n_i T^{-1/2} g_{ff} \, e^{-hv/kT} \, \mathrm{erg\ cm^{-3}\ s^{-1}\ sr^{-1}\ Hz^{-1}}. \tag{2.9}$$

Here, g_{ff} is the free–free gaunt factor, a quantum mechanical correction to the classically derived expression for j_v, that accounts for such effects as those arising from electron–ion collisions with a small impact parameter – collisions in which the electron passes very near the ion – that are not properly described by classical orbits,

$$g_{ff} = \frac{\sqrt{3}}{\pi} \left\{ \frac{\ln(2kT)^{3/2}}{\pi e^2 v m_e^{1/2}} - \frac{5\gamma}{2} \right\} \tag{2.10}$$

where γ is Euler's constant, $\gamma = 0.577$. At radio frequencies one can employ the convenient approximation

$$g_{ff} = 11.962 T_e^{0.15} v^{-0.1} \tag{2.11}$$

so that

$$j_v = 6.507 \times 10^{-38} Z_i^2 n_e n_i T^{-0.35} v^{-0.1} e^{-hv/kT} \, \mathrm{erg \, cm^{-3} \, s^{-1} \, sr^{-1} \, Hz^{-1}} \tag{2.12}$$

We are now in a position to use Kirchhoff's law

$$j_v = \kappa_v B_v(T) \tag{2.13}$$

to relate the bremsstrahlung absorption coefficient κ_v to the emission coefficient determined above. And this is an appropriate use of Kirchhoff's law because, again, H II regions are so very nearly in thermal equilibrium. Thus,

$$\kappa_v = 0.2120 Z_i^2 n_e n_i v^{-2.1} T_e^{-1.35} \, \mathrm{cm^{-1}} \tag{2.14}$$

where the radio-frequency approximation (2.11) has been employed.

2.2.1 *Observations and interpretation of the bremsstrahlung radiation from H II regions*

In order to use observations to infer the physical conditions in an H II region we compute the value of a potential observable in terms of nebular parameters and then compare this expectation with the observations. Radio observations of the continuum emission from H II regions provide us with two observables, the brightness temperature (2.7)

$$T_B = T_e[1 - \exp(-\tau_v)] \tag{2.15}$$

(where we neglect the background contribution), and the flux density S defined as

$$S = \frac{2kv^2}{c^2} \int T_B \, d\Omega \tag{2.16}$$

where the angular integral extends over the solid angle subtended by the nebula. These two observables provide rather different insights: T_B reflects specific conditions along individual lines of sight whereas S provides a measure of the global properties of an H II region.

2.2.1.1 *Brightness temperature*

There are two interesting limits on the radio brightness temperature. The first is when the nebula is optically thick, that is, $\tau_\nu \gg 1$. Since $\kappa_\nu \propto \nu^{-2}$ (2.14) large optical depths are at low frequencies. When the optical depth is very large the observed brightness temperature is the nebular kinetic temperature (2.15). Note that $\tau_\nu \gg 1$ strictly implies that all lines of sight that intersect the nebula are equivalent; each provides information solely on the kinetic temperature of the H II region (which we have implicitly assumed to be constant). No information whatever is available on the density or density distribution within the nebula.

To a single dish radio telescope, temperature is the fundamental observable. This is because in single dish work one compares the intensity in the sky with the intensity of a thermal calibration source whose temperature is known. Thus the sky intensity is converted to a temperature, the so-called *antenna temperature* and, correspondingly, the difference between the measured intensity of an H II region and that of a nearby position on the sky free of radio sources is the antenna temperature T_A of the H II region. But antenna temperature is not equivalent to brightness temperature. Here the distinction is the following: when the radio receiver is looking at the calibration source all the intensity it receives is (black-body) thermal emission, but when it is switched to the sky, intensity is received both in the main telescope beam and in all the sidelobes. We wish to analyze and interpret only the intensity from the main beam. The ratio of the antenna response in the main beam to the total response is called the telescope beam efficiency η_B. This quantity is a function specific to each radio telescope, receiver, and frequency: typical values are $\eta_B \approx 0.70$. By accounting for this correction we can express the *brightness temperature* of an H II region as

$$T_B = \frac{T_A}{\eta_B} \times \begin{cases} 1 & \text{if } \Omega_{\text{source}} > \Omega_{\text{beam}} \\ \Omega_{\text{source}}/\Omega_{\text{beam}} & \text{otherwise} \end{cases} \qquad (2.17)$$

where Ω_{source} and Ω_{beam} are the solid angles subtended by the source and the main telescope beam respectively and T_A is the observed antenna temperature.

Single dish observations of the brightness temperature of optically thick H II regions – which should be a valid measure of the nebular kinetic temperature – show a large scatter with a tendency for values to be somewhat lower than estimates obtained by other techniques. The fact that the temperatures are low is almost certainly a result of the inhomogeneous density structure of real H II regions. Some of the nebular gas, particularly that on the periphery of the nebula, is still optically thin even at very low frequencies.

Since brightness temperature is a quantity which refers to each line of sight through a nebula, and since H II regions are such structured objects, it is preferable to measure T_B with high-enough angular resolution to discriminate, for example, between optically thick and optically thin parts of the nebula. Sufficiently high angular resolution can be obtained with a radio synthesis array. But to an array the fundamental observable is not temperature but rather the flux density or, when the synthesized beam is much smaller than the nebula (which we now intend to be the case), the observable is flux density per synthesized beam S_0. Re-expressing (2.16) in the case where the beam is much smaller than the source, the brightness temperature is related to the observable as

$$T_B(\text{K}) = 1.222 \times 10^6 (v/\text{GHz})^{-2} (\theta_B/\text{arc sec})^{-2} S_0(\text{Jy/beam}) \qquad (2.18)$$

where $\theta_B = \sqrt{(\theta_\alpha \theta_\delta)}$ is the synthesized beam if θ_α and θ_δ are the half-power diameters of a Gaussian beam in right ascension and declination respectively. Synthesis observations of the optically thick parts of H II regions indicate $T_B = T_e \approx 8000\text{--}12\,000$ K, in very good agreement with determinations by other techniques.

If we now consider optically thin nebulae, $\tau_v \ll 1$, the brightness temperature $T_B \approx T_e \tau_v$ or, using (2.14),

$$T_B(\text{K}) = 0.2120 Z_i^2 v^{-2.1} T_e^{-0.35} (n_i n_e l) \qquad (2.19)$$

where c.g.s. units are used throughout. Since this expression depends so weakly on the temperature – the entire range of temperature common to H II regions, $\approx 7000\text{--}12\,000$ K, changes T_B in (2.19) by only ≈ 20 percent – T_B in the optically thin regime can be thought of as a measure of the product $(Z_i^2 n_i n_e l)$ which, for a pure hydrogen nebula is $(n_e^2 l)$ the *emission measure*. Thus, radio maps of the brightness temperature of optically thin H II regions, particularly those made with the high-resolution radio synthesis arrays, provide via (2.19) a reliable measure of the n_e^2 distribution in the nebula. Moreover, since the radio frequency emission, unlike optical radiation, is unaffected by dust either within the H II region or along the line of sight, the radio maps are an accurate and unobscured measure of the extent and distribution of ionized material within the nebula.

As an illustration we present in Fig. 2.1 a 5-GHz map of the Orion nebula as made with the Very Large Array synthesis radio telescope. Here the contours are given in brightness temperature. Since T_B throughout this map is much less than the typical nebular kinetic temperature, $T_e \approx 8500$ K, it is apparent that nearly all of the Orion Nebula is optically thin at 5 GHz. Thus, such an optically thin radio map represents the distribution of emission measure, $n_e^2 l$, in the nebula. Evidently, there exists a great deal of small-scale structure in the emission measure across and within Orion – that is, the density distribution is very inhomogeneous. This is a feature common to most radio maps of galactic H II regions and it reflects the fact that the density distribution within H II regions is quite structured on both large and small scales.

2.2.1.2 *Flux density*

The total flux density from an H II region, defined in equation (2.16), provides fundamental information on the global properties of a nebula. For example, the flux density from an H II region that is optically thin, $\tau_v \ll 1$, along all lines of sight at the frequency of observation is expressed as

$$S_v = \frac{6.51 \times 10^{-15}}{D^2 v^{0.1}} \int \frac{n_e n_i}{T^{0.35}} \, dV \quad (\text{Jy}) \tag{2.20}$$

where D (cm) is the distance to the object, v (Hz) is the observing frequency and the integral is over the total nebular volume (cm^3). If we further presume

Fig. 2.1. The distribution of 4.886 GHz radio brightness in the Orion Nebula. The contours illustrated here are of constant brightness temperature, viz., $T_B = 24, 36, 48, 60, 72, 121, 182, 242, 303, 364, 425, 485,$ and 546 K.

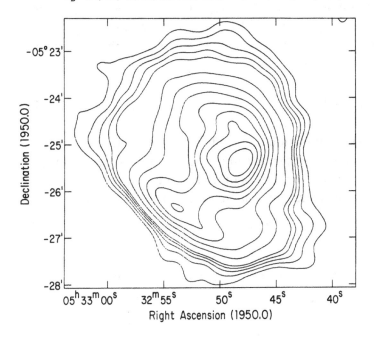

that the nebula is isothermal and homogeneous then an estimate of the mean-square electron density is available from the flux density

$$\langle N_e^2 \rangle = 1.54 \times 10^{14}\, S_\nu(\mathrm{Jy}) D^2 \nu^{0.1} T^{0.35}/V \qquad (2.21)$$

provided that we can estimate the distance to the object and its volume. The latter is usually got from the measured angular size and the distance together with an assumption as to the source geometry. Finally, of course, if we have such estimates for the mean density and the nebular volume we can determine the total mass of ionized gas contained in the H II region,

$$M = 6.94 \times 10^{-74}\, D^3 \theta_s^3 \langle N_e(cm^{-3}) \rangle M_\odot. \qquad (2.22)$$

Here $D(\mathrm{cm})$ is the distance to the nebula and θ_s is the mean angular diameter in arc seconds.

As an example consider the Orion Nebula. Orion is about 500 pc (1.54×10^{21} cm) from the Sun, its flux density at 5 GHz is ≈ 400 Jy and its angular diameter is 2.8. If we assume that the nebula is spherical then its volume is 1.04×10^{54} cm^3, the mean electron density $\langle N_e^2 \rangle^{1/2} \approx 5600$ cm^{-3} (using $T_e = 8500$ K) and the total mass of ionized gas is ≈ 7 solar masses. These are typical numbers for galactic H II regions.

Perhaps most importantly, we can use the total nebular flux density to provide a reasonable estimate of the spectral type of the star that ionizes and excites the nebula – even when that star is totally obscured. Here we note that for an ionization-bounded nebula, irrespective of the nebular geometry or density distribution, the number of stellar Lyman continuum photons produced per second L_c will equal the number absorbed via photoionization within the nebula,

$$L_c = \int n_e n_i (\beta - \beta_1)\, dV \qquad (2.23)$$

where $(\beta - \beta_1)$ is the recombination rate coefficient to all electronic levels less those recombinations that occur directly to the ground state (because a Lyman continuum recombination photon is produced in these latter recombinations). For temperatures characteristic of H II regions

$$\beta - \beta_1 \approx 4.10 \times 10^{-10}\, T^{-0.8}\ \mathrm{cm^3\ s^{-1}}. \qquad (2.24)$$

Combining equations (2.20), (2.23), (2.24) and expanding to first order we can express L_c directly in terms of the observed nebular flux density,

$$L_c = 4.76 \times 10^{48}\, \nu^{0.1} D^2 S T^{-0.45}\ \mathrm{s^{-1}} \qquad (2.25)$$

where in this equation S, D and v are expressed in units of Janskys, kpc and GHz respectively.

Having estimated L_c, the number of stellar Lyman continuum photons/s needed to keep the H II region ionized, we can refer to models of early-type stars to determine which star, or stars, are required to provide that ionization. In Table 2.1 we list the Lyman continuum flux emitted by various main sequence stars of early type together with an illustration of the number of such stars needed to excite a typical 8500 K galactic H II region that is 10 kpc from the Sun and appears as an (optically thin) 1 Jy source at 5 GHz.

2.2.2 *Bremsstrahlung emission from inhomogeneous nebulae*

In much of the previous discussion we considered homogeneous (constant density) nebulae that were either optically thin or optically thick along all lines of sight. This is clearly an idealization. For the more realistic case of an inhomogeneous nebula many of the above equations must be rewritten to take explicit account of the density distribution within each nebula that is to be considered. Although the results of these calculations will be specific to each nebula, we can nevertheless anticipate the general trends.

The spectral shape – the dependence of nebular flux density on frequency – from an H II region that is partially optically thick and partially optically thin will be $S \propto v^\alpha$ where α will be somewhere between the optically thin ($\alpha = -0.1$) and optically thick ($\alpha = +2.0$) limits. For a density distribution that can be described as a power law which decreases with displacement R from a central condensation $N(R) \propto R^{-\gamma}$ we expect that as the density distribution steepens (γ increases) the spectral index α will more nearly approach the optically thick value $\alpha = +2$; conversely, as γ decreases α will tend toward -0.1. In practice we often invert this conclusion and, for a given observed spectral shape α, we infer the nebular radial density distribution.

Observationally, a most important example of nonuniform gas distribution concerns mass loss from a hot, early-type star. Such mass loss

Table 2.1. *Stellar Lyman continuum luminosity*

Spectral type	O5	O6	O7	O8	O9
L_c (photons s^{-1})	4.2×10^{49}	1.2×10^{49}	4.2×10^{48}	2.2×10^{48}	1.2×10^{48}
Stars[a]	1	1	3	5	8

[a] The number of such stars required to ionize and excite an H II region that is 10 kpc from the Sun and appears to us as a 1-Jy radio source at 5 GHz.

driven by a spherical wind can be expressed as $\dot{M} = 4\pi R^2 n\, \mu m_{\mathrm{H}} v$ where v is the wind velocity, m_{H} the mass of the hydrogen atom and μ the mean molecular weight. For constant mass loss rate the radial density distribution is thus $n \propto R^{-2}$ where n, of course, depends on both $\dot{M}$ and v. If we now compute the radio flux density from this wind we find $S \propto v^{0.60}$ as the characteristic spectral shape. The magnitude of the flux density depends on $\dot{M}, v$ and the distance to the star, but the spectral shape is invariant. Observations of many mass loss stars verify this expected spectral dependence (and hence verify that the winds are spherical with approximately constant $\dot{M}$ and v); the magnitude of the radio flux density is subsequently used to estimate the stellar mass loss rate.

2.3   Radio recombination line emission

Every point in an H II region is in approximate ionization balance, meaning that the number of ionizations that occur in a particular volume per unit time is exactly balanced by the number of electron–ion recombinations in that volume in the same time. Although most such recombinations are to the lowest electronic levels, a small fraction, $\approx 10^{-5}$ of the total in hydrogen, are to levels with principal quantum numbers greater than 40. The resulting highly excited atoms decay to the ground state via downward cascade through the electronic levels, each transition of which is accompanied by emission of a photon corresponding to the energy difference of the two levels involved.

The mean radius of a $n > 40$ electronic orbital is so large that the central charge appears to the excited electron as a point. Thus the frequency spectrum of the downward cascade $(n + \Delta n) \rightarrow n$ is hydrogenic,

$$v = Z^2 Rc \left[\frac{1}{n^2} - \frac{1}{(n + \Delta n)^2} \right] \tag{2.26}$$

with the Rydberg constant $R = 109\,737.31(1 + m/M)\,\mathrm{cm}^{-1}$ and where m and M are the electron and nuclear mass respectively. As is evident from (2.26) recombination lines occur throughout the electromagnetic spectrum from radio frequencies to the uv. The lines with $n > 40$ and $\Delta n \ll n$ occur at radio frequencies, and are the focus of this section. The conventional notation is the following: transitions $(n + 1) \rightarrow n$ in hydrogen are referred to as H$n\alpha$ lines; a similar transition in helium is He$n\alpha$; the transition $(n + 2) \rightarrow n$ in hydrogen is H$n\beta$ and so forth. An illustrative list of hydrogen radio lines is given in Table 2.2.

2.3.1 *Recombination line emission and absorption coefficients*

The recombination line emission coefficient $j_L(v)$ analogous, for example, to that for bremsstrahlung (2.9), expresses the energy emitted in the line per unit volume, per unit time, per unit solid angle, per unit frequency. We write $j_L(v)$ for the transition $m \rightarrow n$ as

$$j_L(v) = (A_{mn}/4\pi)N_m h v \phi_v \qquad (2.27)$$

where A_{mn} (s^{-1}) is the Einstein coefficient of spontaneous emission, N_m (cm^{-3}) is the number density of atoms in state m, hv is the energy of the line photon and ϕ_v (Hz^{-1}) is the normalized line profile factor, $\int \phi_v \, dv = 1$, which characterizes the frequency distribution (the shape) of the line.

The absorption coefficient $\kappa_L(v)$ expresses the number of photons removed from the line per unit path length. Here $\kappa_L(v)$ results from interaction of the radiation field with excited nebular atoms: the radiation may either stimulate absorption ($n \rightarrow m$) or it may stimulate emission ($m \rightarrow n$), and because photons produced in the latter case are coherent with the incident radiation this process can be thought of as negative absorption,

$$\kappa_L(v) = (N_n B_{nm} - N_m B_{mn})(hv/4\pi)\phi_v. \qquad (2.28)$$

We can now use the Boltzmann and Saha equations to express the relative atomic populations N_n and N_m in terms of the nebular electron temperature T_e,

$$\frac{N_m}{N_n} = \frac{b_m g_m}{b_n g_n} e^{-(hv/kT_e)} \qquad (2.29)$$

Table 2.2. *Hydrogen radio recombination lines*

Transition	Frequency (GHz)	Separation between adjacent lines (MHz)
H50α	51.071	2918
H75α	15.281	591
H100α	6.478	190
H125α	3.327	78.3
H150α	1.929	38.0
H175α	1.216	20.6
H200α	0.816	12.1
H225α	0.573	7.6
H250α	0.418	5.0

where b_m and b_n are measures of the fractional departure from strict thermodynamic equilibrium,

$$b_n = N_n/N_n^* \tag{2.30}$$

and we use N_n^* to denote the number density of atoms in level n in thermodynamic equilibrium (i.e., as given by the Saha equation); $g_m = 2m^2$ and $g_n = 2n^2$ are the statistical weights of the two levels involved. With these definitions the line absorption coefficient (2.28) becomes

$$\kappa_L(v) = N_n B_{nm} \left(\frac{hv}{4\pi}\right) \phi_v \left[1 - \frac{b_m}{b_n} \exp(-hv/kT_e)\right] \tag{2.31}$$

If we now use the approximation

$$b_m/b_n \approx 1 + d[\ln b_n]/dn \tag{2.32}$$

which is valid so long as $m - n \ll n$ and

$$\exp(-hv/kT_e) \approx 1 - hv/kT \tag{2.33}$$

since $hv/kT \ll 1$, we can express $\kappa_L(v)$ as

$$\kappa_L(v) = \kappa_L^* b_n \beta_n \tag{2.34}$$

with

$$\beta_n = 1 - kT_e \, d(\ln b_n)/dE_m \tag{2.35}$$

and

$$\kappa_L^* = \left(\frac{hv}{kT}\right)\left(\frac{hv}{4\pi}\right) \phi_v N_n^* B_{nm}. \tag{2.36}$$

Written in this way (2.34) the recombination line absorption coefficient is a product of κ_L^*, the line absorption coefficient for a gas in local thermodynamic equilibrium (LTE), b_n the departure of the populations of the atomic levels from LTE and β_n, a term proportional to the first derivative of the b_n with n function which reflects the change in level population with n.

At this point it is convenient to write the Einstein coefficient in terms of the classical oscillator strength f_{mn}

$$B_{nm} = \frac{e^2}{m_e hv} f_{mn} \tag{2.37}$$

and use an approximation for f_{mn}

$$f_{mn} = f(\Delta n)/n = M(\Delta n)(1 + 1.5 \, \Delta n/n) \tag{2.38}$$

which is applicable for $n \gg 1$ and $\Delta n = m - n \ll n$. Here $M(\Delta n) = 0.190\,77$, 0.026 332, 0.008 105 6, and 0.003 491 7 for $\Delta n = 1$, 2, 3, and 4 respectively.

Thus

$$\kappa_L^* = 3.467 \times 10^{-12} N_e N_i T_e^{-5/2} \, \Delta n \left(\frac{f(\Delta n)}{n} \right) \exp \left(\frac{1.579 \times 10^5}{n^2 T_e} \right) \phi_v \text{, cm}^{-1}$$

$$(2.39)$$

where for a Gaussian line the shape function is

$$\phi_v = \left(\frac{4 \ln 2}{\pi} \right)^{1/2} \frac{1}{\Delta v} \exp \left[-4 \ln 2 \left(\frac{v_0 - v}{\Delta v} \right)^2 \right]$$

$$(2.40)$$

and Δv is the full width at half intensity of the line

$$\Delta v = \left(\frac{8 k T_D}{m c^2} \ln 2 \right)^{1/2} v_0$$

$$(2.41)$$

with T_D being the 'Doppler' temperature that characterizes the line. For Stark broadened lines the profile (2.40) must be replaced by a Voigt function.

Finally, the emission coefficient j_L can be written in terms of κ_L by means of Kirchhoff's law:

$$j_L = \kappa_L^* b_n B_v(T_e).$$

$$(2.42)$$

2.3.2 Transfer of recombination line radiation

The specific intensity, I_v, observed along a given line of sight through an H II region is the sum of two terms: (1) the nebular bremsstrahlung continuum intensity I_c, and (2) the intensity of the recombination line radiation I_L; all these quantities are, of course, frequency dependent (the line intensity strongly so). Thus for a homogeneous isothermal H II region,

$$I = \frac{(j_c + j_L)}{(\kappa_c + \kappa_L)} [1 - \exp(-\tau_c - \tau_L)]$$

$$(2.43)$$

where j_c and κ_c are as given by (2.13) and (2.14) respectively, so that

$$I = \left(\frac{\kappa_c + \kappa_L^* b_n}{\kappa_c + \kappa_L^* b_n \beta_n} \right) B_v(T) [1 - \exp(-\tau_c - \tau_L)].$$

$$(2.44)$$

For inhomogeneous nebulae these relations must be written as integrals along the path through the nebula. Note here that at frequencies between recombination lines, i.e., where $\kappa_L = 0$ and hence $\tau_L = 0$, (2.44) reduces to (2.15) as it must.

2.3.3 Approximations to the recombination line intensity

Since equation (2.44) cannot be simply inverted to solve for the nebular parameters N_e and T_e it is often convenient to introduce several approximations:

Approximation (1): The lines and continuum are formed in a plane-parallel, isothermal, homogeneous medium. With this approximation all lines of sight through the nebula are equivalent. We can write (2.44) in terms of a common observable, the ratio of the intensity in the recombination line to the intensity of the adjacent continuum, as

$$\frac{T_L(v)}{T_c} = \left(\frac{\kappa_c + \kappa_L^* b_n}{\kappa_c + \kappa_L^* b_n \beta_n}\right) \frac{[1 - \exp(-\tau_L - \tau_c)]}{[1 - \exp(-\tau_c)]} - 1 \tag{2.45}$$

where we express $I = I_c + I_L$ by means of the Rayleigh–Jeans form of the Planck function as $T = T_c + T_L$ where T_L and T_c are the brightness temperature of the line and (bremsstrahlung) continuum emission respectively.

Approximation (2): All the optical depths, line and continuum, are very small compared with unity. With this approximation we can expand the exponentials in a form first suggested by Goldberg,

$$\frac{T_L}{T_c} = \frac{b_n \kappa_L^*}{\kappa_c} (1 - \beta_n \tau_c / 2). \tag{2.46}$$

For H$n\alpha$ lines from a nebular gas containing one singly ionized helium atom for every ten H$^+$ ions this becomes

$$T_L/T_c = 2.82 \times 10^{-13} b_n v^{1.1} T_e^{-1.15} (v\phi_v)[1 - \beta_n \tau_c / 2]. \tag{2.47}$$

Approximation (3): Recombination lines are formed and transferred under conditions of LTE; in this case $b_n = 1$ *and* $\beta_n = 1$. With this approximation we assume that all atomic states in the nebula are precisely in thermodynamic equilibrium, and since $\tau_c \ll 1$, as in the previous approximation,

$$T_L/T_c = 2.82 \times 10^{-13} v^{1.1} T_e^{-1.15} (v\phi_v). \tag{2.48}$$

The line profile function $(v\phi_v)$ is the sum of Voigt functions and its form is related to the density structure of the nebula through the density dependence of impact broadening. But, since impact broadening has a very strong dependence on principal quantum number (cf. Section 2.3.4), we can observe at a sufficiently high frequency so as to minimize this effect. Hence the following approximation:

Approximation (4): Impact broadening is negligible. With this approximation the line profile function is a gaussian (2.40) and (2.48) becomes

$$T_{\mathrm{L}}/T_{\mathrm{c}} = 6.35 \times 10^3 \, v^{1.1} T_{\mathrm{e}}^{-1.15} \, \Delta v \qquad (2.49)$$

or, solving for T_{e},

$$T_{\mathrm{e}}^* = \left(6.35 \times 10^3 \, v^{1.1} \, \frac{T_{\mathrm{c}}}{T_{\mathrm{L}} \, \Delta v} \right)^{0.87} \qquad (2.50)$$

where Δv is the full line width at half maximum expressed in km s^{-1} and v is the line frequency in GHz. We denote the temperature thus derived as T_{e}^* and although it is commonly called the 'LTE electron temperature' it requires many approximations besides that of local thermodynamic equilibrium.

Use of equation (2.50), together with the implicit additional assumption that $T_{\mathrm{e}} = T_{\mathrm{e}}^*$, has become common in radio astronomy because it allows one to quickly and simply obtain a nebular parameter, the electron temperature, in terms of two observables, the line to continuum ratio $T_{\mathrm{L}}/T_{\mathrm{c}}$ and the half width of the recombination line, Δv. The disadvantage of this technique is that T_{e}^* tends to underestimate T_{e} – radio determinations of T_{e}^* are typically in the range 5000–7000 K whereas optical determinations of T_{e} in the same nebulae tend to cluster closer to 10 000 K. Goldberg resolved this discrepancy by noting that subtle departures from LTE would be reflected as an enhancement of the line intensity T_{L} over the purely LTE value and such an enhancement in (2.50) would result in T_{e}^* being diminished relative to T_{e}.

2.3.4 *Departures from local thermodynamic equilibrium*

As we described in Section 2.1, a gas will be in thermodynamic equilibrium if any net change in the total energy (positive or negative) is shared via collisions among the constituent particles on a time scale short compared with the interval between such changes. For the high-n levels from which the radio recombination lines arise the principal energy change is the energy lost to line radiation, while the rate at which knowledge that a line photon has been radiated away is communicated to neighboring atoms on the collisional time scale $(N_{\mathrm{e}} \sigma v)^{-1}$ (2.2). Since the (classical) cross section of an excited hydrogen atom is proportional to n^4 the collisional time scale is short at large principal quantum number and hence departures from LTE will be smaller at larger n (i.e., b_n will approach unity). On the other hand, at small n, or smaller N_{e}, the collisional time scale is longer, radiative processes increasingly dominate and departures from LTE become large (b_n becomes less than unity).

Quantitatively, to determine the coefficients of departure from thermodynamic equilibrium, b_n, and their derivatives, β_n, it is necessary to determine the population of each atomic level. The method is to equate the number of transitions out of level n per unit volume per unit time with the number of transitions into level n from all other states (including the continuum). Goldberg and his colleagues were among the first to carry out such calculations.

Representative b_n and β_n functions are shown in Fig. 2.2.

Fig. 2.2. The departure from thermodynamic equilibrium (upper figure) and the derivative of the logarithm of this function (lower figure; cf. (2.35)) as a function of principal quantum number. Curves are drawn, and labeled, for nebular electron densities of $N_e = 10^2$, 10^3, and $10^4\,\mathrm{cm}^{-3}$.

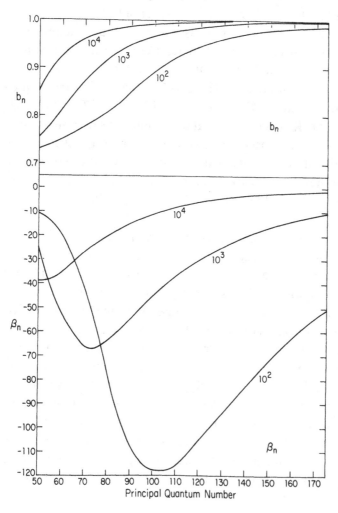

The b_n curve illustrates the features noted previously, viz. the absolute departure of a single level from LTE is not particularly large (less than 20 percent) and this departure of b_n from unity decreases with increasing principal quantum number and increasing nebular density. The quantity β_n which reflects the derivative of the b_n function with n is more surprising because it is negative throughout the radio-frequency regime and can be quite large. As Goldberg noted, this means that the non-LTE recombination line optical depth (2.34) is *negative* and hence the radio lines are weak masers with an intensity that is enhanced over that of the pure LTE case (cf. (2.43) or (2.44)).

The non-LTE enhancement of radio recombination lines has one very important consequence which can be seen as follows. Assume that we observe a plane-parallel, isothermal, homogeneous H II region at a frequency so high that impact broadening is negligible and the line profile factor is Gaussian. In this case (2.47) can be easily modified to describe the correct line to continuum ratio T_L/T_c – or, stated another way, the value of T_e derived from (2.47) and the observables T_L/T_c and Δv in this case would indeed be the correct value of the nebular kinetic temperature. If, however, the data were analyzed with the 'LTE approximation' (2.50) then the value of T_e^* so derived would be related to the correct T_e (2.47) by

$$T_e^* \approx T_e/[b_n(1 - \beta_n \tau_c/2)]^{0.87}. \tag{2.51}$$

As β_n is always negative, the quantity $(1 - \beta_n \tau_c/2) > 1$ and the relationship of T_e^* to the correct T_e depends on the continuum optical depth of the nebula at the frequency of observation. This means that if we observe the radio recombination line emission from two H II regions that have identical temperatures and densities and we use the convenient LTE approximation (2.50) we would inappropriately conclude that the temperature, T_e^*, of the nebula with the greater extent (and hence greater continuum optical depth) is less than that of the other. And, because T_e^* is a function of the continuum optical depth, if we observe an isothermal but inhomogeneous nebula, or a homogeneous, isothermal nebula that is not plane-parallel (i.e., a nebula where various lines of sight through the nebula differ in extent) all variations in τ_c will incorrectly appear as temperature variations if T_e^* is used to estimate nebular temperature.

The 'LTE temperature' T_e^* is a very convenient and useful estimator, but is should be interpreted guardedly.

2.3.5 *Impact broadening*
The large classical diameter of the highly excited atoms that emit

radio recombination lines, $2r \approx 10^{-4} (n/100)^2$ cm, together with the relative proximity of such an excited atom to free charges in its vicinity – the separation between charges being only $\approx 5 \times 10^{-2} (N_e/10^4 \, \mathrm{cm}^{-3})$ cm – implies that the energy levels of the excited atom will be perturbed by the electric field of the nearest-neighbor charge. Most important in this respect are the free nebular electrons which appear to move rapidly with respect to the excited atoms. In the interval $\Delta t \approx 1/2\pi \, \Delta v$ during which an atom makes the transition $(n+1) \to n$ the motion of the nearest free electron will cause the bound radiating electron to see a time-dependent perturbing electric field, the consequence of which is that the energy levels of the excited atom are slightly smeared and less discrete. The radio recombination line emitted between two such smeared levels is thereby broadened and we speak of the process as impact (or collisional) broadening – the 'collision' being the perturbation of the energy levels of an excited atom resulting from the electric field of a free electron.

Impact broadening will become increasingly important as the density of free electrons N_e increases and as the size of the excited atom increases, that is, as n increases. Proper quantum mechanical calculations confirm these expectations and show that the ratio of the impact broadened full width at half maximum Δv^{I} of a radio H$n\alpha$ line to the Doppler broadened Gaussian portion Δv^{D} of the same line in a typical H II region is,

$$\Delta v^{\mathrm{I}}/\Delta v^{\mathrm{D}} \approx 0.14 (n/100)^{7.4} (N_e/10^4 \, \mathrm{cm}^{-3}) \tag{2.52}$$

which increases both with N_e and very rapidly with n. The extreme dependence on n is not observed – in most nebulae the width of H150α lines is nearly the same as the width of H100α lines and certainly not ≈ 3 times larger as expected from (2.52).

The disagreement between observation and expectation in this case is simply attributable to the fact that H II regions are not homogeneous (constant density) objects. Observations of lines at low frequencies (high n) where we expect from (2.52) to observe the effect of impact broadening in dense nebular gas reveal instead the unbroadened line emission from more tenuous material. The very broadened lines, that being broadened are commensurably diminished, become indistinguishable from the instrumental spectral baseline. Furthermore, non-LTE enhancement of the recombination line emission will be particularly important in those nebulae in which we observe dense nebular gas through a foreground of more tenuous nebular material. Of course, if the geometry is reversed and the dense gas is in the foreground no such line enhancement is possible.

Observationally, the principle effect of impact broadening is to preferentially weaken and render undetectable low-frequency recombination lines from dense gas. This means that when one observes at low frequencies (high n) one selectively samples low-density nebular material and the nebular parameters derived from such observations (e.g., T_e) are applicable to that material. Conversely, at high frequencies the line observations are dominated by emission from the most dense gas. Thus, radio recombination line observations made over a wide range of principal quantum number can be used to probe the entire nebular structure, but for their interpretation in terms of physical nebular parameters one must exercise caution in choosing an appropriate model for the nebular density distribution and the large-scale nebular geometry.

2.3.6 *The relative abundance of helium and hydrogen*

One straightforward observational parameter that does not require extensive source modeling for its interpretation is the abundance of helium relative to hydrogen in an H II region. Since the atomic orbitals from which radio recombination lines arise are so extensive the highly excited electron in all atomic species can be treated as hydrogenic. Thus, the physics which describes the emission and transfer of hydrogen radio recombination lines is equally applicable to all atomic species; the observed intensity of the lines from any species X will be related to that of the corresponding hydrogen lines simply by the ratio of the elemental abundances, X/H, weighted by the fractional nebular volume in which the ion X^+ is found. And the recombination lines from X will be displaced in frequency from the hydrogen lines by the reduced mass in the Rydberg equation.

Helium has a cosmic abundance which is approximately one-tenth that of hydrogen by number. We expect, therefore, that helium recombination lines will be detectable in H II regions provided that the exciting star is sufficiently hot to provide an adequate flux of photons with energies greater than the helium ionization threshold (i.e., $h\nu > 24.6\,\text{eV}$). In practice, this latter condition is fulfilled for stars of spectral types earlier than 07. For these stars it is straightforward to show that the helium is once ionized throughout the entire region in which hydrogen is ionized (cf. Chapter 1). In this case the nebular abundance ratio He/H is simply given by the ratio of power in the helium recombination line relative to the power in the corresponding hydrogen recombination line. The utility of such a measurement is that it requires no corrections for extinction or differential excitation and, of course, it is possible to make this determination in nebulae throughout the galaxy.

2.3.7   *Radio recombination lines from atomic carbon*

The first observations of hydrogen and helium radio recombination lines made with high-frequency resolution found an apparent inconsistency in the sense that whereas the hydrogen line profile was exceedingly well described by a Gaussian function the helium line was not. Rather, the helium line profile showed an asymmetry to high frequencies that suggested that the helium line was blended with another line. This so-called 'anomalous' recombination line had a frequency separation relative to the corresponding hydrogen line that implied from (2.26) that it be identified as a recombination line from atomic carbon. The problem with such an identification was that the power in the putative carbon line relative to that in the hydrogen line was a few percent whereas the relative cosmic abundance of carbon is only $[C/H] \approx 10^{-4}$. This inconsistency together with the following observational properties of the carbon recombination line led Dupree and Goldberg to a refined interpretation for the line:

1. The ratio of the carbon recombination line intensity to the hydrogen recombination line intensity varies with n.
2. The radial velocity of the carbon lines $v(C)$ is frequently different from the velocity of the hydrogen or helium lines but $v(C)$ agrees very well with the velocity of molecular line emission seen in the same directions.
3. The width of the carbon recombination lines, 3–10 km s^{-1}, is similar to that of the molecular lines but very much narrower than the width of the nebular hydrogen and helium lines.

The interpretation of the carbon recombination line rests on the recognition that H II regions are frequently found in close proximity to the molecular cloud from which they formed. Radiation longward of the Lyman limit, $\lambda > 912$ Å, radiated by the star which excites the H II region, freely escapes the nebula and ionizes all elements on the face of the adjacent molecular cloud that have ionization potentials less than 13.6 eV. Since carbon is the most abundant of these elements we expect there to be a thin layer of C^+ between the H II region and the molecular cloud proper; radio recombination lines from carbon are emitted at that interface. The thermodynamics of the C^+ region are consequently quite distinct from those of the nebular H and He lines. In particular, the temperature of the C^+ region is much cooler than the H II region, $T(C^+) < 100$ K: the line width is correspondingly smaller, the velocity is that characteristic of the molecular gas and the line intensities reflect conditions in the interface region not those of the nebular gas.

The observed intensity of the carbon radio recombination line depends on the following conditions: the spatial separation between the exciting star and the neutral material; the extinction within the H II region; the density of neutral gas; and the depletion of carbon in the neutral gas. But, in addition, the carbon lines will be stronger if the neutral gas is found along the line of sight to the H II region rather than behind the H II region because in the former case the carbon lines can be enhanced via non-LTE stimulation of the background nebular continuum radiation. Owing to this combination of effects the interpretation of the carbon recombination line emission from each H II region is distinct and model-dependent.

By way of illustration, a typical radio recombination line spectrum from the galactic H II region W3 is shown in Fig. 2.3. Here one can see both the nebular hydrogen and helium lines as well as the carbon line which arises, in this case, from the interface between W3 and its background molecular cloud. The frequency displacement of the helium and carbon lines resulting from their mass difference from hydrogen (and each other) is also evident. The ratio of helium to hydrogen derived from radio spectra such as this is [He/H] ≈ 0.09 for H II regions throughout the galaxy.

2.3.8   *Stimulated radio recombination lines from distant galaxies*

Perhaps the most exciting recent application of radio recombination lines is related to their detection from very distant galaxies. At first glance such a detection appears highly unlikely because the observed flux density in

Fig. 2.3. The spectrum of hydrogen, helium and carbon radio recombination lines from the gaseous nebula W3. Shown here are the 114β recombination lines that occur at 8.6 GHz.

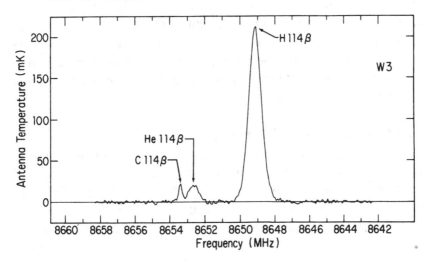

the lines $S_L = \int I \, d\Omega$ (where I is given by equation (2.43)) is diminished as the inverse square of the distance owing to the solid angle term. Given the sensitivity limits of current instrumentation we expect to be able to detect such lines only from the very nearest galaxies. But equation (2.43) accounts only for the spontaneous emission from an H II region. If now we consider the situation where an H II region is found along the line of sight to a distant objects, in which clearly star formation has recently occurred in the nuclei, will be

$$S_0 = S_i \exp(-\tau_c - \tau_L) \tag{2.53}$$

where S_i is the source flux density that would be observed in the absence of the foreground H II region. The effect of the H II region is thus to attenuate the background source flux density by the sum of the H II region's free–free optical depth and the optical depth in the recombination line. But since τ_L is negative the background source flux density will be less attenuated at the frequencies corresponding to the recombination lines in the H II region than at adjacent frequencies. Quantitatively, at the recombination line frequencies the excess (emission) flux density S_L observed from the background source will be

$$S_L = [S_i \exp(-\tau_c)] \exp(|\tau_L|). \tag{2.54}$$

The observed flux density of a radio recombination line from a H II region located along the line of sight to a distant background radio source is related only to the strength of that background source and the optical depth in the line – it is independent of distance to the H II region or to the background source. Such stimulated radio recombination lines have now been detected from regions of ionized gas in galaxies as distant as 300 Mpc; they allow us to compare the density, temperature and extent of these distant regions with those of local H II regions in our Galaxy. As radio instrumentation improves in sensitivity and spectral bandwidth we can expect radio recombination lines to provide an important probe of the physical conditions in many very distant objects.

2.4   Radio observations of H II regions in perspective

At radio frequencies an H II region radiates both recombination line radiation and bremsstrahlung continuum radiation; the total luminosity of the latter being hundreds of times larger than that of the former. This is in marked contrast to the situation at optical wavelengths where the line radiation overwhelms the continuum. Thus at radio frequencies we can expect to obtain information on the spatial distribution of nebular gas independent of any questions regarding chemical abundances or excitation – wherever the gas is ionized we expect, and we measure, bremsstrahlung radiation. Moreover, by comparing the radio continuum and radio recombination line maps we may obtain not only a reliable measure of the distribution of nebular gas but also a reasonable estimate of the electron temperature and the relative abundance of helium; none of these quantities need to be corrected for internal or line of sight dust extinction.

The observations become most useful when radio, infrared and optical data are considered together. Since the optical line observations are so sensitive to excitation conditions (especially to the local electron density) and to extinction by dust one can use the radio maps to minimize one of these uncertainties and then employ the optical lines as probes of the others, e.g., for the nebular ionization structure or the effect of internal nebular dust. Similarly, for those H II regions in which long-wavelength infrared observations are sufficiently complete between ≈ 2–$300 \, \mu m$ one can use the total ir luminosity derived from these data as a measure of the total luminosity of the star (or stars) that excite the nebula. This luminosity together with the Lyman continuum luminosity obtained from the radio flux density via (2.25) uniquely establishes the temperature and hence the nature of the source of nebular excitation.

Thus the utility of radio observations of H II regions is not so much related to what information one can gain from them *per se*, but rather by considering the radio data as one voice in concert with optical and infrared measurements and with adequate theoretical models of nebular ionization and excitation, one can hope to obtain a more consistent and harmonious understanding of the astrophysics of gaseous nebulae.

References

Thermal equilibrium in an H II region is discussed in

Spitzer, L. (1978). *Physical Processes in the Interstellar Medium*. John Wiley & Sons.

The Lyman continuum luminosity of stars of various spectral types is found in

Panagia, N. (1973). *Astron. J.* **78**, 929.

Avedisova, V. S. (1979). *Astron. Zh.* **19**, 965.

The spectral shape of the continuum flux density from an inhomogeneous H II region is given by

Abbott, D. C., Bieging, J. H. & Churchwell, E. (1981). *Astrophys. J.* **250**, 645.

A complete tabulation of radio recombination line frequencies is

Lilley, A. E. & Palmer, P. (1968). *Astrophys. J. Suppl.* **16**, 143.

The formation and transfer of radio line and continuum radiation in an H II region is detailed in

Goldberg, L. (1968). In *Interstellar Ionized Hydrogen*, ed. Y. Terzian. Benjamin & Co.

Brown, R. L., Lockman, F. J. & Knapp, G. R. (1978). *Ann. Rev. Astron. Astrophys.* **16**, 445.

Populations of the electronic levels in an H II region are tabulated in

Salem, M. & Brocklehurst, M. (1979). *Astrophys. J. Suppl.* **39**, 633.

Observations and interpretation of nebular carbon recombination line radiation are discussed in

Dupree, A. K. & Goldberg, L. (1969). *Astrophys. J. Lett.* **158**, L49.

Pankonin, V., Walmsley, C. M., Wilson, T. L. & Thomasson, P. (1977). *Astron. Astrophys.* **57**, 341.

Extragalactic radio recombination line observations are summarized by

Bell, M. B. (1980). In *Radio Recombination Lines*, ed. P. A. Shaver. D. Reidel Publishing Company.

3

Quasars, Seyfert galaxies and active galactic nuclei

DONALD E. OSTERBROCK

3.1 Introduction

Quasistellar radio sources were first recognized as objects at very large distances in 1963, when Schmidt identified several nebular emission lines in the spectrum of the stellar appearing, thirteenth magnitude object 3C 273, and measured its redshift as $z = 0.158$. Greenstein and Matthews soon identified similar lines in 3C 48, with $V = 16.2$, giving $z = 0.367$. It was immediately clear that these highly luminous objects are beacons that can be observed out to the distant reaches of the universe. For many years OQ 172, with $z = 3.53$ and $V = 17.8$, was the most distant object known, until in 1982 an even larger redshift, $z = 3.78$, was measured for PKS 2000-330, a quasar with red magnitude 17.3. However, observations also quickly showed that all quasars and QSOs do not have the same absolute magnitude; like stars, they are spread over an enormous range in luminosity. Hence we can only hope that 'understanding' quasars, through the study of their spectra, will mean not only understanding the strongest energy sources we know in the universe, but also recognizing their absolute magnitudes from their spectra and thus determining their distances.

Long before the first quasars were discovered, galaxies with generally similar emission-line spectra were known. In 1908, Fath, working with a small slitless spectrograph on the Crossley reflector of Lick Observatory, recognized five nebular emission lines (in addition to weak $H\beta$) in the spectrum of NGC 1068, lines we now know as [O II] $\lambda 3727$, [Ne III] $\lambda 3869$, and [O III] $\lambda\lambda 4363, 4959, 5007$. Slipher obtained much better spectrograms of this same object in 1917, and in 1926 Hubble particularly remarked on the planetary-nebula-like emission spectra of NGC 1068, 4051 and 4151. Seventeen years later Seyfert published his important paper, in which he clearly stated that a very small proportion of galaxies, including these three objects, have nuclei with spectra showing many high-ionization emission lines. Invariably, he noted, these galactic nuclei are especially luminous, and the emission lines are wider than the absorption lines in normal galaxies. These properties – broad emission lines covering a wide range of ionization, arising in a bright nucleus – define the class of objects we now call Seyfert galaxies. They are the commonest type of active galactic nuclei.

After World War II, very rapid advances in radio astronomy led to the optical identification of the first radio sources. Among these was Cyg A, identified by Baade and Minkowski with a fifteenth-magnitude galaxy with $z = 0.057$. Its rich emission-line spectrum proved to be very similar to the spectra of Seyfert galaxies. Other identifications of similar objects quickly followed. Most radio galaxies have optical emission lines that arise in their nuclei, and in many of them these emission lines cover a wide range of ionization and are wider than the lines in the spectra of normal galaxies. Thus they are also examples of active galactic nuclei, considerably rarer in space, however, than Seyfert galaxies, which are mostly radio-quiet and are identified by their spectral properties. Table 3.1 contains very approximate space densities of these objects, 'here and now' in the universe, as well as we know them. These numbers of course depend on the ill-defined limits of absolute magnitude for the various groups of objects, and no doubt suffer from varying degrees of incompleteness, but nevertheless are useful for orientation purposes.

Quasars is the short name I reserve for quasistellar radio sources in this chapter, while QSOs, or quasistellar objects, are optically discovered, mostly radio-faint objects. Note that according to the table the relative numbers of quasars and radio galaxies are in the same proportions as the relative numbers of QSOs and Seyfert galaxies. This is only one of the many observational indications that the radio-loud and radio-quiet objects each form physically continuous groups, covering a very wide range of luminosity, the quasars and QSOs representing the rarest but most luminous forms of active galactic nuclei.

Table 3.1. *Approximate space densities here and now*

	Mpc^{-3}
Field galaxies	10^{-1}
Luminous spirals	10^{-2}
Seyfert galaxies	10^{-4}
Radio galaxies	10^{-6}
QSOs	10^{-7}
Quasars	10^{-9}

3.2 Observational aspects

The first Seyfert galaxies were classified, or discovered, on the basis of individual slit spectra obtained in the course of regular programs of spectroscopy of galaxies. Since only a few percent of normal spirals are Seyferts, the total number found by this method in the days of photographic spectroscopy was small indeed. More were added when spectra were obtained of galaxies with unusually bright nuclei – the 'compact galaxies' of Zwicky – many of which turned out to fit the spectroscopic definition of Seyfert galaxies as well. Objective-prism surveys with a Schmidt camera by Markarian and his collaborators turned up many additional Seyfert galaxies. In these surveys galaxies with strong ultraviolet continuous spectra were catalogued; of these, about 10 percent proved to be Seyfert galaxies when individual spectra were obtained. (The rest are mostly galaxies with strong populations of early-type stars, and consequently much ionized gas in the form of H II regions, even in their nuclei. The strongest examples of these objects, in which clearly star formation has recently occurred in the nuclei, are often called 'star-burst' galaxies. The spectrum of one galaxy of this type, NGC 7714, is shown in Fig. 3.1.) Most recently spectral surveys of individual galaxies, with modern, fast, detectors, primarily to measure radial velocities, have turned up more Seyferts. And other objective-prism Schmidt-camera surveys, some based on ultraviolet continua, some on the strongest emission-lines – especially $H\beta$ and [O III] $\lambda\lambda 4949, 5007$ – have turned up even more.

By visual inspection alone, the emission-line spectra of Seyfert galaxies can be divided into two types, a classification scheme first proposed by Khachikian and Weedman. Seyfert 1 galaxies have very broad H I, He I, and He II lines, with total widths ranging from 5×10^3 km/s up to 3×10^4 km/s, while the forbidden emission lines, like [O III] $\lambda\lambda 6300, 6364$, [Ne III] $\lambda\lambda 3869, 3967$, [N II] $\lambda\lambda 6548, 6583$, and [S II] $\lambda\lambda 6716, 6731$, though definitely broader than in normal galaxies, are narrower than the permitted lines, and typically have full widths at half maximum of order 5×10^2 km/s. An example of a Seyfert 1 galaxy is Mrk 1243, whose spectrum is shown in Fig. 3.2. Seyfert 2 galaxies, on the other hand, have both permitted and forbidden lines with essentially the same width, typically about the same as the widths of the forbidden lines in Seyfert 1 galaxies. An example is Akn 347, whose spectrum is shown in Fig. 3.3.

This classification into two main types, Seyfert 1 and 2, may be further subdivided. Some Seyfert galaxies have H I emission-line profiles that are clearly composite, consisting of a broad, Seyfert 1-like component on which a narrower, Seyfert 2-like component is superimposed. An example is Mrk 704, whose spectrum is shown in Fig. 3.4. There is actually a wide range of relative strengths of broad and narrow components of the H I emission lines, from the extreme of those with very strong broad components, which we call Seyfert 1 galaxies, through intermediate objects like Mrk 704, to those with very strong 'narrow' (but still definitely broader than in normal galaxies) components that we call Seyfert 2s. The Seyfert galaxies with clearly intermediate-type H I profiles are generally called Seyfert 1.5 galaxies, to indicate their permitted-line profiles are between Seyfert 1 and Seyfert 2 profiles.

In radio galaxies the radio emission typically comes from two large, widely separated lobes, on opposite ends of a diameter that passes through the central galaxy. In most cases no optical emission has been detected at the positions of the radio lobes, although in a very few objects detection of very

Fig. 3.1. Spectral scans of NGC 7714, a narrow-emission-line or 'star-burst' galaxy, in which the gas is photoionized by ultraviolet radiation from hot stars in the nucleus.

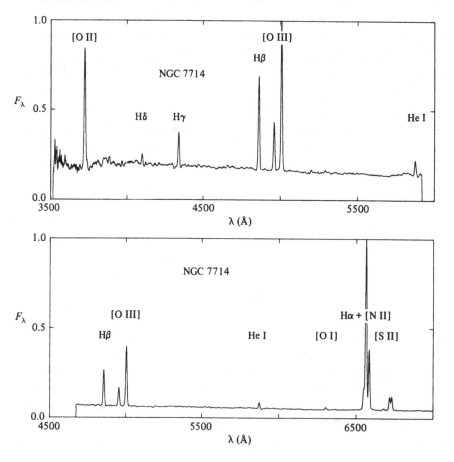

faint optical radiation, in the form of knots or condensations, has been claimed. Very frequently there is a weak, compact (flat radio-frequency spectrum) radio source at the position of the nucleus of the galaxy. All the optical radiation (emission lines plus featureless continuum, to be discussed later) comes from this active galactic nucleus.

The emission-line spectra of radio galaxies again can be classified into two types, more or less similar to the two types of Seyfert galaxies. One type, exemplified by 3C 390.3, whose spectrum is shown in Fig. 3.5, has very broad permitted emission lines, and narrower forbidden lines; this type, the radio-loud equivalent of a Seyfert 1 optical spectrum, is called a broad-line radio galaxy. The other type, with permitted and forbidden lines of similar width, analogous to a Seyfert 2 optical spectrum, is called a narrow-line radio galaxy. An example is Cyg A = 3C 405, shown in Fig. 3.6. Of course, in radio

Fig. 3.2. Spectral scans of Mrk 1243, a Seyfert 1 galaxy. This, like all the other spectral scans shown in this chapter, was taken with the image-dissector scanner on the 120-inch Shane telescope of Lick Observatory.

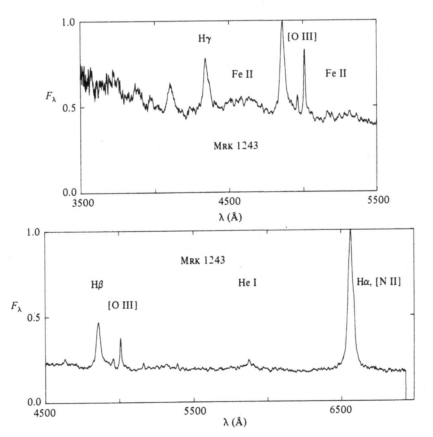

galaxies, as in Seyfert galaxies, the lines we call 'narrow' are narrow only in comparison with the extremely broad permitted H I, He I, and He II lines of Seyfert 1s and broad-line radio galaxies; even the 'narrow' lines in active galactic nuclei are noticeably broader than the emission lines of 'normal' galaxies, that arise in H II regions and complexes.

Although the optical emission-line spectra of radio and Seyfert galaxies are generally similar, there are some real differences between them. Typical Seyfert 1 galaxies have in their spectra broad, permitted emission lines, not only of H I, He I and He II, but also Fe II. The strongest Fe II lines are those of multiplets (35), (37), (38), (41), (42), (46), arising from radiative transitions from the excited energy terms of odd parity to lower even terms, as shown in

Fig. 3.3. Spectral scans of Akn 347, a Seyfert 2 galaxy. Note several of the strongest absorption features in the integrated stellar spectrum are marked *below* the spectrum.

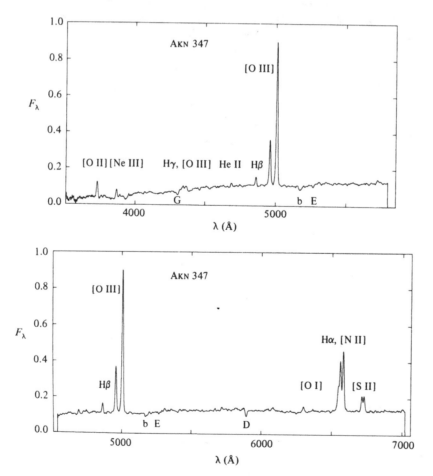

Fig. 3.7. Similar Fe II emission lines are not observed in planetary nebulae or H II regions, though they are observed in many Be stars, T Tauri stars and novae. Many of these lines are clustered in two large 'clumps' or 'bands', as can be seen in the spectrum of Mrk 1243 in Fig. 3.2. The relative strengths of these Fe II features are more or less the same in all Seyfert 1 and Seyfert 1.5 galaxies; their overall strength varies from object to object but they are present in observable strength in nearly all galaxies of these types. On the other hand, they are extremely weak, if present at all, in typical broad-line

Fig. 3.4. Spectral scans of Mrk 704, a Seyfert 1.5 galaxy. This, like all the other spectral scans shown in this chapter, is shown in relative energy-flux units per unit wavelength versus wavelength *in the rest system of the emitting object.*

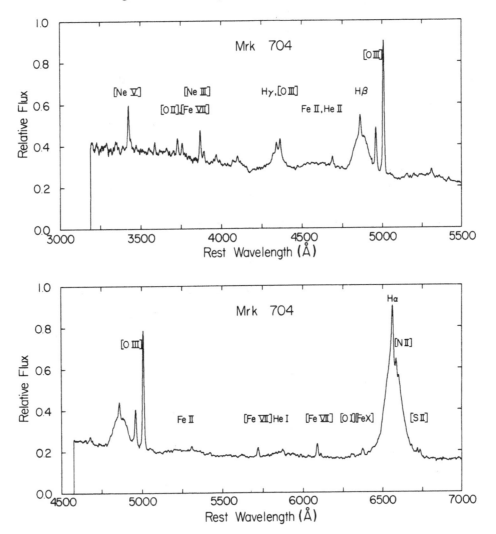

radio galaxies. For instance, they cannot be seen in the spectrum of 3C 390.3 in Fig. 3.5. This clearly represents a difference in the physical conditions between these two types of active galactic nuclei.

Another, more subtle difference, is that the ratio of fluxes in the broad Balmer emission lines Hα and Hβ (which we shall write simply Hα/Hβ) is, on the average, larger in the broad-line radio galaxies, typically ≈6±, than in

Fig. 3.5. Spectral scan of 3C 390.3, a broad-line radio galaxy. The spectral resolution, as in all the other scans shown in this chapter, is approximately 10 Å (full width at half maximum).

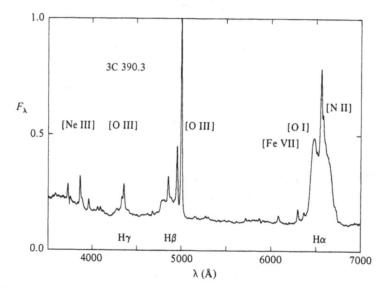

Fig. 3.6. Spectral scan of Cyg A = 3C 405, a narrow-line radio galaxy.

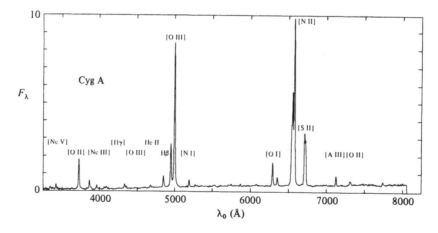

Seyfert 1 galaxies, in which typically $H\alpha/H\beta \approx 3.5\pm$. Finally, the broad-line radio galaxies typically have composite H I profiles – they are actually more like Seyfert 1.5 profiles then Seyfert 1s – and their broad-line profiles tend to be more flat topped, and to have more irregularities and structure, than the more nearly triangular, smoother broad-line profiles of Seyfert 1s and 1.5s. All of these are tendencies – specific objects can be found that contradict nearly every one of them – but they indicate that, on the average, radio-loud and radio-quiet broad emission line objects differ in their optical properties. The same is not true for the narrow-line objects; in spite of close study we know of no consistent differences between the optical emission-line spectra of narrow-line radio galaxies and Seyfert 2 galaxies.

Probably even more significant than the spectral differences between Seyfert and radio galaxies are the morphological differences. Nearly all Seyfert galaxies that are close enough to be resolved and classified by form are spiral galaxies. Many of them are more or less distorted, but still basically

Fig. 3.7. Schematic energy-level diagram of Fe II, with strongest observed multiplets in Seyfert 1 galaxies indicated.

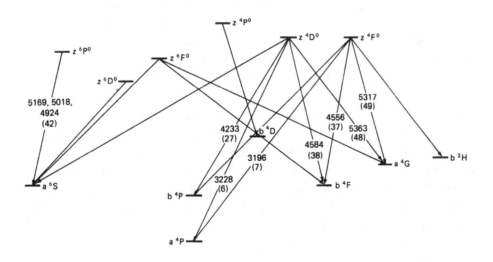

Fe II

spirals. Most of them are closer to Sb, than to Sa or Sc. Many of them are barred spirals, or have a companion galaxy. On the other hand, few if any of the radio galaxies are spirals. Most of the narrow-line radio galaxies that are close enough to be classified are cD, D or E galaxies. Nearly all of the broad-line radio galaxies are N galaxies, systems with brilliant star-like nuclei containing most of the luminosity of the system, but with faint, nebulous envelopes visible. They are thus nearly quasistellar, but not quite.

Spiral galaxies certainly contain more interstellar matter outside their nuclei than elliptical or 'giant elliptical' (cD and D) galaxies, and are more condensed to the principal plane defined by the overall angular-momentum vector of the galaxy, than the latter. We shall discuss physical models of active galactic nuclei later in this chapter, but for the present a good working hypothesis is that the difference between a Seyfert galaxy and a radio galaxy may be more in the environment – a flattened, rotating, interstellar-matter rich one, or a more nearly spherical, interstellar-matter free one – than in the energy source within the nucleus itself.

All Seyfert galaxies show a featureless continuum in their spectra, in addition to the emission lines and the integrated stellar absorption-line spectrum of a more or less normal galaxy. It comes from a tiny, unresolved object, the 'engine' or visible appearance of the central energy source, within the nucleus. In typical Seyfert 1 galaxies the featureless continuum is very strong, much stronger than the integrated stellar absorption-line spectrum, while in typical Seyfert 2 galaxies the featureless continuum is considerably fainter, and the galaxy spectrum can easily be seen. Thus its broad emission lines go with a strong featureless continuum, an observational fact that must be significant in undertaking the physics of active galactic nuclei. We shall return to this point at the end of the chapter. For the present discussion, however, it has the effect that the nuclei of Seyfert 1 galaxies are generally more luminous than the nuclei of Seyfert 2 galaxies, at least partly because of the extra contribution of the strong featureless continua of the Seyfert 1s. Furthermore, this additional contribution makes the nuclei so bright that a typical Seyfert 1 galaxy, *as a whole*, is more luminous than a typical Seyfert 2 galaxy. According to the most recent observational determination, the luminosity function of Seyfert 2 galaxies has a weak maximum near $M_B = -20$, while the luminosity function of Seyfert 1s has a similar maximum near $M_B = -21$.

Furthermore, in the past few years increasing observational evidence has accumulated that many QSOs, originally defined as 'quasistellar' in appearance, in fact turn out, if examined closely, to have faint 'fuzz', that is, low surface-brightness, small-diameter, nebulosity surrounding them. In many cases the faint nebulosity has approximately the angular size and apparent magnitude expected for a more or less normal spiral galaxy at the distance indicated by the redshift of the QSO. Thus again the indications are that QSOs represent the most luminous active galactic nuclei, so bright that the galaxies in which they lie are too faint with respect to the nuclei to be detected in many cases, and only bright enough to be detected with advanced instrumentation in others. This interpretation seems further confirmed by the fact that at the brightest absolute magnitudes, $M_B \approx -24$ or so, the luminosity function of Seyfert 1 galaxies fits smoothly on the luminosity function of QSOs. Above about $M_B \approx -22$ there are no galaxies that are not Seyferts.

Note also that there are practically no known QSOs with narrow permitted and forbidden emission lines, analogous to Seyfert 2 galaxies. This fits in well with the other observational statements we have been discussing – if the featureless continuum is so bright that the active galactic nucleus completely dominates the total light, and the rest of the galaxy cannot be seen (or can only be seen with difficulty), broad permitted emission lines of the Seyfert 1 type are almost certain to be present. Completely analogously, but with much less complete statistics, quasars seem to be the high-luminosity extension of broad-line radio galaxies that are also almost invariable N galaxies.

3.3   Narrow-line region

The 'narrow' lines of Seyfert 2 and narrow-line radio galaxies are more or less the same as those of H II regions and planetary nebulae, except that in a typical active galactic nucleus the range of ionization is very great, extending from [O I] and [S II] through [N II], [O III], and [Ne III] to [Ne V], and often [Fe VII] and [Fe X]. Lines of H I, He I, and He II are moderately strong, but weaker than the strongest forbidden lines. Clearly the physical situation is that of a low-density, highly ionized gas, with more or less 'normal' abundances. The well-developed diagnostics of nebular spectroscopy may be used to analyze it further. They depend on accurate spectrophotometric measurements of the relative strengths of the emission lines.

The relative strengths of the lower Balmer lines, Hα, Hβ, and Hγ, have been measured in many narrow-line radio galaxies. An example is Cyg A, for which the observational data are listed in Table 3.2. The observed Balmer-line strengths do not fit the recombination decrement, the predicted emission-line spectrum resulting from ionization and subsequent recombination and downward radiative transitions, characteristic of an ionized gas. These predicted ratios of strengths of emission lines, which we shall write in the rest of this chapter in abbreviated form as Hα/Hβ, Hγ/Hβ, etc., are nearly independent of density and temperature over a wide range of nebular and quasinebular conditions. The observed Balmer decrement is steeper than the calculated decrement. It is natural to suppose that, as in H II regions and planetary nebulae, the emitted spectrum is modified by interstellar extinction (or reddening) resulting from dust, both within the active galaxy and within our Galaxy. The amount of extinction calculated

Table 3.2. *Observed and calculated relative line fluxes in Cygnus A*

Ion	λ	Relative fluxes		Photo-ionization model	Crab Nebula
		Measured	Corrected		
[Ne V]	3346	0.14	0.38	0.12	—
[Ne V]	3426	0.36	0.95	0.34	0.27
[O II]	3727	2.44	5.00	0.24	12.6
[Ne III]	3868	0.66	1.23	0.53	1.90
[Ne III]	3969	0.22	0.40	0.16	—
[S II]	4071	0.14	0.23	—	—
Hδ	4101	0.17	0.28	0.26	0.31
Hγ	4340	0.32	0.46	0.47	0.61
[O III]	4363	0.16	0.21	0.19	0.19
He I	4471	≤0.07	≤0.09	0.02	0.28
He II	4686	0.25	0.28	0.18	0.68
Hβ	4861	1.00	1.00	1.00	1.00
[O III]	4959	4.08	3.88	6.3	3.92
[O III]	5007	13.11	12.30	18.1	11.92
[N I]	5199	0.40	0.32	—	—
[Fe XIV]	5303	≤0.10	≤0.08	0.01	—
[Fe VII]	5720	≤0.10	≤0.06	0.03	—
[N II]	5755	0.14	0.09	—	0.11
He I	5876	0.13	0.08	0.06	0.79
[Fe VII]	6087	≤0.07	≤0.04	0.04	—
[O I]	6300	2.10	1.10	1.24	1.20
[O I]	6364	0.69	0.35	0.41	0.33
[Fe X]	6375	0.10	0.05	0.07	—
[N II]	6548	3.94	1.90	0.29	1.36
Hα	6563	6.61	3.08	2.85	3.16
[N II] ·	6583	13.07	6.15	0.86	4.10
[S II]	6716	3.65	1.66	—⎫	9.24
[S II]	6731	3.29	1.51	—⎭	
[A III]	7136	0.64	0.25	—	—
[O II]	7324	0.35	0.13	—	0.50
[A III]	7751	0.13	0.043	—	—

from the Hα/Hβ ratio on this hypothesis generally agrees fairly well with the amount of extinction calculated from Hγ/Hβ. For instance, in Cyg A the amount of extinction derived in this way is $E_{B-V} = 0.69 \pm 0.04$. We can then suppose, as a first crude approximation, that the amount of extinction derived from the H I lines in this way is characteristic for the entire narrow-line spectrum, and correct all the observed relative narrow-line strengths for it. Corrected intensities for Cyg A calculated in this way, are also listed in Table 3.2. (The amount of extinction in Cyg A is relatively large in comparison with most other active galactic nuclei, but this is no doubt due to its position near the plane of our Galaxy; from observations of elliptical galaxies near it in the sky about half the total extinction arises within Cyg A, and the other half from interstellar matter in our Galaxy.)

There is no doubt of the presence of dust in active galaxies, and that it causes extinction. Quantitative statements about it, however, are less definite than they may at first seem, for they are all based on the assumption that the optical properties of the dust in these objects are identical with the optical properties, derived from observations, of dust in the plane of our Galaxy within a thousand parsecs of the Sun. Much further observational and theoretical work will be necessary before we fully understand the dust in active galactic nuclei. No doubt the distribution of dust within a nucleus, and within a galaxy, is patchy and irregular; the assignment of a single representative value of the extinction is clearly an extreme simplification.

With this understanding, once the intensities of the other lines have been corrected, they may be used to derive diagnostic information on the properties of the ionized gas. The most directly useful ratio is [O III] $(\lambda 4959 + \lambda 5007)/\lambda 4363$ which, for instance, $= 77$ in Cyg A. As is well known in nebular astrophysics, this ratio depends strongly on temperature in the low-density $(N_e < 10^4 \, \text{cm}^{-3})$ regime, and on both T and N_e at higher densities. The observed ratio in Cyg A corresponds to a mean $T = (1.5 \pm 0.1) \times 10^4 \, \text{K}$ in the [O III] emitting region, or lower temperatures if N_e is higher. Many other narrow-line radio and Seyfert 2 galaxies have similar ratios, indicating temperatures in the range $1-2 \times 10^4 \, \text{K}$. This is strong observational evidence that the main energy input to the ionized gas is by photoionization. The only other type of input we know, the conversion of kinetic energy into heat – 'shock wave heating' or 'collisional heating' or 'cloud–cloud collisions' – results in collisional ionization and a relation between temperature and ionization. The [O III] lines, in particular, would be radiated mostly at $T > 3 \times 10^4 \, \text{K}$ and would be expected to indicate a much higher representative temperature than $1.5 \times 10^4 \, \text{K}$, and a much lower $(\lambda 4959 + \lambda 5007)/\lambda 4363$ ratio, as in the Cygnus Loop, an old supernova remnant that is the best-studied example of shock heating. On the other hand, under photoionization conditions there is no direct relationship between temperature and ionization, but the thermostatic effect of the radiative cooling by collisionally excited line radiation, which goes up very rapidly with increasing temperature, tends to keep $T \approx 1\text{–}2 \times 10^4$ K over a wide range of input ultraviolet spectra and fluxes.

In some Seyfert 2 and narrow-line radio galaxies, the [O III] ($\lambda 4959 + \lambda 5007$)/$\lambda 4363$ ratio is smaller, indicating higher T (up to 5×10^4 K in some cases) if $N_e < 10^4 \, \text{cm}^{-3}$, or alternatively higher densities (up to $N_e \approx 10^7 \, \text{cm}^{-3}$ in some cases) if $T \approx 1\text{–}2 \times 10^4$ K. It seems more plausible that one physical energy-input mechanism is going on in all (or most of) the active-galactic nuclei and, since it cannot be collisional input in many of them (such as Cyg A), the process is presumably photoionization. Another diagnostic ratio, [N II] ($\lambda 6548 + \lambda 6538$)/$\lambda 5755$, though not observed in so many objects, also gives $T \approx 1\text{–}2 \times 10^4$ K in those in which it is seen. This, however, does not discriminate between photoionization and collisional heating, because N^+ is a lower stage of ionization than O^{++}, and the [N II] lines can be radiated at these temperatures under either situation.

The most unambiguous density indicator observable in Seyfert 2 and narrow-line radio galaxies is [S II] $\lambda 6717/\lambda 6731$, which indicates $N_e \approx 10^{3.5\pm1} \, \text{cm}^{-3}$ in many objects. [O II] $\lambda 3729/\lambda 3726$, also a good density diagnostic in H II regions and planetary nebulae, is not useful in active galactic nuclei because their 'narrow' emission lines are almost always broad enough so that this doublet is blended into a single unresolved feature. The ionization potential of S^0, 10.4 eV, is less than that of H^0, 13.6 eV, so S may be ionized to S^+ in an extensive volume outside the H^+ zone, and the electron density derived from the [S II] lines is probably particularly unrepresentative of the average density within the ionized region where most of the other emission lines arise.

Although the temperature-sensitive line-ratio measurements strongly suggest photoionization as the energy-input mechanism, it is clear that photoionization by hot stars, as in H II regions and planetary nebulae, will not give the emission-line spectra observed in Seyfert 2 galaxies and narrow-line radio galaxies. Hot stars will not produce the wide range of ionization, with low stages such as [O I] and [S II], and high stages such as [Ne V] and [Fe VII], strong in comparison with [O III], [N II], and [Ne III]. What is required is photoionization by a much 'harder' spectrum, extending even further into the ultraviolet than the spectra of the central stars of planetary nebulae, some of which have effective temperatures up to 2×10^5 K. The copious high-energy photons ($h\nu > 100$ eV) of such a hard spectrum will produce high ionization (up to Ne^{+4}, Fe^{+6}, Fe^{+9}, etc.) near the central source, and will also produce a long, partly-ionized 'transition zone', in which these photons have a long mean free path. In such a zone H^0 and H^+, O^0 and O^+, and S^+ can all coexist with an appreciable electron density, and [O I] and [S II] lines can be produced by collisional excitation.

Now in fact there is a featureless continuum observed in essentially every active galactic nucleus, as stated above. The form of this spectrum, in the observed optical region, approximately fits a power law,

$$L_\nu = C\nu^{-\alpha}. \tag{3.1}$$

Typically $\alpha \approx 1$–2; for instance in Cyg A the observed featureless continuum has $\alpha = 3.8$, or if it is corrected for the same amount of extinction as determined from the reddening of the Balmer decrement, $\alpha = 1.6$. If this spectrum continues with the same power-law form to high energies, it can qualitatively explain the observed Cyg A line spectrum. Furthermore, quantitatively, a model nebula, calculated with 'normal' abundances of the elements and an input power-law spectrum with $\alpha = 1.2$ does fit, in an approximate way, the observed Cyg A emission-line spectrum, as shown in Table 3.2. The agreement is by no means perfect, but the wide range of ionization is clearly exhibited. Adjustments of the assumed abundances, density, and form of the featureless continuum and its extrapolation to high energies would probably improve somewhat the agreement, although no doubt the fit would not be perfect.

The true situation in the ionized gas is no doubt more complicated than can be represented in any simplified model. In particular, every observed nebula shows complicated density structure, often with large-scale gradients, and always with small fine structure-condensations, filaments, density fluctuations and the like. It would be surprising indeed if Cyg A did not also have similar fine structure. Instead of trying unsuccessfully to model these complications, we can instead compare the Cyg A observed emission-line spectrum to that of NGC 1952, the Crab Nebula, which is known to be photoionized by an ultraviolet continuum resulting from synchrotron radiation, approximately following a power law with $\alpha \approx 1.5$. This comparison, also listed in Table 3.2, shows even better agreement, except for the fact that the He I and He II lines are stronger in the Crab Nebula than in Cyg A, because of the known high abundance of helium in the supernova remnant.

Another test of the photoionization idea is that the total number of ionizing photons available from the central source should be at least large enough to balance the total number of recombinations in the ionized gas, which in turn is related to the total number of $H\beta$ photons emitted by the gas. This is the basis of the Zanstra method for determining the effective temperatures of the 'exciting' (ionizing) stars of planetary nebulae. For a power-law source of the form (3.1), this condition may be written

$$\int_{v_0}^{\infty} \frac{L_v}{hv} \, dv \geqslant \alpha_B N_e N_p V, \tag{3.2}$$

$$L_{H\beta} = \alpha_{H\beta}^{\text{eff}} h v_{H\beta} N_e N_p V. \tag{3.3}$$

Expressing the $H\beta$ line luminosity in terms of W_0, its equivalent width (in wavelength units) with respect to the luminosity in the featureless continuum,

$$L_{H\beta} = L_v(\lambda 4861) \frac{dv}{d\lambda} W_0(H\beta), \tag{3.4}$$

this becomes

$$W_0(H\beta) \lesssim \frac{\lambda_{H\beta}}{\alpha} \frac{\alpha_{H\beta}^{\text{eff}}}{\alpha_B} \left(\frac{v_0}{v_{H\beta}}\right)^{-\alpha} = \frac{568}{\alpha} (5.33)^{-\alpha}. \tag{3.5}$$

In the case of Cyg A, the observed equivalent width observed with a slit $2.7'' \times 4''$ in area, projected on the sky, is $W_0(H\beta) = 39$ Å (in the rest frame of Cyg A). The continuum, however, is diluted by the integrated stellar absorption line spectrum of the galaxy. From analysis of the spectrum, the fraction of the observed continuum at $\lambda 4800$ that results from the featureless continuum is approximately $f_{FC} = 0.6$, the remainder, $f_G = 0.4$, is galaxy spectrum. Thus the corrected equivalent width of $H\beta$, expressed in terms of the featureless continuum, is $W_0(H\beta) = 65$ Å, and the Zanstra condition (3.4) is fulfilled with $\alpha < 1.2$.

Similar semi-quantitative agreement exists between the observed spectra of essentially all Seyfert 2 and narrow-line radio galaxies that have been analyzed in detail, and the predictions of the photoionization picture. There seems little doubt that a hard photon spectrum, extending to X-ray energies, is the primary energy-input mechanism to the observed gas in Seyfert 2 and narrow-line radio galaxies.

Furthermore, the *narrow* emission-line spectra of Seyfert 1, Seyfert 1.5, and broad-line radio galaxies are generally similar to the spectra of Seyfert 2s and narrow-line radio galaxies. It seems most likely that in these objects also the energy input is photoionization by high-energy photons. The one known systematic difference is that, on the average (although not in every single object), the ionization goes up to a higher level in the narrow-line regions of Seyfert 1 galaxies than in Seyfert 2s; this may indicate that in Seyfert 2s there is a cut-off or downturn in the ionizing spectrum, at energies of perhaps $h\nu \approx 10^2$ eV, that is not present, or occurs at higher energies, in Seyfert 1s.

The abundances of the elements seem to be more or less normal in active galactic nuclei. The abundance ratio He/H ≈ 0.1 is especially well determined because it involves ratios of intensity of two lines (for instance Hβ and He I $\lambda 5876$) formed mainly by recombination, which have only small (and nearly identical) temperature dependences. Abundance ratios such as O/H, Ne/H, etc., are less certain for two reasons. One is the exponential temperature dependence of the emission coefficients of collisionally excited lines such as [O III] $\lambda 5007$, [Ne III] $\lambda 3869$, etc., and the other is the large correction for unobserved stages of ionization, such as O^{+3} (the only lines of [O IV] are in the far infrared and in the satellite ultraviolet), Ne^+ (until [Ne II] $\lambda 12.8\,\mu$ can be observed in active galactic nuclei), etc. Nevertheless, to the apparent accuracy of the available observational data and its analysis by these standard nebular methods, there are no surprises. O, Ne, N and (from ultraviolet measurements described below) C are the most abundant elements after H and He. They have approximately standard relative abundances to one another. All of them may be somewhat below 'normal' abundance with respect to H, but, if so, only by a factor of three or so. The abundances of the heavier elements, Ar, S, Si, seem to be approximately normal with respect to O. One interesting result is that Fe is observed, in apparently approximately normal abundance, in many active galactic nuclei. In planetary nebulae, on the contrary, Fe is observed to be below normal (stellar) abundance in the gas; the interpretation is that in planetaries most of the Fe is locked up in dust grains. Evidently this removal of Fe from the ionized gas by dust is less complete in the observed narrow-line regions of active galactic nuclei.

Model calculations of the temperature and ionization distributions in assumed model active galactic nuclei may be carried out by the methods originally developed for model planetary nebulae or H II regions, substituting an assumed power-law input spectrum (or a more complicated spectrum extending to high energies) for the assumed high-temperature model stellar atmosphere spectra used in the nebular cases. However, there are some additional complications in the model active galactic nuclei calculations, which of course reflect complications in the actual physical processes that occur in these objects. They result from the high-energy photons that are so important in active galactic nuclei, but that do not exist in typical planetary nebulae and H II regions.

One is that at very high energies ($h\nu \gtrsim 2$ keV) Compton scattering by initially bound electrons in H^0 becomes more important than the ordinary electric-dipole bound-free process, so an additional Compton scattering term

$$a_{cv} = \frac{8\pi}{3} \tau_0^2 \left(1 - \frac{3}{4} \frac{m_0 c^2 v_0}{hv^2}\right) = 0.66 \times 10^{-24} \text{ cm}^2 \left(1 - \frac{3}{4} \frac{m_0 c^2 v_0}{hv^2}\right), \quad (3.6)$$

must be added to the standard 1^2S cross section formula.

More importantly, at high energies photoionization of heavy ions can dominate the opacity, rather than photoionization of H^0, He^0, and He^+, as at low energies. Furthermore, photoionization not only of the outer valence electron, but also of inner-shell electrons can occur. Thus, for instance O^{+2} ions can be photoionized by photons with energy $hv > 55$ eV in the standard way

$$O^{+2}(1s^2 2s^2 2p^2\ ^3P) + hv \rightarrow O^{+3}(1s^2 2s^2 2p\ ^2P^0) + e. \quad (3.7)$$

For photons with energies $hv > 64$ eV, inner shell ionization

$$O^{+2}(1s^2 2s^2 2p^2\ ^3P) + hv \rightarrow O^{+3}(1s^2 2s 2p^2\ ^4P) + e, \quad (3.8)$$

for instance can also occur, while for photons with energy $hv > 570$ eV,

$$O^{+2}(1s^2 2s^2 2p^2\ ^3P) + hv \rightarrow O^{+3}(1s 2s^2 2p^2\ ^2P) + e \quad (3.9)$$

can also. In this process the higher excited level decays by a radiationless Auger transition, releasing a second electron

$$O^{+3}(1s 2s^2 2p^2\ ^2P) \rightarrow O^{+4}(1s^2 2s^2\ ^1S) + e. \quad (3.10)$$

These Auger processes thus couple stages of ionization which differ by two electrons, such as O^{+2} and O^{+4}, in contrast to normal photoionization, which only couples stages differing by one electron.

From high-energy photons, processes like (3.7) and (3.8) produce high-energy photoelectrons; for instance, a 750-eV photon will yield electrons with kinetic energy 695 eV and 686 eV by these processes, respectively. The Auger electron produced in reaction (3.10) has an energy between 438 eV and 460 eV. These high-energy electrons lose part of their energy in Coulomb collisions, directly heating the gas, part by collisional ionization, yielding secondary electrons, and part by exciting bound levels, especially of H^0, leading to line radiation.

All these effects can be taken into account quantitatively, and are, in modern model calculations of the structure, ionization, and temperature distribution in active galactic nuclei, and their resulting line radiation. The aim of such calculations is to vary the assumed input parameters – such as abundances of the elements, density or density distribution, geometrical situation, and assumed photoionizing spectrum – and compare the resulting calculated line strengths with observational data. If there is disagreement, that assumed model can be rejected. Ideally, in the end all but one set of assumptions and assumed parameters might be rejected, and the one remaining possibility whose predictions did not disagree with the measurements would be the correct description of the object. We are still far from that situation. Instead, at present, we have approximate (but far from complete) agreement with some models, which we therefore feel are somewhere in the direction of the true physical structure. The most important missing part of the picture, of course, is a physical model of the central energy source. The power-law photoionizing spectra (or more complicated broken power laws with cut-offs) that go into active-galaxy models are analogous to the black-body input spectra that were assumed for the central stars of planetary nebulae before we understood their evolution, internal structure, compositions and atmospheres, and could calculate, at least approximately, the resulting ionizing spectrum on a physical basis.

One further result is of interest here, concerning the Balmer decrement in the narrow-line regions of active galactic nuclei. In a completely ionized gas, which is a good approximation for H II regions and planetary nebulae, the main process by which the hydrogen lines are emitted are by recaptures to the excited states

$$H^+ + e \rightarrow H^0(n'l') + h\nu, \tag{3.11}$$

followed by downward radiative transitions

$$H^0(n'l') \rightarrow H^0(nl) + h\nu_{n'n}. \tag{3.12}$$

In active galactic nuclei, however, the harder photoionizing spectrum results in a larger 'transition zone', or partly ionized region, in which neutral H^0 coexists with ionized H^+ and free electrons. In this zone collisional excitation

$$H^0(1s) + e \rightarrow H^0(n'l') + e \tag{3.13}$$

is also important. This is particularly so because high-energy electrons, produced by photoionization by high-energy photons and by Auger processes, are especially effective in heating the gas, which therefore often has a somewhat higher equilibrium temperature than the gas in traditional nebulae.

Under these collisional excitation conditions, Hα is particularly strongly enhanced. The result is that the Balmer decrement is steepened, and Hα/H$\beta \approx 3.1$ is a better approximation than the recombination value

Hα/H$\beta \approx 2.8$, as both calculations and analysis of the available observational data indicate. The correct value should of course be used in future determinations of the amount of extinction from Balmer-line measurements. The higher lines are less affected by collisional excitation (because of their larger excitation energies), and the recombination ratio Hγ/H$\beta \approx 0.47$ is probably a sufficient approximation at present.

3.4   Broad-line region

The characteristic feature of the spectra of Seyfert 1, 1.5, and broad-line radio galaxies is their broad H I emission lines, as shown in Figs. 3.2, 3.4, and 3.5. Usually broad He I can also be seen, considerably weaker than H I, and Fe II almost always as well, in the Seyfert 1s and 1.5s, but not in the broad-line radio galaxies. All these are permitted lines. None of the forbidden lines have similar broad profiles. The only interpretation known is that these broad lines arise in a region in which the density is so high that all the levels of abundant ions that give rise to the forbidden lines are collisionally deexcited. A more nearly correct form of this statement is that the actual electron density in the broad-line region is considerably larger than the critical densities of all these levels, so that the lines these levels emit are weaker, in the ratio N_c/N_e, than they would be with respect to Hβ (for instance), at the same temperature and ionization but at low density. The quantitative limit is thus quite vague, but roughly $N_e \gtrsim 10^8\,\mathrm{cm}^{-3}$ seems plausible.

Although no lines in the optical region can be used to set an upper limit to the density, C III] $\lambda 1909$ has been observed with a broad profile, similar to the H I profiles, in the ultraviolet spectra of several Seyfert 1 and 1.5 galaxies, and a few broad-line radio galaxies, as well as in the redshifted spectra of many QSOs and quasars. Thus, in these specific objects, and presumably in all the observed broad-line regions, $N_e \lesssim 10^{10}\,\mathrm{cm}^{-3}$, the approximate critical density for C III $2s2p\ ^3P_1^0$. An intermediate value, $N_e \approx 10^9\,\mathrm{cm}^{-3}$, may therefore be adopted as a very rough mean for the observed broad-line regions.

There is little direct information in the temperature. There are no direct diagnostics for it in the H I, He I, and He II lines. The presence of Fe II indicates that $T \lesssim 4 \times 10^4\mathrm{K}$, for at higher temperatures it would be nearly completely collisionally ionized to Fe III, even with no contribution from photoionization. For orientation, $T \approx 10^4\,\mathrm{K}$ is a satisfactory round figure. Although the Balmer decrements show that other processes besides simple recombination must be involved, probably the recombination emission coefficient gives a crude idea of the amount of ionized gas in the broad-line region. If we visualize it as spherical, the volume, mass and radius are defined by the equations

$$L(H\beta) = N_e N_p \alpha_{H\beta} h\nu_{H\beta} V\varepsilon, \tag{3.14}$$

$$M = (N_p m_p + N_{He} m_{He}) V\varepsilon, \tag{3.15}$$

$$V = \frac{4\pi}{3} r^3, \tag{3.16}$$

where ε is the volume filling factor. To a sufficiently good approximation $N_{He} = 0.1 N_p$ and $N_e = N_p + 1.5 N_{He}$. The most luminous nuclei of Seyfert 1 and broad-line radio galaxies have $L(H\beta) \approx 10^9 L_\odot$, which gives $M \approx 36 M_\odot$ $(19^9/N_e)$ and $r = 0.015\varepsilon^{-1/3} (10^9/N_e)^{2/3}$ pc. The sizes and masses of the broad-line regions are thus extremely small; for instance if we imagine a filling factor $\varepsilon \approx 0.1$, more extreme than observed in planetary nebulae and H II regions, and $N_e \approx 10^9$ cm^{-3}, then $r \approx 0.03$ pc ≈ 0.1 light year. No broad-line region has been resolved with an earth-based telescope, and this simplified calculation predicts none will be resolved even with the Space Telescope (although, of course, the nearest candidates should certainly be tried).

However, the broad $H\beta$ and $H\alpha$ profiles, and fluxes, have been observed to vary on time scales as short as a month or so (≈ 0.1 year) in a significant fraction of Seyfert 1 galaxies. This agrees with the light-travel times across these objects, according to the approximate calculation above. Faster variations should not be expected unless the broad-line regions were smaller.

A similar calculation for Seyfert 2 galaxies, or for the narrow-line regions of Seyfert 1 galaxies, of which the most luminous have $L(H\beta) \approx 2 \times 10^8 L_\odot$, gives $M \approx 7 \times 10^5 (10^4/N_e) M_\odot$, and $r \approx 20\varepsilon^{-1/3} (10^4/N_e)^{2/3}$ pc. Thus, for an assumed $\varepsilon \approx 0.1$, $r \approx 40$ pc, and in fact a few of the nearest Seyfert 2 galaxies narrow-line regions apparently have been resolved, with diameters of order 10^2 pc. Correspondingly, there is no well-observed case of narrow-line variation in active galactic nuclei.

The energy-input mechanism from the central source to the small, dense broad-line region is not so clear as in the case of the narrow-line region. Most probably it is photoionization, as in the narrow-line region. The most convincing evidence for this is that the $H\alpha$ or $H\beta$ luminosities of narrow-line and broad-line objects, Seyfert 2s, 1.5s, 1s, and QSOs, are, to a good approximation, proportional to the featureless continuum luminosity. Put another way, the equivalent width of the total $H\beta$ emission line, broad plus narrow components if both are present, or either if only one is present, expressed in terms of the featureless continuum, is approximately the same for all active galactic nuclei. This, of course, is just what is expected from (3.4) if photoionization is effective in all active galactic nuclei, whether or not they have dense, broad-line gas close around their central sources. This argument of course does not prove photoionization is the input mechanism; it is merely consistent with it. The fact that the narrow components of the H I emission lines are not stronger than observed in Seyfert 1.5 galaxies with strong featureless continua would be hard to interpret on any other basis.

The large widths of the Balmer emission lines (and the other permitted lines) in Seyfert 1, 1.5, and broad-line radio galaxies can only be interpreted as Doppler shifts due to mass motions. No other mechanism is known that seems remotely possible under the physical conditions in these objects. All the H I lines in a given active galactic nucleus have approximately the same width and profile, but there are differences at the 10-percent level. He I $\lambda5876$, the strongest and best observed emission line of this atom, is often but not always wider than the H I lines. He II $\lambda4686$ is difficult to observe because it is badly blended with several Fe II multiplets but, in the objects in which they are weak, $\lambda4686$ seems definitely broader than the H I lines. The Fe II lines have the same widths as H I, or perhaps 75 percent of the width of the H I lines. All these trends are understandable if the velocities in the broad-line region represent motion in the gravitational field of a central object, and hence decrease outward. Rotation, radial flow, and turbulence have all been proposed as the dominant form of motion. Understanding the velocity field is one of the major problems to be solved to understand active galactic nuclei.

In the Seyfert 1 and broad-line radio galaxies, the ratios of intensities of the broad Balmer emission lines, $H\alpha/H\beta$ and $H\gamma/H\beta$ do not agree with the recombination decrement, modified by interstellar extinction. They cluster very roughly about the 'reddening line', the locus of points in an $H\alpha/H\beta$, $H\gamma/H\beta$ diagram along which the recombination point is moved by varying amounts of extinction. The natural interpretation is that, although reddening is important, the emission coefficients of the Balmer lines are also affected by density and optical depth effects. Because of the high density in the broad-line region, the optical depth in the H I $L\alpha$ resonance line is very large even within a single 'cloud' or 'density condensation' in which the velocity gradient is not large. Resonance-line scattering thus maintains a sizable population in the excited $2\,^2P$ term of H I, the amount depending critically on the optical depth and on the absorption of $L\alpha$ photons by dust. These excited H^0 atoms can in turn absorb Balmer-line photons themselves, leading to fluorescence or scattering, thus modifying the 'emitted' Balmer decrement, as well as the Paschen decrement, Brackett decrement, etc. Also, collisional excitations from $2\,^2P$ to $n = 3, 4, 5, \ldots$ have only low thresholds and large cross sections. They further modify the emitted decrements. The physical processes are all well understood theoretically, but linking them all correctly, particularly in a physically realistic model with density and temperature variations included, makes for very complicated computer programs indeed. Yet, ultimately, it should be possible from detailed analyses in this way of the broad H I, He I and He II lines and their profiles, to understand the physical conditions in the ionized gas, and perhaps even in the dust as well, within the broad-line regions of active galactic nuclei.

The origin of the broad Fe II lines is probably even more complicated, and is far from being understood. These lines are not observed at all in planetary nebula and H II regions; thus, their excitation must depend on some of the unique properties of active galactic nuclei, probably the combination of high electron densities and strong radiation fields. The ionization potential of Fe I is only 7.9 eV, and the bulk of the Fe II emission probably comes from the large, partly ionized 'transition' zones outside the fully ionized H II zones. Fe II has an extremely complicated energy-level diagram, with many metastable levels within a few volts of the ground $a\,^6D$ term. The densities in the broad-line region are higher than the critical densities for all these Fe II metastable levels, and they therefore have approximately Boltzmann thermal populations, although the forbidden lines they emit are for this very reason too weak to be observable. Collisional excitation and/or resonance fluorescence, resulting from the strong, near-ultraviolet featureless continuum, then populates the higher Fe II terms which are connected with the lower terms by permitted radiative transitions. It is clear from the observed relative intensities that the optical depths in all the permitted ultraviolet Fe II lines are quite large, and therefore a correct and reasonably efficient treatment of the radiative-transfer problem is absolutely essential. Understanding the Fe II lines, and thus why they are strong in radio-quiet Seyfert 1 active galactic nuclei of spirals, but weak in radio-loud active galactic nuclei of N galaxies, will undoubtedly be extremely important in deciphering the nature of these objects.

3.5 High-energy photons

Most Seyfert 1 galaxies have [Ne V] $\lambda\lambda 3346, 3426$ and [Fe VII] $\lambda\lambda 5721, 6087$ emission lines in their spectra, indicating photoionization by photons with $h\nu \gtrsim 100$ eV. Many also show [Fe X] $\lambda 6375$ (the 'red coronal line') and [Fe XI] $\lambda 7892$ emission. Two galaxies, III Zw 77 and Tololo 0193-383, even have the [Fe XIV] $\lambda 5303$ (the 'green coronal line') in emission. The mechanism of ionization to these high stages is not certain, but their correlations with one another and with [Fe VII] make photoionization (rather than collisional ionization, as in the solar corona), the most promising possibility. If so, [Fe X], [Fe XI], and [Fe XIV] indicate the presence of copious supplies of photons with $h\nu > 234$, 262, and 361 eV, respectively. These are in the range of soft X-ray energies.

Correspondingly, every bright Seyfert 1 or broad-line radio galaxy is an X-ray source, and nearly every highly luminous X-ray galaxy is a Seyfert 1, broad-line radio galaxy, quasar, or QSO. The X-ray spectra of these objects are evidently extensions, to keV or even MeV energies, of the optical featureless continuum. The continuum does not follow a single power law over this entire range, however; the best observational information and indirect evidence together lead to a composite spectrum of the approximate form

$$L_\nu = C\nu^{-\alpha} \quad \nu \leqslant \nu_B$$
$$L_\nu = D\nu^{-\beta} \quad \nu \geqslant \nu_B,$$

with a break or turn-up at $h\nu_B \approx 1$ keV, and $\alpha > 1 > \beta$.

Measurements show that there is a very good correlation between the X-ray luminosity and the luminosity in the *broad* emission component of Hα. Several X-ray galaxies, originally described as Seyfert 2 galaxies or narrow-emission-line galaxies on the basis of early spectrograms, turned out with high signal-to-noise-level data to have weak, but definitely present, broad Hα emission. The lowest (X-ray)-luminosity known example of this correlation at the present writing is M 81, which has strong narrow emission lines of Hα and [N II] $\lambda\lambda6548$, 6583, plus a very weak Hα emission component with a width ≈ 4500 km/s. Knowledge of this weak broad component led to an attempt to detect M 81 as an X-ray source; the measurement was successful and showed that it fits the correlation well.

The X-ray fluxes from some Seyfert 1 nuclei have been observed to vary appreciably on time scales as short as 10^3 sec. This indicates that the X-ray emitting regions in them are very small, with light-crossing times not greatly larger than this. Variations in the optical featureless continuum are not observed with so short a time scale. Evidently, the X-rays are emitted in a smaller volume (according to the models we shall discuss, closer to the center) than the optical continuum-photons.

3.6   Ultraviolet spectra

With earth-bound telescopes we cannot observe the ultraviolet spectra of galaxies because of the very strong absorption of air (specifically

O_3) at $\lambda < 2950$. However, many quasars have sufficiently large redshifts so that their ultraviolet spectra are shifted into the observable region. Furthermore, the International Ultraviolet Explorer satellite, with its 18-inch aperture, ultraviolet-transmitting telescope, has provided data in the last few years on the short-wavelength spectra of many of the brighter Seyfert galaxies, and of a few radio galaxies and low-redshift quasars. The results obtained in these two ways agree well. The ultraviolet spectra of Seyfert galaxies, radio galaxies, quasars and QSOs are similar, to the extent we have been able to observe them.

Practically all of the ultraviolet spectral information we have on active galactic nuclei refers to Seyfert 1s and broad-line radio galaxies; the narrow-line objects nearly all are too faint in the ultraviolet to have been observed effectively with the IUE to date. One of the strongest ultraviolet emission lines in all active galactic nuclei is $L\alpha$ $\lambda1216$, the strongest resonance line of H I. It has a broad profile, similar to the H I Balmer lines, but often strongly affected by self-absorption, and by absorption of clouds containing H^0 along the line of sight between us and the nucleus. (The study of the absorption components of $L\alpha$, from the observed wavelength of the $L\alpha$ emission line, down to the rest wavelength of $L\alpha$ in the reference system of our Galaxy, gives important information on gas around active galactic nuclei, as well as in intergalactic space, but is outside the scope of this short chapter.)

Typically, $L\beta$ and the higher Lyman emission lines are much fainter, indicating strong resonance fluorescence (conversion to $L\alpha$ plus Balmer, Paschen, Brackett, etc., photons) as expected in these optically thick lines. The ratio $L\alpha/H\beta$ deviates greatly from the recombination value, indicating the great importance of the collisional and line-transfer effects mentioned in Section 3.4, as well as absorption of $L\alpha$ by dust. This absorption is enhanced, of course, by resonance-line scattering of the $L\alpha$ photons by H^0, which greatly increases their path lengths within the cloud or nucleus in which they were emitted. Collisional excitation of $L\alpha$ is also important, particularly within the partly ionized transition zones. Though many calculations of these effects have been made, the real situation in the dense, broad-line region of active galactic nuclei is so complicated that much work still remains to be done before we shall understand completely the H I spectrum.

In the ultraviolet spectral region, as in the optical, most of the emission lines are collisionally excited lines of abundant ions, with excited levels within a few volts of the ground level. Some of these are forbidden lines, such as the [Ne IV] $2p^2$ ^4S–$2p^2$ ^2D $\lambda2423$ blend, completely analogous to the collisionally excited forbidden [O II] $\lambda3727$ blend observed in the optical region. Others are permitted blends, such as Mg II $3s$ ^2S–$3p$ ^2P $\lambda2798$, and

C IV $2s$ ^2S–$2p$ ^2P $\lambda1549$ (see Fig. 3.8). There are no strong analogues of these lines in the optical region, because of the energy-level structure of the abundant ions that occur in active galactic nuclei (and in gaseous nebulae), but there are several such allowed transitions with $h\nu > 5.5$ eV in the ultraviolet. Also, there are quite a few semiforbidden (or intercombination) lines, such as the C III] $2s^2$ ^1S–$2s2p$ ^3P $\lambda1909$ and O III] $2p^2$ ^3P–$2s2p$ ^5S $\lambda1665$ blends shown in Fig. 3.9. Most of these lines are permitted by the general electric-dipole selection rules ($\Delta J = 0$, ± 1, but $0 \nleftrightarrow 0$, and parity changes), but are forbidden in strict Russell–Saunders coupling because $\Delta S = 1$. However, because of the small breakdown of LS coupling (typically of order 10^{-2} in the amplitude of the wave function for abundant light ions), these transitions occur with probabilities roughly intermediate (logarithmically) between normal permitted and forbidden lines. An exception is the [C III] $2s^2$ $^1S_0$–$2s2p$ $^3P_2$ $\lambda1907$ component, which is strictly forbidden as an electric-dipole transition, but occurs as a permitted magnetic-quadrupole transition. Its transition probability is thus very small, $A = 5.2 \times 10^{-3}$ s^{-1}, and it is collisionally deexcited below its critical density $N_c = 4 \times 10^5$ cm^{-3}. All the collisionally excited ultraviolet lines, combined with optical lines of the same ions, provide additional diagnostic information on the temperatures and densities in the ionized gas.

Fig. 3.8. Schematic energy-level diagrams of Mg II and C IV, with strong ultraviolet emission lines indicated.

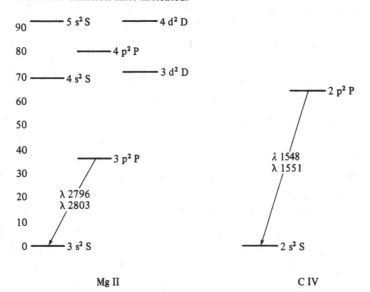

3.7 Physical models

Because of limitations of space, this chapter is necessarily chiefly devoted to the spectroscopic methods for analyzing the observed plasma in the nuclei of quasars, Seyfert galaxies and active galactic nuclei. Many studies have been made toward an understanding of these objects, but there is still much more to be learned. No doubt the basic ideas and methods described in this chapter will be more useful to the reader than details of our present knowledge of active galactic nuclei.

Yet, a few words are in order on the general picture we have of these objects. The luminosity of a typical active galactic nucleus, $\approx 10^{12} L_{\odot}$, is much larger than can be accounted for as a result of hydrogen burning (or, indeed, from any nuclear burning in the stars in the nucleus), typically totalling $\approx 10^9 M_{\odot}$. (In contrast, the star-burst galaxies mentioned in Section 3.2 can be understood in these terms.) Therefore, the mechanism of energy release in quasars, Seyfert galaxies, and other galactic nuclei is almost certain to be gravitational. The most plausible picture consistent with our knowledge of physics as we understand it is that there is a black hole, typically of $M \approx 10^8 M_{\odot}$, in the nucleus of a typical active galactic nucleus. Around it is an accretion disk, in which matter spirals into the black hole, releasing energy by friction into heat, which is then radiated away in the form of photons. The temperature increases inward toward the center of the accretion disk, so the radiation emitted is the sum of thermal (crudely black-body, actually more nearly like stellar) spectra emitted over a wide range of

Fig. 3.9. Schematic energy-level diagrams of C III and O III, with strong semiforbidden ultraviolet emission lines indicated. For O III the well-known optical forbidden lines are also indicated, by dashed lines.

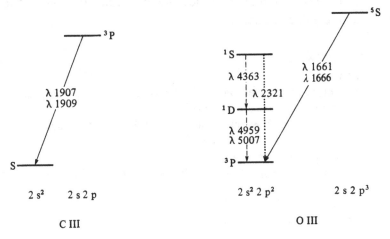

temperatures. The radius of a $10^8 M_\odot$ black hole is $\approx 10^{-5}$ pc; the accretion disk, though much larger than this, is smaller than the broad-line emission region in which it is immersed. Mass is fed from the broad-line region into the accretion disk and by instabilities in the opposite direction.

No doubt near the central black hole many instabilities came into play. It is clear that in quasars and radio galaxies, very-high-energy, relativistic plasmas are emitted from the central region near the black hole, quite often in two oppositely directed jets. It seems natural to assume that they are perpendicular to the plane of the accretion disk. There is some evidence, in a few cases, that they are perpendicular to the plane of the parent galaxy.

Probably in the radio-quiet QSOs and Seyfert galaxies, similar physical processes occur. Perhaps they do not become quasars or radio galaxies because the relativistic jets cannot escape freely to large distances, but are stopped by interstellar matter within the galaxy. There are several indications, mostly indirect but partly directly observational, that the jets that occur in radio-quiet objects are not perpendicular to the plane of their parent galaxies. This suggests that the accretion disks are not always coplanar with the galaxies of which they form a (very small) part.

The question of which galaxies form active galactic nuclei, how often, and why, on this picture is closely connected with the conditions under which mass, with (on a galactic scale) very nearly zero angular momentum, is delivered at their centers. This question is far from understood. More and more evidence is accumulating that many active galactic nuclei are fed by the results of tidal interactions or gravitational collisions with other galaxies. Many Seyfert galaxies are barred spirals, and evidently a non-circularly symmetric gravitational field can help in directing mass toward the center. Probably other mechanisms, not yet ever glimpsed, are important in still other cases. Much work remains for the student of quasars, Seyfert galaxies, and other active galactic nuclei.

References

Section 1

Many of the early papers on quasistellar radio sources, including those mentioned in the text, are reprinted in

Robinson, I., Schild, A. & Schucking, E. L. (1965). *Quasistellar Sources and Gravitational Collapse*. University of Chicago Press.

An important early monograph is

Burbidge, G. & Burbidge, M. (1967). *Quasistellar Objects*. W. H. Freeman & Co.

A brief selection of papers is

Seyfert, C. K. (1943). *Astrophys. J.* **97**, 28–40.

Baade, W. & Minkowski, R. (1954). *Astrophys. J.* **119**, 206–14.

Osterbrock, D. E. (1982). In *Extragalactic Radio Sources*, ed. D. S. Heeschen & C. M. Wade, pp. 369–71 (Table 1). D. Reidel Publishing Company.

Wampler, E. J., Robinson, L. B., Baldwin, J. A. & Burbidge, E. M. (1973). *Nature*, **243**, 336–7 (OQ 172).

Peterson, B. A., Savage, A., Jauncey, D. L. & Wright, A. E. (1982). *Astrophys. J.* **260**, L27–29 (PKS 2000-330).

Section 2

Four reviews that give references to many of the original papers are

Weedman, D. W. (1977). *Ann. Rev. Astron. Astrophys.* **15**, 69–95.

Osterbrock, D. E. (1979). *Astron. J.* **84**, 901–9.

Sargent, W. L. W. (1980). In *Scientific Research with the Space Telescope*, ed. M. S. Longair & J. W. Warner, pp. 197–214. U.S. Government Printing Office.

Osterbrock, D. E. (1984). *Quart. J. Roy. Astron. Soc.* **25**, 1.

The best discussion of luminosity functions, both optical and radio, is

Meurs, E. J. A. (1982). *The Seyfert Galaxy Population*. Leiden, Ph.D. Thesis.

Section 3

The data for Cyg A, the example discussed in this section, are from

Osterbrock, D. E. & Miller, J. S. (1975). *Astrophys. J.* **197**, 535–44.

Osterbrock, D. E. (1983). *Pub. Astron. Soc. Pacific*, **95**, 12–17.

The diagnostic methods, taken over from studies of gaseous nebulae, are discussed in Chapter 1. Several papers that give some of the concepts used in the model calculations are

Aldrovandi, S. M. V. (1981). *Astron. Astrophys.* **97**, 122–7.

Davidson, K. (1977). *Astrophys. J.* **218**, 20–32.

Davidson, K. & Netzer, H. (1979). *Rev. Mod. Phys.* **51**, 715–66.

Ferland, G. J. & Truran, J. W. (1981). *Astrophys. J.* **244**, 1022–32.

Halpern, J. P. & Grindlay, J. E. (1980). *Astrophys. J.* **242**, 1041–45.

Shuder, J. M. & MacAlpine, G. M. (1979). *Astrophys. J.* **230**, 348–59.

Section 4

The references cited for Section 3 are all applicable here. In addition, for Fe II see

Phillips, M. M. (1978). *Astrophys. J.* **226**, 736–52.

Grandi, S. A. (1981). *Astrophys. J.* **251**, 451–64.

A few of the most recent papers on the problems of the H I spectrum under high-density, large optical-depth conditions are

Canfield, R. C. & Puetter, R. C. (1981). *Astrophys. J.* **243**, 390–403.

Collin-Souffrin, S., Dumont, S. & Tully, J. (1982). *Astron. Astrophys.* **106**, 362–74.

Drake, S. A. & Ulrich, R. K. (1980). *Astrophys. J. Suppl.* **42**, 351–83.

Kwan, J. & Krolik, J. H. (1981). *Astrophys. J.* **250**, 478–507.

Mathews, W. G., Blumenthal, G. R. & Grandi, S. A. (1980). *Astrophys. J.* **23**, 971–85.

Weishert, J. C., Shields, G. A. & Tarter, C. B. (1981). *Astrophys. J.* **245**, 406–15.

Section 5

The two Seyfert 1 galaxies with the highest ionization lines are discussed by

Osterbrock, D. E. (1981). *Astrophys. J.* **246**, 696–707 (III Zw 77).

Fosbury, R. A. E. & Sansome, A. E. (1983). *Mon. Not. Roy. Astron Soc.* **204**, 1231–6 (Tololo 0109-383).

A few of the many excellent papers on X-ray measurements of Seyfert galaxies; starting with two early reviews that give many references

Culhane, J. L. (1978). *Quart. J. Roy. Astron. Soc.* **19**, 1–37.

Wilson, A. S. (1979). *Proc. Roy. Soc. London A*, **366**, 461–89.

Maccaraco, T., Perola, G. C. & Elvis, M. (1982). *Astrophys. J.* **257**, 47–55.

Elvis, M. & Speybroeck, L. (1982). *Astrophys. J.* **257**, L51–5 (M 81).

Mushotzky, R. F. (1982). *Astrophys. J.* **256**, 92–102.

Section 6

A selection of observational data

Baldwin, J. A. (1975). *Astrophys. J.* **201**, 26–41 (QSOs and quasars).

Wills, B. J., Netzer, H., Uomoto, A. K. & Wills, D. (1980). *Astrophys. J.* **237**, 319–25 (QSOs).

Wu, C. C., Boggess, A. & Gull, T. R. (1982). *Astrophys. J.* **266**, 28–40 (Seyfert 1 galaxies).

Section 7

Very good expositions of this picture are in

Rees, M. J. (1977). *Ann. N.Y. Acad. Sci.* **302**, 613–36.

Rees, M. J. (1978). *Observatory*, **98**, 210–23.

Some ideas on the tipped disk model are given by

Osterbrock, D. E. (1978). *Proc. Nat. Acad. Sci.* **75**, 540–44.

Tohline, J. E. & Osterbrock, D. E. (1982). *Astrophys. J.* **252**, L49–52.

4

Chemical abundances

L. H. ALLER

4.1 Introduction

Stellar spectra differ widely. Lines of atomic hydrogen are strong in the spectrum of the bright star Sirius, relatively weak in the solar spectrum, while lines of iron and other metals are strong in the solar spectrum, weak or absent in the spectrum of Sirius. Do such differences in the relative strengths of spectral lines reflect differences in relative abundances? Can we conclude that the atmosphere of Sirius is made up largely of hydrogen while the solar atmosphere consists largely of metal atoms?

Since the early 1920s astronomers have understood that the most conspicuous differences between stellar spectra arise from differences in the temperature of the atmospheric layers where the spectral lines are formed, rather than from differences in the relative abundances of the chemical elements. To absorb light at the frequency of one of the Balmer lines, a hydrogen atom must be in its first excited state. The fraction of hydrogen atoms in this state depends on the temperature (and, much more weakly, on the pressure). At the temperature of the solar photosphere (the visible layer of the Sun's atmosphere), nearly all of the hydrogen atoms are in the ground state, where they can absorb lines of the ultraviolet Lyman series. The spectrum of Sirius is formed at a temperature of around 10 000 K. A much larger (though still numerically small) fraction of the hydrogen atoms is in the first excited state at this temperature. At still higher temperatures hydrogen is almost fully ionized, so the relative population of the first excited state is again very low. Metallic lines are weak in Sirius because metallic atoms are almost wholly ionized at the temperatures that prevail there, and most of the metal ions cannot absorb light at visible wavelengths.

These qualitative considerations are made quantitative by the *theory of ionization and excitation equilibrium*. This theory, formulated by M. N. Saha in 1922, is a special case of the statistical theory of chemical equilibrium. For the benefit of our readers who are chemists, the following remarks are included to show the close relationship between physical chemistry and the interpretation of stellar atmospheres. Consider a gas composed of particles A, B, C, ..., Z. (For example, A might stand for the hydrogen molecule, B for the hydrogen atom, C for the proton, D for the electron.) We may symbolize a chemical reaction by the formula

$$n_A A + n_B B + \cdots + n_Z Z = 0.$$

Here n_A denotes the number of particles of type A that *disappear* in the reaction. Thus if n_A is a negative number, $-n_A$ particles of type A *appear* in the reaction. Examples:

$$2H - H_2 = 0,$$

$$p + e^- - H = 0.$$

The first equation represents the formation of a hydrogen molecule from two hydrogen atoms; the second, the formation of a hydrogen atom from a proton and an electron. The condition for thermodynamic equilibrium when the temperature and pressure of the gas are specified is that the *Gibbs free energy* should have its smallest possible value. In an ideal gas the Gibbs free energy is a sum of contributions from the various kinds of particles present. It is shown in books on statistical thermodynamics that the contribution to the free energy from a given set of reactants is a minimum if the following condition is satisfied:

$$n_A \mu_A + n_B \mu_B + \cdots + n_Z \mu_Z = 0,$$

where the coefficients n_A, n_B, ... are the coefficients in the formula that symbolizes the reaction and the quantities μ_A, μ_B, etc., are *chemical potentials*. The chemical potential of a reactant depends on the temperature, the pressure, and the reactant's energy spectrum (see any standard book on statistical thermodynamics, e.g., Landau & Lifshitz: *Statistical Physics*).

Every possible reaction yields a relation among the corresponding free energies, and these relations determine the partial pressures (and hence the relative concentrations) of the various particles.

The ionization equation enables us to relate the number of atoms of any chemical element in the ith stage of ionization N_i to the number in the $(i+1)$st stage, N_{i+1}. If N_ε is the electron density, T is the absolute temperature, X_i is the ionization potential in going from the ith to the $(i+1)$st ionization state, $U_i(T)$ and $U_{i+1}(T)$ are the partition functions for the ith and $(i+1)$st stages, while m, k, and h have their usual meanings, then

$$\frac{N_{i+1}N_{\varepsilon}}{N_i} = \frac{(2\pi m k T)^{3/2}}{h^3} \frac{2U_{i+i}(T)}{U_i(T)} e^{-X_i/kT}.$$

Analogously one can write the dissociation equation for a molecule, e.g., $O + H \leftrightarrows OH$

$$\frac{n(O)n(H)}{n(OH)} = K(OH)$$

where the reaction 'constant' K depends strongly on T, the dissociation potential D_{OH}, and the internal degrees of freedom in the molecule (see, e.g., p. 131 in Aller, *Atmospheres of the Sun and Stars*, 2nd edition, 1963).

The theory of ionization and excitation equilibrium and of molecular dissociation enabled astrophysicists to show that most, though not all, of the conspicuous systematic difference between stellar spectra result from differences in the temperature of the layers where the spectra are produced. It also provided a basis for quantitative chemical analyses of individual stellar atmospheres. The strength of a given iron line (say) is proportional to (1) the number density of iron atoms at the level where the line is produced, (2) the fraction of iron atoms capable of absorbing that line, and (3) an 'intrinsic' factor, the transition probability. The theory of ionization and excitation equilibrium tells us the second factor – the fraction of iron atoms in a given atomic state at a given temperature and pressure. Theoretical calculations or laboratory experiments, or a combination of the two, allow us to estimate the third factor, the transition probability – at least in favorable cases. With estimates of the second and third factors in hand, we can then infer the value of the first factor – the number density of iron atoms – from appropriately reduced measurements of the strength of the spectral line (see below).

Even in the 1920s, however, it was clear that not all differences between stellar spectra could be attributed to differences in ionization and excitation. The spectra of cool stars (spectra containing molecular bands) fall into four distinct classes, labelled M, N, R, and S. The spectra of M stars feature bands of TiO (titanium oxide). In S stars, bands of zirconium oxide replace those of titanium oxide. Both kinds of bands are absent in R and N stars, which feature bands of carbon compounds such as CN and C_2. These differences cannot be caused by differences in the temperature of the layers where the spectra are formed, because the stars have closely similar continuous spectra. (The continuous spectrum of a star approximates that of a black body, and thus indicates the temperature of the layers in which the spectrum is formed.) R. N. Curtiss suggested that the differences between the M stars on the one hand and the R and N types on the other could be understood in terms of differences in the O/C ratio. In cool stars where the O abundance exceeds that of C, formation of CO removes nearly all the carbon, and oxygen is left free to form moelcules such as TiO whose bands are characteristic of M-type spectra. In stars where carbon predominates, all the O is locked up in CO. The excess carbon forms compounds such as CN, C_2, CN, etc., whose lines appear in the spectra of stars. Similarly, differences in chemical composition had to be invoked to explain the prominence of ZrO bands in the spectra of S stars.

At the other end of the temperature range are the hot, seemingly hydrogen-deficient Wolf–Rayet (WR) stars. Two sequences exist. One has broad emission lines of C, O, and He, the other has broad emission lines of N and He, with a weak trace of C. Quantitative analysis in 1941 at Harvard indicated that in the C–O WR sequence, $n(C)/n(O) \approx 3$, $n(He)/n(O) \approx 50$, while in the N WR sequence, $n(He)/n(N) \gtrsim 20$, $n(C)/n(O) \approx 3$, and $n(C)/n(N) \approx 0.05$. The last three estimates imply that the helium–oxygen ratio in the N sequence is of the order of 1000 or more, 20 times larger than in the C–O sequence. This difference and the differences between the spectra of the two sequences can be understood in terms of the operation of the C–N–O cycle under differing conditions of temperature and density.

By the late 1940s and early 1950s the reality of differences in chemical composition for differences between stars had been accepted for these 'classical' examples, and also for differences between stars belonging to the spectral classes, B, A, F, G, and K. Differences arise not only from nuclear processes in stellar interiors (as in WR stars, cool giants, and cool supergiants), but also from differences in initial chemical composition. Presumably, the chemical composition of the interstellar medium in which stars formed varied with time or with location in the Galaxy, or both.

4.2   Importance of composition studies

There are at least five reasons why composition studies are important in astronomy, apart from their intrinsic interest.

(1) The relative abundances of the chemical elements contain information about the processes by which the elements were formed. H, He, deuterium (^2H), and some Li perhaps, were made in the Big Bang. Be, B, and most of the Li on the Earth and meteorites appear to have been produced by spallation of cosmic rays on grains from which the solar system was formed. All other elements must have been cooked in stars, some in relatively quiescent processes, others in violent supernova events.

(2) The rising importance of chemistry as a tool in the study of the solar system and its evolution, thanks largely to the insights of H. C. Urey, requires us to know solar (especially volatile) elemental abundances.

(3) Chemical composition studies of processed material, as found in the envelopes of highly evolved stars and in planetary nebulae, provide checks on our concepts of stellar evolution and element building.

(4) Some people have speculated that the Sun may have accreted metal-rich material early in its history. Consequently, the chemical composition of the Sun's outer layers may not reflect that of the core (apart from differences caused by thermonuclear reactions). This speculation bears on the problem of the Sun's unexpectedly low 'neutrino luminosity' – i.e., the discrepancy between the predicted rate of neutrino emission, calculated on the assumption of an initially homogeneous chemical composition, and the measured rate. The most straightforward interpretation of the measured rate requires the region where the neutrinos are produced, the Sun's central core, to have a much lower metal abundance than the visible layers. To test this idea, theoretical stellar models with accurate chemical compositions as inputs are needed. A comparison of predicted and observed free periods of oscillation of the Sun may help settle this question.

(5) Chemical-composition studies also influence concepts of the evolution of galaxies. For example, the metal/hydrogen ratio varies from one part of a galaxy to another, possibly because rates of star formation, stellar masses, stellar lifetimes, and the rates of nucleosynthesis, may vary from point to point. The element-building rate may also vary with time. Thus chemical compositions, intrinsic colors, and energy distributions of distant galaxies may differ from those in our neighborhood.

4.3 The Sun and the primordial solar system composition (PSSC)

The Sun was the first star to be analyzed qualitatively, but for more than sixty years after the invention of the spectroscope, quantitative analysis had to await the development of modern atomic physics and radiation theory. In the heyday of the universal-cosmic-composition hypothesis, the Sun was proposed as a basic standard since it could be analyzed better than other stars. Knowledge of the solar composition (originally only an astronomer's concern) became crucial when it was recognized that chemistry would be important for understanding our solar system. Several lines of evidence are relevant, in varying degrees to the primordial solar system composition (PSSC).

(1) Easily accessible rocks of the Earth's crust unfortunately give a distorted view of the chemical make-up of the stuff from which the Earth was formed. Siderophile elements such as Fe, Co, Ni, and Au tend to favor the Earth's interior, lithophilic elements like Si, Mg, and Al tend to concentrate in its crust, while chalchophile elements associate with sulphur. Although the composition of the crustal rocks is biased in favor of lithophile elements, their analysis nevertheless yields important abundance information, particularly isotope ratios (since elements tend to separate by chemical processes that do not distinguish between different isotopes); and also information about the relative abundances of lanthanides, which have similar chemistries.

(2) The rubble-strewn lunar surface reflects a complex history of meteorite and asteroid bombardment and tells us little of the PSSC.

(3) Planetary atmospheres can provide data on isotope ratios such as $^{12}C/^{13}C$. Those of Jupiter and Saturn are composed mostly of H and He, but we are not yet able to subject them to quantitative analysis.

(4) Young comets probably reflect the PSSC, but no samples are available. Their spectra exhibit effects of selective resonance excitation by a Doppler-shifted solar spectrum whose wavelength variation is far from smooth.

(5) Although achondritic and iron or stony-iron meteorites show severe fractionation effects, the chondrites, the most numerous of meteorites in falls, offer important data. The carbonaceous chondrites in particular (which show the smallest effect of thermal and chemical processing) are the best sample of the PSSC, insofar as nonvolatile constituents are concerned. A remarkable correlation exists between their chemical composition and that of the solar photosphere (see Table 4.2).

The Sun offers several avenues for abundance studies: (a) the solar photosphere, (b) the solar corona, (c) the 'normal' solar wind, and (d) high-energy solar particles (HESP) (solar cosmic rays). Sources (b), (c), and (d) are far removed from local thermodynamic equilibrium (LTE) and give abundance data on a few plentiful elements. Abundance measurements for some elements can be made in the solar photosphere, in the transition zone between the photosphere and the corona, in the corona proper, in the normal wind, and in HESP. A comparison of the results shows that care must be taken in the interpretation of the data. Most attention has been paid to the photospheric spectrum where the greatest number of elements is represented under conditions amenable to precise theoretical analysis. In other stars, we also normally work with photospheres, although the IUE (International Ultraviolet Explorer) has given data on stellar chromospheres and coronae.

Stellar photospheres have been analyzed by a number of methods. Table 4.1 gives a brief summary of procedures which range from use of simple eye estimates to intricate procedures.

The first quantitative analysis of the solar photosphere was made by H. N. Russell in 1928. He calibrated the eye estimates of intensity in the Rowland atlas of the solar spectrum by using relative line strengths in multiplets. This early approach gave a broad, generally satisfactory picture of the solar composition, except for a poor metal/H ratio. Eye estimates of line intensities have been replaced by measurements of equivalent widths and line profiles.

The equivalent width, W_λ, of a line is the amount of energy it removes from the spectrum, expressed in Ångstroms of the local continuum. The quantity of energy that the line removes from the continuous spectrum is the product of the equivalent width and the (interpolated) intensity of the continuum at the wavelength of the line. (The precise definition is given in column 1 of Table 4.1.)

The equivalent width of a very weak absorption line is proportional to the number of absorbing atoms in the line of sight: doubling the number of absorbing atoms doubles the energy they extract from the beam. But as the number of absorbing atoms increases, the amount of energy they extract from the beam begins to increase more slowly, because atoms close to the source partially 'shadow' atoms more distant from the source. More precisely, the probability that an atom will absorb a photon is proportional to the intensity of the incident beam, and the intensity of the beam, at a wavelength within an absorption line, decreases with distance from the source. A fuller discussion of the curve of growth is given in Section 4.4 below.

The relationship between a line's equivalent width and some convenient function of the product of the number of absorbing atoms and the line's *f*-value (a number proportional to the transition probability), is called the *curve of growth* (see Fig. 4.1). If we know the transition probability or *f*-value and have constructed a curve of growth, we can deduce the number of atoms in the line of sight from the measured equivalent width.

A more modern approach is to measure the actual profile of a spectrum line and represent it by a number of parameters, one of which is the

Table 4.1. *Some methods of analysis of stellar spectra*

Type of data	Theory or method	Application	Authors
A. Eye estimates of line intensities			
1. Stellar spectra	Purely empirical.	Test of ionization theory.	C. H. Payne (1925) D. H. Menzel (1925)
2. Dark line solar spectrum	Calibration of Rowland intensities with theoretical relative atomic line strengths.	Chemical composition of solar photosphere.	H. N. Russell (1928)
3. Eclipse data	Russell method applied to emission lines.	Chemical composition and physical state of solar chromosphere.	D. H. Menzel (1931)

B. Equivalent widths $W_\lambda = \int \left(\frac{I_c - I_L}{I_c}\right) d\lambda$ *where* I_c = *intensity in continuum,* I_L = *residual intensity in the line. The integral is taken over the line.*

Type of data	Theory or method	Application	Authors
	Curve of growth.		
	1. Simple theory of spectral line formation in a Schuster–Schwarzschild (SS) model.	$T_{\text{EXCITATION}}$ and $\langle P_\epsilon \rangle$ in stellar atmospheres, chemical compositions.	Minnaert (1930) C. W. Allen (1934) D. H. Menzel (1936) and A. Unsöld (1938) gave correct formal theory. M. Wrubel (1948, 1951) precise formal theory. See text
	2. Refined theory based on SS or ME model with precise solutions of transfer equations.		
	3. Model atmosphere T_{eff} and $\log g$ from accurate measurements of continuum flux and H-line profiles.		

C. *Line profiles* 1. Individual profiles of single lines	Line-broadening theory including Doppler, radiative, and collisional broadening. 1. Schuster–Schwarzschild or Milne–Eddington (ME) model. 2. Model atmosphere.	Elemental abundances.	A. Unsöld (1928)
D. *Detailed* $I(\lambda)$ *over an extended region, possibly containing many overlapping lines*	Spectral synthesis requires detailed line broadening theory and model atmosphere, $P_\theta(\tau_\lambda)$, etc., with kinematical structure $\xi(\tau_\lambda)$ and $\Xi(\tau_\lambda)$	Elemental abundances, empirical f-values and damping constants constraints on model atmosphere	See text
E. *Line profiles, frequently requiring data from several spectral regions*	1. Non LTE approach requires analysis of entire stellar atmospheric structure, radiation field and detailed atomic parameters 2. Same as above but with an atmosphere not in hydrostatic equilibrium	Elemental abundances, particularly in early-type stars	See text

abundance we wish to determine. Suppose we observe the radiation in a line from a point on the solar disk at which the ray to the observer makes an angle θ with the outward normal. We observe the residual intensity:

$$r_\lambda(\theta) = I_\lambda(0, \theta)/I_\lambda^c(0, \theta), \tag{4.1}$$

where I_λ^c is the interpolated continuum intensity at wavelength λ. $I_\lambda(0, \theta)$, the intensity in the line, is given by:

$$I_\lambda(0, \theta) = \int_0^\infty S_\lambda(t_\lambda) \exp(-t_\lambda \sec \theta) \sec \theta \, dt_\lambda. \tag{4.2}$$

$S_\lambda(t_\lambda)$, the source function, measures the emissivity of the material and

$$dt_\lambda = (\kappa_\lambda + l_\lambda)\rho \, dh \tag{4.3}$$

is the element of optical depth in the line. Here κ_λ is the coefficient of continuous absorption: the line absorption coefficient is

$$l_\lambda = n_{r,s}\alpha_\lambda = n_{r,s}\alpha_0 H(a, v), \tag{4.4}$$

where $n_{r,s}$ is the number of atoms of the species in question (e.g., Ca, C, Fe) in state s of the rth stage of ionization capable of absorbing the line in question. The absorption coefficient at the line center is related to the f-value by the formula

$$\alpha_0 = (\pi \varepsilon^2/mc) f/\Delta v_0 \sqrt{\pi}. \tag{4.5}$$

Here Δv_0 is a measure of the Doppler broadening. The line broadening function $H(a, v)$ handles the combined effects of Doppler broadening and radiation and collisional damping (see, e.g., Aller (1963), p. 323, or Mihalas (1978), p. 279). H requires a broadening function different from $H(a, v)$. The continuous absorption coefficient, κ_λ, is due primarily to processes involving hydrogen (notably the H^- ion in the Sun and atomic hydrogen in hot stars).

To use the preceding formalism to estimate the quantity $n_{r,s}$, the number of absorbing atoms, we need to know the source function S_λ in formula (4.2). In the simplest situation the source function is equal to the source function of a black body (the Planck function) at the local temperature. This black-body source function is denoted by B_λ, and the condition $S_\lambda = B_\lambda$ is called *local thermodynamic equilibrium* (LTE). In LTE

$$I_\lambda^c(0, \theta) = \int_0^\infty B_\lambda(T_\lambda) \exp(-\tau_\lambda \sec \theta) \sec \theta \, d\tau_\lambda, \tag{4.6}$$

where $d\tau_\lambda = \kappa_\lambda \rho \, dh$. Thus

$$dt_\lambda = (1 + \eta_\lambda) \, d\tau_\lambda, \quad \eta_\lambda = l_\lambda/\kappa_\lambda \tag{4.7}$$

and T_λ is the temperature of the layer at which radiation of wavelength λ originates.

Local thermodynamic equilibrium seems to be a good approximation for most photospheric lines in solar-type stars. $B_\lambda(\tau_\lambda)$ and $B_\lambda(t_\lambda)$ both increase monotonically with depth. Absorption lines are formed because of this atmospheric temperature gradient.

To calculate $r_\lambda(\theta)$ we need a model of the solar atmosphere. That is, we need to know the variation of temperature, pressure, and absorption coefficients with optical depth τ_λ. The model itself depends on the chemical composition, so it must be determined by an iterative procedure: One constructs a model atmosphere using plausible values of the chemical abundances, uses this model to deduce values of the chemical abundances from measured line profiles, constructs a new model incorporating these abundances, and so on, until the derived abundances are satisfactorily close to those used in the model. In practice this procedure converges rapidly.

In the Sun one can determine $T(\tau_\lambda)$ and also check the wavelength dependence of the theoretical κ_λ by measuring the brightness of the solar disk at different angular distances from its center. The angle θ between the line of sight and the outward normal to the photosphere is related to the angular distance r between the center of the disk and the point of observation by the formula $\sin \theta = r/R$, where R is the Sun's angular radius. In practice, astronomers measure the 'limb-darkening' function, $I_\lambda^c(0, \theta)/I_\lambda^c(0, 0)$ and intensity distribution at the center of the solar disk $I_\lambda^c(0, 0)$.

Because stars other than the Sun do not have resolved disks, we must use a different method to determine the variation of temperature with the quantity τ_λ, the optical depth at wavelength λ. This variation is determined by the requirement that the energy flux in the star – the rate at which radiant energy crosses a given spherical surface – must be the same at all depths; if it were not, energy would accumulate in some layers and be depleted in others. And physical depth is related in a known way (if the absorption coefficient is known) to optical depth at any given wavelength. Of course this condition also applies to the Sun. If energy is transported by convection in any stratum, then it is clear that it is the sum of convective and radiative flux that must be constant.

Each proposed solar atmospheric model yields different numerical relationships and can lead to appreciably different predictions of line profiles and equivalent widths. With temperature and pressure given as functions of optical depth at some standard wavelength, we calculate $n_{n,s}/n_E$ (where n_E is the number of atoms of element E per gram of stellar material). Given $H(a, v)$ and the f-value, we can predict the line profiles for different assumed values of n_E and obtain the abundance n_E/n_H from the best fit.

A critical quantity in making the fit is the ratio l_λ/κ_λ which determines for each point in the line the relationship between t_λ and τ_λ. It depends on presumably known factors affecting line broadening (molecular and turbulent motions, radiative and collisional damping) and also on the ratio n_E/n_H. Of crucial importance, of course, is a knowledge of the f-values. For rare elements, we need to know them for only the strongest lines, since only these, if any, appear in the solar spectrum. For an abundant metal such as iron, however, the intensities of strong lines depend not only on the f-value but also on the so-called damping constant. The damping constant determines the width of the line profile at zero temperature – the 'natural' line width. Every spectral line has a finite width, related to the lifetime of the initial level of the transition that gives rise to it by the uncertainty relation

$$\Delta v\, \Delta t \approx 1.$$

The lifetime of an atomic state is determined by the rate of radiative transitions from that state and the rate of collisionally induced transitions. The damping constant is in fact the sum of damping constants associated with the two kinds of processes: $\Gamma = \Gamma_{rad} + \Gamma_{coll}$. In solar-type stars, the collisional contribution dominates $\Gamma_{coll} \gg \Gamma_{rad}$. Since Γ_{coll} is usually poorly known, we are forced to employ weak lines whose f-values often are hard to determine. Thus, until recently, inaccurate f-value measurements seriously hindered abundance work. As f-values have improved, the chief source of uncertainty for an abundant metal usually now lies in the solar atmospheric model. From this cause we can get an error $\Delta \log N_E/N_H \approx 0.2$.

Unfortunately, abundance determinations for noble gases and halogens remain poor. At the low photospheric temperatures of the Sun, lines of the noble gases whose wavelengths lie in the visible region cannot be excited. Ar and Ne are observed in the solar corona and wind; He is observed in the solar wind and occasionally in prominences, but under conditions very far removed from local thermodynamic equilibrium.

Reviews of solar abundances have been published by a number of writers, e.g., Withbroe (1971), Ross & Aller (1976), Hauge & Engvold (1977), and Holweger (1979). Cameron (1982) and Anders & Ebihara (1982) give tables of solar system abundances. Table 4.2 lists elements for which significant new solar abundance determinations have been made since the reviews of 1976 and 1977. In the second column we give $\log N$(element) on the 'astrophysical' scale, i.e., $\log N(H) = 12.00$ from the compilation by Ross and Aller (1976). The next columns give new determinations and references. The column headed 'adopted values' gives suggested values as of 1985. The next column gives elemental abundances for CI carbonaceous chondrites (Wasson 1985) on the scale $\log N(\text{Si}) = 7.64$.

First, consider elements such as Si and metals whose absorption lines appear in the photospheric spectrum. For boron, Kohl *et al.* (1977) used a spectrum synthesis method to find a $\log N(\text{B})$ and a B/Be abundance ratio that agrees well with predictions of light nuclide formation by galactic cosmic-ray spallation. With newer *f*-values, Lambert & Luck (1978) revised abundances of Na, Mg, Al, Si, P, S, K, and Ca to get values differing only slightly from those given by Ross and Aller (1976).

Abundances of Fe-peak elements deduced from ionic lines agree well with those found from neutral atomic lines, used generally heretofore (Biémont 1978). The iron abundance itself is a classical problem on which much effort has been expended. Accurate measurements of *f*-values by the Oxford group permit a new assessment of the problem; the error in the iron abundance arises from inaccuracies in equivalent widths, damping constants, and model atmospheres. Blackwell & Shallis (1979) conclude that 'until an acceptable model atmosphere is available, the spectroscopic abundance of iron derived from the Fe I spectrum will remain uncertain by at least 0.2 dex'.

Improved determinations (primarily due to better *f*-values) have also been made for several relatively rare elements, including Nb, Ru, Sn, Pr, Sm, Er, Hf, Os, and Ir. Tin disagrees with the chondritic values by a factor of two, the differences for other elements are smaller. For example, with the new Ir *f*-values measured by Ramanujam & Andersen (1978), the $\log Nf$ value found by Drake and Aller yields identical solar and chondritic values.

In summary, it seems that, for the metals, solar and type CI chondritic values agree well (Table 4.2). Discordances can be attributed to limitations in solar photospheric models, to line blending, erors in W_λ, *f*-values, and Γ_{coll}s. Thus, for the metals we might regard the abundance problem as essentially 'solved' and invert the problem – using spectral line data to improve atmospheric models, investigate turbulence, etc.

Abundance determinations are difficult for many nonmetals, such as the noble gases. When we must rely on the solar corona and solar wind, problems pertaining to systematic separations of different atomic species, diffusion, selective excitation and acceleration mechanisms, etc., become serious. Fortunately, the abundances of some of these elements can be found from both coronal and dark-line photospheric spectra. Often, bad discordances appear. To what physical mechanisms can these differences be attributed? One suggestion is that complicated effects loosely called 'diffusion' can modify the element/hydrogen ratio in the corona as compared with its photospheric value. Another possibility is that a selective

Table 4.2. *Abundances of some elements in the Sun*

(Data are listed for those elements for which improved data have become available since 1976.)

$$\log \frac{N(\text{element})}{N(H)} + 12.00$$

	Ross & Aller (1976)	More recent measurements and references	Adopt	C-I chondrites, Wasson (1983)	Δ (CI-0)
He	10.8±0.20	11.07 (1)	11.0:		
Li	1.0±0.1	1.16±0.10 (30)	1.16	2.42	1.2
Be	1.15±0.2			1.54	0.39
B	<2.1±0.2	2.6±0.03 (2)	2.6	3.03	0.43
C	8.62±0.12	8.67 (3)	8.66		
N	7.94±0.15	7.99 (3)	7.98		
O	8.84±0.07	8.92±0.035 (3) 8.91±0.01 (28, 31) 8.84 (32)	8.91		
Ne	7.57±0.12	8.05 (4)	8.05		
Na	6.28±0.15	6.32 (5)	6.31	6.39	+0.08
Mg	7.60±0.15	7.62 (5)	7.62	7.67	+0.05
Al	6.52±0.12	6.49 (5)	6.51	6.57	+0.06
Si	7.65±0.08	7.63 (5) 7.55 (23)	7.64	7.64	
P	5.50±0.15	5.45 (5)	5.48	5.59	+0.11
S	7.2±0.15	7.23 (5)	7.23	7.33	+0.10
Ar	6.0±0.2	6.57 (4)	6.57		
K	5.16±0.10	5.12 (5)	5.15	5.22	+0.07
Ca	6.36±0.10	6.34 (5) 6.32 (6) 6.30 − 6.36 (19)	6.34	6.43	+0.09
Sc	3.04±0.07	3.08±0.15 (7) 3.08±0.03 (24)	3.08	3.18	+0.10
Ti	5.05±0.12	4.99±0.16 (7) 4.98±0.15 (8) 5.08 (16)	5.06	5.01	−0.05
V	4.02±0.15	4.14±0.17 (7)	4.11	4.11	+0.00
Cr	5.71±0.14	5.88±0.15 (7) 5.64 (15)	5.76	5.77	+0.01
Mn	5.42±0.16	5.4±0.2 (7)	5.42	5.61	+0.19
Fe	7.50±0.08	7.56 − 7.66 (9) 7.46 − 7.65 (10) 7.57 (17) 7.63±0.05 (33)	7.59	7.58	−0.01
Co	4.90±0.18	4.98±0.38 (7) 4.92±0.08 (18)	4.93	5.00	+0.07
Ni	6.28±0.09	6.28±0.39 (7) 6.22±0.13 (22)	6.27	6.33	+0.06
Zn	4.45±0.15	4.60±0.08 (21)	4.58	4.75	+0.17
Yt	2.10±0.25	2.71±0.15 (25) 2.24±0.03 (26)	2.24	2.28	+0.04
Zr	2.75±0.15	2.56±0.05 (20)	2.66	2.69	+0.03
Nb	1.9±0.2	1.42±0.05 (34)	1.42	1.53	+0.11

Mo	2.16±0.2	1.92±0.05 (25)	1.92	2.05	+0.13
Ru	1.83±0.4	1.84±0.07 (29)	1.84	1.91	+0.07
Sn	2.0±0.4	1.85±0.3 (11)	1.94	2.23	+0.29
Pr	0.66±0.15	0.71±0.08 (12)	0.71	0.89	+0.18
Sm	0.72±0.3	0.80±0.11 (27)	0.80	1.06	+0.26
Er	0.76±0.4	0.93±0.06 (35)	0.93	1.05	+0.12
Hf	0.8±0.1	0.88±0.8 (13)	0.88	0.90	+0.02
Os	0.7±0.2	1.45±0.10 (36)	1.45	1.48	+0.03
Ir	0.85±0.2	1.45±0.2 (14)	1.45	1.45	0.00

(1) Cameron, A. G. W. (1982). Elemental and nuclidic abundances in the Solar System. In *Essays on Nuclear Astrophysics*, ed. C. A. Barnes *et al.* Cambridge University Press.
(2) Kohl, J. L., Parkinson, W. H. & Withbroe, G. L. (1977). *Astrophys. J. Lett.* **212**, L101.
(3) Lambert, D. L. (1978). *Mon. Not. Roy. Astron. Soc.* **182**, 249.
(4) Dietrich, W. F. & Simpson, J. A. (1979). *Astrophys. J.* **231**, L91. See also Geiss, J., *et al.* (1969). *NASA Spec. Publ.* (214), p. 183.
(5) Lambert, D. L. & Luck, R. E. (1978). *Mon. Not. Roy. Astron. Soc.* **183**, 79.
(6) Ayres, T. R. & Testerman, L. (1978). *Solar Physics*, **60**, 19.
(7) Biémont, E. (1978). *Mon. Not. Roy. Astron. Soc.* **184**, 683.
(8) Whaling, W., Scalo, J. M. & Testerman, L. (1977). *Astrophys. J.* **212**, 581.
(9) Hauge, O. & Engvold, O. (1977). Report No. 39, Inst. Theo. Astrophys. (Oslo).
(10) Blackwell, D. E. & Shallis, M. J. (1979). *Mon. Not. Roy. Astron. Soc.* **186**, 673.
(11) Allen, M. S. (1978). *Astrophys. J.* **219**, 307.
(12) Biémont, E., Grevesse, N. & Hauge, O. (1979). *Solar Physics*, **61**, 17.
(13) Andersen, T., Peterson, P. & Hauge, O. (1976). *Solar Physics*, **49**, 211.
(14) Ramannujam, P. S. & Andersen, T. (1978). *Astrophys. J.* **226**, 1171.
(15) Biémont, E., Grevesse, N. & Huber, M. E. (1978). *Astron. Astrophys.* **67**, 87.
(16) Blackwell, D. E., Shallis, M. J. & Simmons, G. J. (1982). *Mon. Not. Roy. Astron. Soc.* **199**, 33, 37.
(17) Gurtovenko, E. A. & Kostik, R. I. (1981). *Astron. Astrophys. Suppl.* **46**, 239.
(18) Cardon, B. L., Smith, P. L., Scalo, J. M., Testerman, L. & Withbroe, G. L. (1982). *Astrophys. J.* **260**, 395.
(19) Smith, G. (1981). *Astrophys. J.* **103**, 351.
(20) Biémont, E., Hannaford, P. & Lowe, P. M. (1981). *Astrophys. J.* **248**, 867.
(21) Biémont, E. & Godefroid, M. (1980). *Astron. Astrophys.* **84**, 361.
(22) Biémont, E., Grevesse, N., Huber, M. E. & Sanderman, R. J. (1980). *Astron. Astrophys.* **87**, 242.
(23) Becker, U., Zimmerman, P. & Holweger, H. (1980). *Geochim. Cosmochim. Acta*, **44**, 2145.
(24) Youssef, N. H. (1979). *J. Astron. Soc. Egypt*, **1**, 96.
(25) Biémont, E. (1983). *Astrophys. J.* **275**, 889.
(26) Hannaford, P., Lowe, R. M., Grevesse, N., Biémont, E. & Whaling, W. (1982). *Astrophys. J.* **261**, 736.
(27) Saffman, L. & Whaling, W. (1979). *J. Quant. Spec. Rad. Trans.* **21**, 93.
(28) Grevesse, N., Sauval, A. J. & van Dishoeck, E. F. (1984). *Astron. Astrophys.* **141**, 10.
(29) Biémont, E., Grevesse, N., Kwiatkovsky, M. & Zimmerman, P. (1984). *Astron. Astrophys.* **131**, 364.
(30) Steenbock, W. & Holweger, H. (1984). *Astron. Astrophys.* **130**, 319.
(31) Sauval, A. J., Grevesse, N., Brandt, J. W., Stokes, G. M. & Zander, R. (1900). *Astrophys. J.* **282**, 330.
(32) Goldman, A., Murcray, D. G., Lambert, D. L. & Dominy, F. (1983). *Mon. Not. Roy. Astron. Soc.* **203**, 767.
(33) Rutten, R. J. & van der Zalm, E. B. J. (1984). *Astron. Astrophys. Suppl.* **56**, 143.
(34) Hannaford, P., Lowe, R. M., Biémont, E. & Grevesse, N. (1985). *Astron. Astrophys.* **143**, 447.
(35) Biémont, E. & Youssef, N. Y. (1984). *Astron. Astrophys.* **140**, 177.
(36) Kwiatkowski, M., Zimmerman, P., Biémont, E. & Grevesse, N. (1984). *Astron. Astrophys.* **135**, 59.

acceleration mechanism may enhance the coronal abundance of one group of elements with respect to another. Thus, J. P. Meyer concluded that diffusion effects are actually unimportant. He found a tendency for atoms with low first ionization potentials (FIP) ($5 < \text{FIP} < 8.5$ eV) to have essentially photospheric abundances in the corona, solar wind, or solar cosmic rays while elements of higher FIP ($10 < \text{FIP} < 22$ eV), e.g., O, N, and Ar, are underabundant by a factor of 4–6. These elements have a normal abundance ratio among themselves. Meyer (1981) suggests that, in the cooler chromosphere, metals are almost completely ionized, while O, N, and the noble gases remain neutral. Particles are drawn into the corona by a process that favors the charged ones by a factor of about 4:1. Thus the compositions of the corona and both the 'normal' and fast solar winds are biased against elements of high FIP.

Galactic cosmic rays show a similar abundance anomaly, which suggests that the material comes from solar-type stars and is later accelerated from MeV to GeV energies in the interstellar medium (ISM). Fortunately, a partial correction for this effect seems possible. Some elements of moderately high FIP. such as C, N, and O are found in both photosphere and corona. (Note that the new $\log N$ [element] values obtained by Lambert (1978) are systematically higher than those proposed by Ross and Aller by 0.05 for C and N to 0.08 for oxygen.) Ne is not significantly more depleted than C, N, and O in the solar corona. Thus, to get the Ne abundance we may use the solar wind Ne/O ratio. I have given the highest weight to the measurement by Dietrich & Simpson (1979), viz., $n(\text{Ne})/n(\text{O}) = 0.14$ and have adopted $\log N(\text{Ne}) = 8.05$. Then a Ne/Ar ratio of 30 was chosen from solar wind measurements by Geiss *et al.* (1969). The final values selected seem to be in harmony with those given by J. P. Meyer and with the values quoted for the interstellar medium.

For He the situation is puzzling. The solar wind value, He/H ≈ 0.05, is lower than the 'cosmic' value of 0.10. Analyses of solar prominence observations are uncertain. For the present, I recommend a value of 0.11.

The next-to-last column of Table 4.2 lists abundances normalized to $\log N(\text{Si}) = 7.64$ as derived from CI (or C–I) chondritic meteorites (Wasson (1985)). The last column gives $\log(\text{CI} - \text{Sun})$. We notice (as have many other investigators, e.g., Cameron, Anders, Wasson, and Holweger) that there is a close agreement between chondritic and photospheric abundances for S, Si, and all metals for which good f-values are known. Larger discrepancies found for Sn, Pr, and Sm are probably due to severe blends; that for Mn may reflect poor f-values or the influence of hyperfine structure. Helium remains the outstanding embarrassment among solar abundances.

4.4 Stellar abundance determinations, curve of growth

Astronomers were quick to apply developments in spectroscopy and atomic physics to spectral analyses of stars. The earliest, necessarily qualitative, efforts provided a check on ionization theory (see Table 4.1). Real progress began with Minnaert's invention of the curve of growth (COG) and subsequent refinements by Menzel (1936), Unsöld (1938) and others. The COG basically relates the equivalent width W_λ of an absorption line to the number N of atoms acting to produce it (see Fig. 4.1). The curve owes its characteristic shape to the nature of spectral-line broadening. When a line is very weak all absorption occurs in the core where the shape of the absorption coefficient is governed by Doppler motion. Initially, W_λ is proportional to Nf, but as the line core saturates, the curve flattens, varying eventually roughly as $\sqrt{(\log Nf)}$. When Nf is very large, heavy absorption occurs far out in the wings of the line and W_λ now increases as $\sqrt{(Nf\Gamma)}$, where Γ is the damping constant discussed in Section 4.3. Every COG has this general shape but the details depend on: (a) the value of the microturbulent eddy velocity ξ_{turb}, which determines the vertical position of the COG, (b) the value of Γ which fixes the damping portion of the curve, and (c) the adopted structure of the atmosphere. Early COGs assumed a Schuster–Schwarzschild or SS model – a photosphere radiating a pure continuous

Fig. 4.1. Curve of growth. We plot log $Wc/\lambda V$ against log X_0, W is the equivalent width in the same units as the wavelength, λ, $V = \sqrt{(2kT/M) + \xi^2}$, where ξ is the most probable turbulent velocity, and M is the mass of the atom. In the simple Schuster–Schwarzschild (SS) atmospheric model the lines are formed in a thin layer overlying a photosphere which emits a continuum. $X_0 = N_{r,s}\alpha_0$ where $N_{r,s}$ is the number of atoms lying above the atmosphere and capable of absorbing the line; α_0 is defined in (4.5).

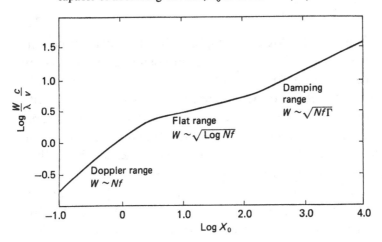

spectrum over which lay a stratum that produced the dark lines. Theoretical curves were also calculated for Milne–Eddington (ME) models in which l_λ/κ_λ is constant with depth. For an example, see, e.g., Aller (1961).

In modern work we employ a theoretical model atmosphere and predict line profiles from which we then compute W_λ. The model atmosphere is specified by the effective temperature, T_{eff}, the logarithm of the gravitational acceleration in the photosphere, log g, and the stellar chemical composition, particularly the He/H and metal/H ratios. (The effective temperature of a star is the temperature of a black body with the same radiated power. It is determined by the star's luminosity and surface area. The gravitational acceleration enters into the condition of hydrostatic equilibrium; it determines the density stratification in the atmosphere.) We modify the input abundance ratios, n(element)/n(H), until a suitable fit between the assumed and derived abundance ratios is obtained.

The 'best' value of ξ_{turb} is that which gives the same value of the elemental abundance for lines of that element of different intensities.

How is the model atmosphere for a given star chosen? The investigator usually starts with a family of model atmospheres that have approximately the expected metal/H ratio. From the unreddened color of the star, its spectrum, and its luminosity class, one can estimate preliminary values of T_{eff} and log g. Consider, e.g., a star in the range B2 to F5. The H line profiles are sensitive primarily to log g while the energy distribution, $F(\lambda)$, depends on both T_{eff} and log g. Hence, we can isolate a small region in the T–log g plane in which the star must lie. We represent the W_λs for a small grid of models enclosing this area. For each grid point, defined by (T_{eff}, log g) we derive abundances and interpolate them for the most probable value (T_{eff}^0, log g^0). The abundances, $A = \log N_E/N_H + 12$ may then be expressed in the form:

$$A = A_0 + \alpha \, \Delta_t^* + \beta \, \Delta \log(g) \tag{4.8}$$

where $\Delta_t^* = (T - T_0)/T_0$ and $\Delta \log(g) = \log(g/g_0)$. This formulation permits one to see quickly how slight changes in model–atmosphere parameters can modify abundances when improved estimates of T_{eff} and log g are found.

We may expect this method to work well for main sequence stars with known energy distributions. We may be pressing our luck when we apply these models to giants and supergiants where the assumptions underlying model–atmosphere calculations are less secure. Nevertheless, the method does appear promising. As an example, consider an A 2Ib supergiant, G 421, in the Large Magellanic Cloud. This star was observed by B. J. O'Mara, J. E. Ross, and Bruce Peterson with the Anglo–Australian telescope; the data were analyzed by a curve of growth program written by Ross. Kurucz (1979) published a grid of LTE model atmospheres in the range $5500 < T_{eff} < 50\,000$ and log g values ranging from those appropriate to main sequence stars to objects that approached the stability limit. Kurucz also predicted Balmer line profiles, emergent fluxes $F(\lambda)$, and UBV and uvby colors. The observed energy distributions and H line profiles for G 421 indicate $T \approx 8000$ K and log $g = 1.5$. See Figs. 4.2 and 4.3. Although the data are preliminary and the scatter is large, it seems clear that metals of the iron group, viz., Sc, Ti, Cr, and Fe, are depleted by factors of between three and six as compared with the Sun.†

For the Sun and cooler stars, and for all stars in the UV ($\lambda < 3000$), line crowding and overlapping or profiles can become severe. Frequently, the

† A more rigorous procedure might entail an iterative process, i.e., one would get a first approximation to abundances in these Magellanic Cloud stars and then compute new models to insure a proper dependence of T, P_g, and P_ε on τ_0. Alas, uncertainties in relevant atmospheric theory (e.g., is the atmosphere in hydrostatic equilibrium?) as well as data limitations due to inadequate observing time render such refinements premature.

Fig. 4.2. Energy distribution in the supergiant G 421 in the Large Magellanic Cloud (LMC). Note close agreement between the predictions of the Kurucz theory (indicated by circles) and observations by Ross, O'Mara, and Peterson with the Anglo–Australian Telescope.

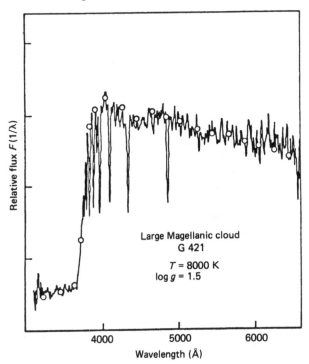

faint lines of a target rare element fall on the wings of a strong line of some abundant element such as Fe. The COG is of limited use here. What we must do is represent, i.e., synthesize, at least a limited spectral region in detail. We need a good model atmosphere, with appropriate values of the r.m.s. microturbulent and macroturbulent velocities ξ and Ξ. (In microturbulence the mean free path of a photon is long compared with the size of a single turbulent element while in macroturbulence the opposite is true.) For solar spectral synthesis work, $\xi = 0.8\,\mathrm{km\,s^{-1}}$ and $\Xi = 1.9\,\mathrm{km\,s^{-1}}$ have been proposed. In principle, both ξ and Ξ can vary with optical depth. For a spectral synthesis calculation, we need, for each line in the spectral domain, values of λ_0 (wavelength of the line center, Γ_{rad}, and Γ_{coll}) as a function of gas pressure. We use the product Nf as a free parameter, and proceed by trial and error. The Γ-values can be adjusted only within narrow limits, sometimes small corrections to the assumed central wavelength are needed, and sometimes we find evidence of additional lines not apparent from an

Fig. 4.3. Profile of Hδ in G 421 in the LMC. Theoretical profiles (indicated by solid lines) are taken from Kurucz' calculations.

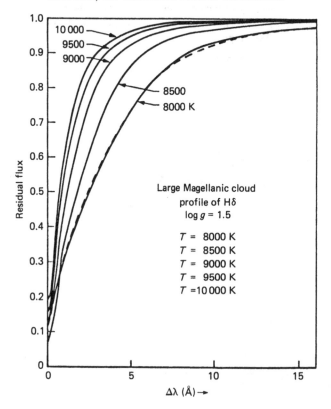

inspection of the spectrum. Unidentified lines (which abound even in the Sun) are troublesome.

High-speed computers have popularized spectral synthesis procedures. One can calculate both the broadband and detailed energy distribution and allow for a finite resolution, and thus construct a 'convolved' spectrum that can be compared directly with an observed spectrogram or scan. A variation of the method is to take observed scans of stars of presumably known chemical compositions (e.g., as derived from COGs) rather than using purely theoretical spectra. These are convolved with scan slot widths, filter transmissivities, etc., to give resultant spectra and colors that can then be used to interpret observations of stars of unknown composition. For late-type stars with prominent molecular bands, synthetic spectra are clearly necessary. One example is the work by Carbon *et al.* (1982) to get C and N abundances in giant stars of the metal-poor globular cluster M92. They matched the observed spectra of some 45 stars brighter than $M_v = -2$ with synthetic spectra calculated with laboratory *f*-values and model atmospheres given by Gustafsson *et al.* (1975). Another example is the extensive work by D. L. Lambert *et al.* on late-type stars, used to assess predictions of nucleogenesis theory.

An important application of spectrum synthesis is to the integrated light of stellar systems such as distant globular clusters and galaxies. Assuming a luminosity–spectral class function, and adding together contributions from sets of stars, one proceeds by iteration. For each set of stars (of defined values T_{eff}, M_v, and metal/H) the observed spectrum and energy distribution are presumed known. The resultant composite spectrum must also be convolved with the random space motions of individual stars (often ≈ 180–$300 \, km \, s^{-1}$). The method requires a sufficient supply of observed basic stellar spectra of relevant T_{eff}, M_v, and metal/H groups. In principle, one can adjust individual elemental abundances to fit the observations well. Clearly to extract information from the light of globular clusters and galaxies, we need as long a spectral range as practicable. Infrared radiation is contributed almost entirely by giants and supergiants. It depends not only on temperature but also on metallicity, since, the greater the metal/H ratio, the redder the color of the giant branch. The observations typically involve low spectral resolutions, usually V, K, and R colors, but can be useful in establishing important parameters such as the metal/H ratio. Comparisons between predicted and observed spectra can yield information about the metal/H ratio, place limits on the dwarf/giant ratios, and establish deficiencies or excesses of elements such as C or N (see, e.g., Kraft (1979) and Freeman & Norris (1981)). One can also study radial abundance gradients in galaxies from radial gradients of absorption features (see, e.g., Cohen (1979)).

4.5   Breakdown of LTE and hydrostatic equilibria in stellar atmospheres

A popular assumption is that in (4.2) we can set $S_\lambda = B_\lambda$ or $S_v = B_v$, i.e., postulate LTE, but for some lines in the Sun and in many hot stars this assumption is not justified and we must use more elaborate treatments (see Mihalas, *Stellar Atmospheres*, 1978). The basic relationship is:

$$S_v = \frac{j_v}{k_v} = \frac{2hv^3}{c^2} \left[\frac{b_{n'}}{b_n} e^{hv/kT} - 1 \right]^{-1} = \frac{2hv^3}{c^2} \left[\frac{N_{n'} g_n}{N_n g_{n'}} - 1 \right]^{-1} \tag{4.9}$$

where j_v and k_v are respectively the emission and absorption coefficients, and n, n′ label the upper and lower states respectively of the transition that gives rise to radiation of frequency v. For a gas in thermodynamic equilibrium, the coefficients b_n are all unity, and this relation expresses Kirchhoff's law: in thermodynamic equilibrium the ratio between the emission and absorption coefficients is the intensity of black-body (Planck) radiation. For a gas that is in statistical equilibrium, but not in thermodynamic equilibrium, the coefficients b_n differ from unity. The relative populations of two levels is given by:

$$N_n / N_{n'} = (b_n / b_{n'})(g_n / g_{n'}) \exp(-\chi_{nn'}/kT), \tag{4.10}$$

where g_n and $g_{n'}$ are the statistical weights of the two levels n and n′ and $\chi_{nn'}$ is the energy difference between them. Here T is the temperature defined by the local Maxwellian velocity distribution of the electrons. Although deviations from LTE may be severe, the electronic velocity distribution will still follow closely a Maxwellian law.

Not only do the level populations no longer follow the Boltzmann law, but the non-LTE ionization equilibrium may depart far from that predicted by the Saha equation. Further, S_v may depart appreciably from B_v. One obtains the populations in the excited levels by solving the equations that express the conditions of statistical equilibrium: the average population of each atomic or molecular level must remain constant in time. In gaseous nebulae, where this problem was first successfully addressed, there is a simplifying feature: a gaseous nebula is nearly transparent, except in resonance transitions, so the radiation incident at any given interior point is often simply the dilute radiation of the exciting star. In a stellar photosphere the situation is much more complicated. To predict line intensities, one must calculate the radiation field in the lines as well as in the continuum and evaluate the steady state population of each relevant excited level throughout the pertinent atmospheric strata.

The resultant predictions often depart from those based on LTE theory.

Non-LTE effects are particularly severe in stars earlier than about B3, and can enormously complicate abundance determinations. To mention one example, P. Dumont tried to derive the nitrogen abundance in certain O-type stars from N III lines. These transitions had often been employed in LTE analyses, but a detailed non-LTE treatment showed the W_λs to be insensitive to abundance and very sensitive to details of the radiation field and atmospheric structure.

Lines of He I and He II in O-type stars often tend to be stronger than LTE theory would predict for a fixed T_{eff}. Thus, a temperature calibration of spectral class deduced from LTE theory may be seriously in error. To obtain abundances of elements like C, N, and O in early-type stars, both accurate profiles and careful non-LTE analyses are necessary.

Yet another complication appears in the atmospheres of early-type stars – and many others as well – a breakdown of the condition of hydrostatic equilibrium. Ultraviolet observations of the spectra of O- and B-type stars show resonance lines with P Cygni-type profiles (an emission line on whose violet wing there appears an absorption feature). The emission line is produced in an envelope that surrounds the main body of the star where the continuous spectrum arises. The absorption component is due to material in the line of sight and its violet displacement indicates that the shell is expanding. Thus there is a substantial, optically thick wind blowing from the star. Much, if not all, of the observed spectrum may be produced in this rapidly expanding envelope which is certainly not in hydrostatic equilibrium. So added to the complexities of non-LTE, we must discard conventional model atmospheres and learn how to develop dynamical models that will satisfy the available observational data.

4.6 Abundances in gaseous nebulae

The elements He, C, N, O, and Ne are important for stellar structure and as tracers of element building in stars. Unlike stars, which now can be analyzed only in our Galaxy or in the Magellanic Clouds, gaseous nebulae in remote galaxies can be studied. In particular, we can study the N/H and O/H ratios as a function of position in a galaxy.

The spectra of gaseous nebulae are produced by plasmas under extreme non-LTE conditions. Fortunately, the line-excitation mechanisms appear to be well understood. Bowen & Wyse's (1939) semi-quantitative pioneering studies of chemical compositions of gaseous nebulae showed a basic similarity between stellar and nebular abundances. Improvements in the theories of collisional and radiative processes in gaseous nebulae (largely at Harvard 1937–41) permitted a realistic attack on the problem.

Some nebular spectral lines are produced by photoionization followed by recapture and cascade (H, He, and weak permitted lines of C, N, O, and Ne). Strong lines of ions of C, N, O, and Ne, etc., are produced by collisional excitation followed by radiative decay. Thus, in order to interpret the observed data we need certain atomic parameters: coefficients of photoionization (a_v), recombination coefficients (α), spontaneous-transition probabilities (A), collision strength parameters for electron excitation (Ω), and charge exchange cross sections (σ_x).

We can find the 'plasma diagnostics' (T_ε, N_ε) (T_ε = electron temperature, N_ε = electron density) from certain line ratios; radio-frequency and optical data involving H lines and continuum are also valuable. For nebular plasma diagnostics we normally use ions of $2p^n$ and $3p^n$ configurations. Now p^2 and p^4 configurations give ^3P, ^1D, and ^1S terms. We compare the auroral-type to nebular-type transitions, i.e., the ratio $I(^1S - {}^1D)/I(^1D - {}^3P) = f_1(N_\varepsilon, T_\varepsilon)$. At typical nebular densities $100 < N_\varepsilon < 2 \times 10^4$ electrons/cm^3, this ratio depends primarily on T_ε and only slightly on N_ε (except for [N II]). In a p^3 configuration, the ratio of the two nebular-type transitions, $I(^2D_{3/2} - {}^4S_{3/2})/I(^2D_{5/2} - {}^4S_{3/2}) = r_1(N_\varepsilon, T_\varepsilon)$, depends strongly on N_ε, and weakly on T_ε. We can also use the ratio of auroral-to-nebular-type transitions, $I(^2P - {}^2D)/I(^2D - {}^2P) = r_2(N_\varepsilon, T_\varepsilon)$, which is sensitive to both N_ε and T_ε. In a given nebula, the observed intensity ratios define curves, f_1, r_1, and r_2. If the nebula were homogeneous and the atomic parameters and the observed intensities were all accurate, the curves would intersect at a point that would fix a definite (N_ε, T_ε). In practice, they often do not, as is shown by the diagnostic diagram for IC 2165 (Fig. 4.4). We may then need to represent the nebula by a two-zone model, characterized by T_ε(low) and T_ε(high).

Next we calculate the ionic concentration. For recombination lines, we compare I_L (the intensity of a given line) with $I(H\beta)$, and assume that $I_L/I(H\beta) = E(\lambda_L)/E(H\beta)$. The emission per unit volume of an ion X^i in I_L is $E(\lambda_L) = \alpha_{jj'} N(X^{i+1}) h_{\nu_{jj'}}$ where $\alpha_{jj'}$ is the effective recombination coefficient for ions that reach the upper level j of the transition jj' corresponding to λ_L and then cascade to j'. Similarly, for hydrogen, we have that $E(H\beta) = N_\varepsilon N(H^+) E^0_{4,2} \times 10^{-25}$ where $E^0_{4,2}(T_\varepsilon)$ is a factor of the order of unity ($E^0_{4,2} = 2.22$, 1.24, and 0.66 at $T_\varepsilon = 5000$, 10 000, and 20 000, respectively). Note that recombination lines of an ion X^i give the ionic concentration of ions X^{i+1}. Thus, values of $I(\lambda 4267)$ would be expected to give $N(C^{++})$; e.g., $N(C^{++})/N(H^+) = 0.109[T_\varepsilon/10^4]^{0.14}[I(4267)/I(H\beta)]$. Actually, there is some controversy whether this line is formed purely by a simple recombination process. At nebular densities we can neglect the N_ε dependence.

For collisionally excited lines, the problem is often more complicated. We set up the equations of statistical equilibrium for each ground configuration level and may include other low-lying levels. We have equations of the form:

$$\sum_{i \neq j} N_i(X^k) N_e q_{ij} + \sum_{i>j} N_i(X^k) A_{ij} = N_j(X^k) \left\{ \sum_{j>i} A_{ji} + N_e \sum_{j \neq i} q_{ji} \right\} \quad (4.11)$$

$$q_{ij} = 8.63 \times 10^{-6} \frac{\Omega_{ij}}{\tilde{\omega}_i \sqrt{T_\varepsilon}} e^{-\chi_{ij}/kT_\varepsilon}, \quad q_{ji} = 8.63 \times 10^{-6} \frac{\Omega_{ij}}{\tilde{\omega}_j \sqrt{T_\varepsilon}}, \quad E_j > E_i.$$
$$(4.12)$$

The emission per unit volume in a line jj' is $E(\lambda_{jj'}) = N_j A_{jj'} h\nu_{jj'}$. Hence, from $I(\lambda_{jj'})/I(H\beta)$ we can solve for $N_j(X^k)/N(H^+)$, where X denotes the ion under consideration. We easily obtain $N(X^k) = \sum N_i(X^k)$. In p^3 configurations

Fig. 4.4. Diagnostic diagram for IC 2165. Note that the ratios of the nebular-to-auroral (N/A) lines of the p^2 and p^4 configurations of [N II], [O III], and [Ar III] run nearly horizontally. Thus, these ratios are sensitive to T_ε and insensitive to N_ε (except for [N II] at $N_\varepsilon \gtrsim 10^4$. The p^3 doublet line ratio $I(^2D_{3/2} - {}^4S_{3/2})/I(^2D_{5/2} - {}^4S_{3/2})$ depends mostly on N_ε, while the $\lambda7320/\lambda3727$ [O II] ratio is sensitive to both N_ε and T_ε. Dotted lines indicate uncertain ratios. The scatter is due to errors in Is, atomic parameters, and $(N_\varepsilon, T_\varepsilon)$ variations. Thus, {[S II], [N II]}, {[O II], [O III], [Ne III]} and {[Ne IV], [Ne V]} may originate in zones with $T_\varepsilon \approx 9000$, 13 500, and 15 000, respectively.

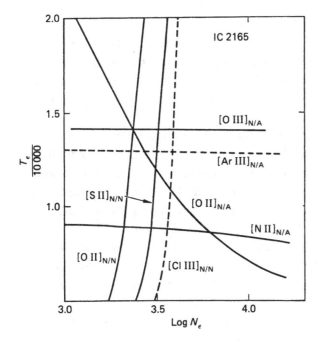

$N_1(X^k)$ will very nearly equal $N(X^k)$. Thus, we obtain $N(X^k)/N(H^+)$, e.g., $N(O^+)/N(H^+)$, $N(O^{++})/N(H^+)$, $N(N^+)/N(H^+)$, etc., for all ions whose lines are measured. To find $N(\text{element})/N(H^+)$, we must allow for the unobserved ionization stages. This step is one of the most difficult in the analysis. Various extrapolation formulae have been proposed. Elementary considerations of ionization patterns suggest that for nebulae of low-to-moderate excitation, one can use approximations such as: $n(O) = n(O^+) + n(O^{++})$, $n(N^+)/n(N) = n(O^+)/n(O)$, $n(Ne^{++})/n(Ne) = n(O^{++})/n(O)$. For ions of Ar, Cl, and S, various extrapolation recipes have been proposed. They are empirical, some are nonphysical, and none is really satisfactory.

Alternative procedures are possible. They entail calculating theoretical models to represent the observed intensities. A spherically symmetrical model is usually assumed, although it is possible to introduce shells of nonuniform density, etc. One adopts a functional dependence $N_H(r)$, a nebular radius R_N, a nebular chemical composition, an exciting star of radius R_s, and emergent energy flux $F_s(v)$. One introduces the relevant physics, the atomic coefficients a_v, Ω, A, σ_x, and α (including dielectronic recombination). The model program then produces the radiation field at each grid point, solves for the energy balance, the ionization equilibrium, and for the emissivity in each line. Adjustable parameters include the chemical composition, the energy distribution, $F_s(v)$, and the truncation radius. Is the nebula radiation or ionization bounded? The objective is to fit the observed intensities as closely as possible. Usually, an exact match is impossible. Few planetary nebulae are spherically symmetrical; most have irregular condensations and may contain dust. Galactic and extragalactic H II regions are usually very irregular, dusty, and often are excited by several imbedded stars. The choice of a set of 'observed intensities' is important. One should use the flux as integrated over the nebula if this is what theory predicts.

Ideally, the abundances given by the model should be most reliable. In practice, definitive – fully satisfactory – models are often not obtainable.

We may use the theoretical models in another way: as an interpolation device for ionic concentrations. We iterate the model until it gives the correct level of excitation, e.g., represents the line ratios such as $(4949 + 5007)/3727$ [O III]/[O II], 4686/5876 (He II/He I), 3425/3868 [Ne V]/[Ne III], and then fit as many of the other intensities as we can. With such a model we can get T_e in the inner hotter zones, where no diagnostic lines are produced. In particular, we can estimate the ionization correction factor $ICF = N(\text{element})/\sum N(X^k)$. Here the summation is taken over all observed ions, k. ICFs afford a systematic way for handling nonobserved ionization stages of elements such as N, Na, K, and Ca.

Table 4.3 illustrates the procedure. Successive columns list the ion, the wavelength, λ, of the lines used, the intensity, I, on the scale, $I(\text{H}\beta) = 100$, $N(X^k)$ as calculated from the observed Is, the sum $\sum N(X^k)$, the ICF, and $n(\text{element})/n(\text{H})$ as obtained from previous columns, and from the model that best represents the data. For most elements the agreement is reasonably good – within 10–15 percent. In this analysis we have used lines originating in both optical and near-uv regions. The measured uv C and N line intensities probably contain contributions from the central star. Consequently, care and judgment must be used in any interpretation of such data.

Whatever method is used, there are inherent limitations in the accuracy of the results:

    (a) The input atomic data depend mostly on theoretical calculations for which there are few empirical checks. It is sometimes difficult to get good A- and Ω-values.

    (b) Observational data are often incomplete. Frequently, the intensities are measured only in a limited region of the nebula. Ideally, one should have calibrated isophotic contours for all important lines.

    (c) Irregularities in the density distribution are revealed by knots and filamentary structures. Such nebulae are poor targets for most theoretical models. Also, approximate methods are affected by fluctuations in N_ε and T_ε. More elaborate models may help eventually.

In H II regions, items (b) and (c) give even more trouble than in planetaries. Distinct filamentary structures are seen in galactic and Magellanic Cloud H II regions, but not in more distant galaxies. Also, giant H II regions are often excited by several stars. Although theoretical models have been used, empirical procedures have generally been preferred.

Assuming that we can get good elemental atomic abundances for the gas phase, to what extent do such numbers reveal overall abundances? Refractory metals such as Fe and Ca are locked up in grains. Comparisons of absorbing clouds in the interstellar medium with atmospheres of involved stars show that many metals are thus depleted. Presumably they are locked up in grains. Carbon easily forms solid particles and CO molecules, but presumably most of the O and N, and certainly the noble gases, are unaffected.

Planetary nebulae, which are the ejecta of highly evolved stars show effects of nuclear processing in advanced stages of stellar evolution. Some are enhanced in nitrogen. Others are carbon-rich, while yet others appear to have compositions reflecting that of the interstellar medium in which progenitor stars were formed (see Table 4.4).

Table 4.3. *Analysis of IC 2165*

Ion	λ	I	$N(X^k)$	$\sum N(X^k)$	ICF	N(element) Direct	N(element) Model
He I			0.053	0.104	1.0	0.104	0.098
He II			0.051				
C III	1908	1345	4.7 (−4)	6.1 (−4)	1.36	8.3 (−4)	9.2 (−4)
C IV	1549	1010	1.4 (−4)				
N II	6583	39.4	8.82(−6)	1.03(−4)	1.04	1.07(−4)	8.6 (−5)
N III	1747	26.2	3.86(−5)				
N IV	1487	50.4	4.35(−5)				
N V	1240	40.9	1.18(−5)				
O II	3727	49	9.35(−6)	1.79(−4)	1.45	2.6 (−4)	2.9 (−4)
O III	4959, 5007	1550	1.70(−4)				
Ne III	3868	98	3.29(−5)	8.79(−5)	1.01	8.9 (−5)	8.4 (−5)
Ne IV	4725, 2422	1.2, 113	4.2 (−5)				
Ne V	3346, 3476	79	1.3 (−5)				
Na IV	3241	7.39	1.56(−7)		2.87	4.5 (−7)	6.2 (−7)
S II	6717, 6730		2.1 (−7)	1.86(−6)	1.39	2.6 (−6)	3.3 (−6)
S III	6312	5.3	7.1 (−7)				
S IV	10.5μ	0.9, 34.5	9.4 (−7)				
Cl III	5517, 5537	0.56	4.79(−8)				

CIV	7530, 8045	1.08	3.64 (−8)	8.43 (−8)	1.68	1.4 (−7)	1.25 (−7)
Ar III	7135	8.67	3.8 (−7)				
Ar IV	4740	6.16	1.12 (−6)	1.59 (−6)	1.24	2.0 (−6)	2.6 (−6)
Ar V	6435, 7005	2.0	0.92 (−7)				
K IV	6101	0.21	2.54 (−6)		2.35	6.0 (−8)	6.0 (−8)
Ca V	5309	0.086	1.91 (−8)		4.7	9.0 (−8)	1.2 (−7)

Compare with Aller, L. H. and Czyak, S. J. (1983) *Astrophys. J. Suppl.* **51**, 211 (Table 25); see also Marionni, P. A. and Harrington, J. P. (1981) in *The Universe at Ultraviolet Wavelengths*, ed. R. D. Chapman (NASA Conf. Publ. 2171) p. 633.

4.7   Summaries of some further perspectives on problems of elemental abundances

We have emphasized spectroscopic procedures for analyses of stars, nebulae, and the Sun. The central problems include the solar chemical composition and the primordial solar-system composition. Our astrophysical insights on the compositions of stars, nebulae, and the ISM are being supplemented in a dramatic way by ingenious advances in analytical chemistry applied to subtle clues provided by meteorites. The concept of a vast cloud of chemically homogeneous dust and gas that condensed into the present solar system is being replaced by a much more complex picture. The early solar nebula was not uniformly mixed. Impressive evidence for the existence of short-lived radioactive nuclides indicates that the extended cloud destined to produce the solar system was seeded with nuclides made in

Table 4.4. *Compositions of planetary nebulae*

	Mean	High-excitation objects		LMC (P40)	SMC (N2)	Fornax
		N-rich	C-rich			
He	1.1 (11)	1.35(11)	1.07(11)	1.05(11)	1.07(11)	1.2(11)
C	7.7 (8)	5.2 (8)	9.8 (8)	5.01(8)	4.0 (8)	9 (8)
N	1.8 (8)	6.3 (8)	1.4 (8)	3.1 (7)	2.6 (7)	2.5(7)
O	4.4 (8)	4.5 (8)	4.9 (8)	2.14(8)	2.0 (8)	2.4(8)
F	4.0 (4)					
Ne	1.07(8)	1.1 (8)	1.1 (8)	4.3 (7)	3.4 (7)	3.7(7)
Na	1.5 (6)					
S	1.0 (7)	6.2 (8)	1.2 (7)	2.6 (6)	2.3 (6)	5 (6)
Cl	1.65(5)	2.4 (5)	1.85(5)			
Ar	2.7 (6)	5.4 (6)	2.9 (6)	1.55(6)	1.1 (6)	8.0(5)
K	9.0 (4)	8.5 (4)	1.1 (4)			
Ca	1.07(5)	1.5 (5)	8.5 (5)			

Notes: The abundances are given on scale $n(H) = 1 \times 10^{12}$. The figure in (...) indicates the power of 10 by which the entry is to be multiplied. Thus 7.7(8) means 7.7×10^8. The 'mean' values in the second column refer to data for 47 planetaries; 'N-rich' and 'C-rich' refer to high-excitation objects. Values in columns 2, 3, and 4 are from Aller & Czyzak (1983). *Astrophys. J. Suppl.* **51**, 240. The sample of 47 is somewhat biased towards objects of high surface brightness since one of the program objectives was to probe abundances of rarer elements. Results are quoted from Maran *et al.* (1984). *Astrophys. J.* **280**, 615, for one planetary nebula out of each of the Large and Small Magellanic Clouds and the Fornax system. These objects were observed with the IUE. All nebular data were analyzed with the aid of theoretical models, mostly used as interpolation devices.

stars not more than a few million years before the solar system was formed. For example, ^{22}Ne is carried in carbon grains and also in the minerals, spinel and apatite, in certain meteorites. It originates from ^{22}Na, whose half-life is 2.6 years. This radioactive sodium was trapped in grains formed in ejecta of a nova or supernova, soon after the element-building event occurred! Distributions of Xe isotopes differing substantially from standard atmospheric Xe appear in different types of meteorite grains. All were created within stars, ejected, and condensed onto grains that were later incorporated into meteorites. In one type of grain, Anders and his co-workers found almost exclusively Xe isotopes produced by the slow-neutron (s) process in red giants. This material was ejected and later condensed into particles that were destined to be locked into meteorites. Deuterium-rich polymers are found in CI chondrites. The 1H/2H ratio is notably higher than in 'normal' solar system material. These polymers were obviously built up from deuterium-rich molecules that formed in the ISM. Thus, there is clear evidence that interstellar material has been incorporated in the solar system. In the primordial solar system there may have been appreciable differences between at least some of the condensed material and the enveloping cool gas.

The relevance of abundance studies to concepts of stellar structure and evolution is well illustrated in studies of globular star clusters (see, e.g., Kraft (1979)). It was once believed that all stars in a cluster were formed simultaneously with identical initial chemical compositions. The most luminous observed stars, i.e., the giants, have evolved away from the main sequence relatively recently, i.e., over a time-interval short compared with the age of clusters – generally 10^{10}–2×10^{10} years. Studies of chemical compositions ought to cast light on hypotheses of stellar evolution, particularly element-building processes and convection patterns. The practical difficulty is that globular clusters are so remote that even the brightest stars are very faint. Few have been studied yet by accurate methods, but recent technical advances will enable us to reach many additional stars, particularly in the nearest globulars such as ω Centauri and 47 Tucanae. Because the stars are so faint, many observations have been restricted to broad-band pass, U, B, V colors, although narrow-band pass filters are also used. Comparisons with conventional analyses of individual stars provide calibration. Although such color measurements have yielded valuable clues to stellar chemical compositions, many results are more tantalizing than definitive. There is no adequate substitute for the spectroscopic method.

The He abundance is another important parameter. Direct measurement of hot stars on the horizontal branch show He/H ratios an order of magnitude lower than that predicted by the primordial-fireball hypothesis. Possibly He has settled out in the gravitational field of the star. Indirect He abundance estimates may be derived from evolutionary considerations and from pulsation theory for RR Lyrae stars; these lead to He/H ratios between 0.07 and 0.085.

In metal-deficient stars, reduced line blanketing in the blue and ultraviolet produces energy distributions corresponding to an earlier spectral class. If we compare the integrated-light spectral class of a globular cluster with its energy-distribution spectral class (as measured with a scanner) we would infer that globular clusters differ among themselves and from the Sun in metal/H ratio (see Table 4.5). Integrated radiation is of limited usefulness. For meaningful quantitative data we must study individual stars. Thus, [Fe/H] ranges from −0.6 for 47 Tuc to about −2.2 for M92.† Different techniques yield different results; for M92, a crude (U–B) color method gives −1.6, while curve of growth and spectra synthesis methods give −2.4 and −2.2, respectively.

Some surprising results have emerged. Briefly, stars at presumably the same evolutionary stage often show discernible differences, some of which could be attributed to nucleogenesis processes, others to differing initial chemical compositions. The great ages of the present giant stars require stellar masses less than the solar mass. For such stars, conventional evolutionary theory predicts that surface chemical compositions are modified only late in their evolution, when only abundances of C, N, and O

† We use the notation $[X/H] = \log\{N(X)/n(H)\}_{star} - \log\{N(X)/n(H)\}_{Sun}$.

Table 4.5. *Integrated-flux distributions of spectral classes for globular clusters*

Globular cluster	Energy distribution equivalent spectral class	Spectral class	[Fe/H]
47 Tuc	G2 III	G3	−0.6
NGC 362	G4 V	F8	−1.0
NGC 2808	G2 III	F7	−0.8
ω Cen	G4 V	F8	−0.8 → −2.2
NGC 6752	G4 V	F6	−1.7

Data in columns 2 and 3 (Faulkner & Aller (1964)), *IAU Symp. No. 20*, p. 368. Metal deficiencies from Kraft (1979).

(involved in the CNO cycle) are changed. Actually, differences in the strength of CN bands are found in the giant branch almost back to the main sequence, in flagrant contradiction to predictions. Some cluster stars show depletion of atmospheric C and wide variations in CH, CN, and NH bands at the same evolutionary stage. In ω Cen and M13, Fe-peak and light-metal abundance variations are observed from star to star. There is no way to make iron in a star of solar mass. The most likely explanation seems to be that part of the original material from which these low-mass stars were formed got seeded with debris from supernovae and massive first-generation stars. The presence of barium stars shows neutron sources occur in rather low mass objects.

Abundance variations in galaxies as a function of distance from the center have been studied often (see, e.g., Searle, Shields, Dufour, the Peimberts, Pagel *et al.*). Intensity variations in [O II] and [O III] lines with distance from the center of M33 were noted as early as 1942. In many spiral galaxies, the metal/H ratio declines with distance from the center. Sometimes, as in M31, the inner spiral arms are dusty, while in the outer arms a gaseous component predominates.

The effect differs from galaxy to galaxy. It illustrates differences in star formation rates. In our Galaxy, Shaver *et al.* (1983) applied radio-frequency and optical-spectroscopy methods to H II regions between 3.5 to 13.7 kpc from the galactic center. They found a good correlation between $T\varepsilon$ and the $([O\ II] + [O\ III])/H\beta$ ratio and established the abundances of C, N, Ne, S, and Ar. Although $n(\mathrm{He}+)/n(\mathrm{H}+)$ showed no gradient, $d \log[N(\mathrm{element})/N(\mathrm{H})/dr_{\mathrm{gal}} = -0.7$ and -0.09 (kpc)$^{-1}$ for O and N, respectively.

Some other galaxies offer the great advantage that they are seen in plan. An example is M33, a galaxy much smaller than our own, where the O gradient, -0.13 per kpc, is steeper. The method used here was to calculate models of H II regions to obtain ICFs, especially for elements such as Ne, S, and Ar.

Application of model nebular methods to the giant H II regions in M101 suggests that several exciting stars are involved. Table 4.6 compares abundances in the Sun and the Orion Nebula (representative galactic sources) with the Large and Small Magellanic Clouds. In the SMC, especially, element building went more slowly than in the Galaxy, particularly for C, N, and O, which can be made in low-mass stars. Depletions for S and Ar are less severe, suggesting that many stars of high mass were formed early in the history of these systems. A comparison of $n(\mathrm{element})/n(0)$ ratios in several sources shows that, although element-building proceeds at differing speeds in different systems, Ne, S, and Ar seem to be manufactured at similar rates with respect to O in a diversity of sources (see Table 4.7).

Abundance studies have turned out to be very important in connection with the whole subject of stellar structure and evolution. Compositions of envelopes of interacting and cataclysmic binaries, including classical novae,

Table 4.6. *Comparison of Solar and Orion nebular composition with Magellanic Clouds*

	Sun	Orion Nebula	SMC	LMC
He	11.0:	11.0	10.92	10.93
C	8.66	8.46	7.16	7.90
N	7.98	7.48	6.52	6.98
O	8.91	8.60	8.07	8.41
Ne	8.05	7.79	7.48	7.73
S	7.23	7.12	6.50	6.96
Ar	6.57	6.27	5.81	6.24

Notes: The abundances for the Orion Nebula are taken from work of the Peimberts (1977) as revised and extended by Dufour, Shields & Talbot (1982). Abundance of C in the Magellanic Cloud is from Dufour *et al.*; for the other elements we tabulate mean values from the investigation by Dufour *et al.*, and by Aller, Keyes & Czyzak (1979). See also Pagel *et al.* (1978).

Table 4.7. *Abundances with respect to O compared for H II regions, the Sun and planetaries*

	N	Ne	S	Ar
M33	−1.25	−0.77	−1.55	−2.40
M101	−1.07	−0.74	−1.68	−2.47
LMC	−1.43	−0.68	−1.45	−2.17
SMC	−1.55	−0.59	−1.57	−2.26
Orion	−1.12	−0.81	−1.48	−2.33
Sun	−0.93	−0.86	−1.68	−2.34
Mean of planetaries	−0.38	−0.61	−1.64	−2.21

Notes: Data for LMC, SMC, Orion are from references cited in Table 4.6; solar data are from Table 4.2, M33 data from Kwitter & Aller (1981), the M101 data are from measurements by Sedwick & Aller (1981), *Nat. Acad. Sci. USA*, **78**, 1994, analyzed by theoretical methods. Somewhat smaller N/O and Ne/O ratios were found by Rayo, Peimbert & Torres-Peimbert (1982), *Astrophys. J.* **255**, 1.

promise to be of great help in the construction of evolutionary scenarios. Analyses of supernova remnants such as Cas A indicate that these nebulosities arise in different parts of the exploding star; each carries a composition appropriate to the region when it arose.

Acknowledgment is made to NSF grant AST 83-12384 to UCLA for partial support of some of the research results reported here. Professor Wasson kindly supplied his list of C I abundances in advance of publication.

General references

Earlier work on stellar atmospheres, theories of curve of growth, etc., are reviewed in

The Sun (1950). Ed. L. Goldberg. Chicago: Univ. of Chicago Press.

Unsöld, A. (1955). *Physik der Sternatmosphären*. Berlin: Springer.

Stellar atmospheres, *Stars and Stellar Systems*, Vol. 6 (1960). Ed. J. L. Greenstein. Chicago: Univ. of Chicago Press.

Aller, L. H. (1963). *Atmospheres of Sun and Stars*. New York: Ronald Press Co.

Abundances of the Elements (1961). New York: Interscience.

More recent developments are covered in

Jefferies, J. T. (1968). *Spectral Line Formation*. Waltham, MA: Blaisdell Publishers.

Cowley, C. R. (1970). *Stellar Atmospheres*. San Francisco: Freeman.

Mihalas, D. (1978). *Stellar Atmospheres*. San Francisco: Freeman.

For an account with emphasis on observational techniques, see

Gray, D. F. (1976). *Observation and Analysis of Stellar Photospheres*. New York: Wiley.

Reviews of Solar Abundances: A comprehensive survey was made by

Goldberg, L., Müller, E. A. & Aller, L. H. (1960). *Astrophys. J. Suppl.* **5**, 1.

See also:

Ross, J. E. & Aller, L. H. (1976). *Science*, **191**, 1223.

Holweger, H. (1979). *Liège Coll.*, **22**, 119.

Solar System Abundances of Elements and Meteorites:

Grevesse, N. (1984). *Physica Scripta*, **78**, 49.

Anders, E. & Ebihara, M. (1982). *Geochim. Cosmochim. Acta*, **46**, 2363.

Wasson, J. (1985). *Meteorites. Their Record of Early Solar System History*. San Francisco: Freeman.

Cameron, A. G. W. (1982). In *Essays on Nuclear Astrophysics*, ed. C. A. Barnes *et al*. Cambridge University Press.

See, e.g.

Anders, E. (1981). *Proc. Roy. Soc. London A*, **374**, 207 (also Lewis, R. S. and Anders, E. (1983). In *Scientific American*, **249**, 66).

Gaseous Nebulae:

Seaton, M. J. (1960). *Rept. Prog. Phys.*, **23**, 313.

Aller, L. H. & Liller, W. (1968). Nebulae and interstellar matter, in *Stars and Stellar Systems*, Vol. 7. Chicago: Univ. of Chicago Press.

Osterbrock, D. E. (1974). *Astrophysics of Gaseous Nebulae.* San Francisco: Freeman Co.

Aller, L. H. (1984). *Physics of Thermal Gaseous Nebulae.* Dordrecht, Holland: Reidel Publishers.

Pottasch, S. R. (1900). *Planetary Nebulae.* Dordrecht, Holland: Reidel Publishers.

Kaler, J. B. (1985). *Ann. Rev. Astron. Astrophys.*, p. 89, and references therein cited.

Stellar atmospheres: the literature is so vast that any attempt to list references in the alloted space would be superficial and unsatisfactory. An important topic we have not discussed is the effect of diffusion on atmospheric compositions: See:

Vauclair, S. & G. (1982). *Ann. Rev. Astron. Astrophys.* **20**, 37.

Atmospheric models of late type stars are discussed by

Carbon, D. F. (1979). *Ann. Rev. Astron. Astrophys.* **17**, 513.

Halo stars (extreme popl. type II) are discussed by

Spite, M. & F. (1985). *Ann. Rev. Astron. Astrophys.* **23**, 225.

Further articles on stellar atmosphere topics may be found in IAU symposium volumes, e.g. Nr. 26 (1966), *Abundance Determinations in Stellar Spectra.*

Globular Clusters:

See I.A.K. Symposium No. 85 (1980) and

Kraft, R. P. (1979). *Ann. Rev. Astron. Astrophys.* **17**, 309.

Specific references

Aller, L., Keyes, C. & Czyak, S. (1979). *Proc. Nat. Acad. Sci.* **76**, 1525.

Bowen, I. S. & Wyse, A. B. (1939). *Lick Obs. Bull.* **19**, 1.

Carbon, D. F. *et al.* (1982). *Astrophys. J. Suppl.* **49**, 207.

Cohen, J. G. (1979). *Astrophys. J.* **228**, 405.

Dufour, R. J. (1975). *Astrophys. J.* **195**, 315.

Dufour, R. J. & Harlow, W. V. (1977). *Astrophys. J.* **216**, 706.

Dufour, R. J., Shields, G. & Talbot, R. (1982). *Astrophys. J.* **252**, 461.

Freeman, K. C. & Norris, J. (1981). *Ann. Rev. Astron. Astrophys.* **19**, 319.

Holweger, H. & Müller, E. P. (1974). *Solar Physics,* **39**, 19.

Kurucz, R. L. (1979). *Astrophys. J. Suppl.* **40**, 1.

Kwitter, K. B. & Aller, L. H. (1981). *Mon. Not. Roy. Astron. Soc.* **195**, 939.

Menzel, D. H. (1936). *Astrophys. J.* **84**, 462.

Meyer, J. P. (1981). *International Cosmic Ray Conference,* **3**, 149, 145.

Pagel, B. E. J., Edmunds, M. G., Fosbury, R. & Webster, B. L. (1978). *Mon. Not. Roy. Astron. Soc.* **189**, 569.

Peimbert, M. & Torres-Peimbert, S. (1977). *Mon. Not. Roy. Astron. Soc.* **179**, 217.

Peimbert, M., Serrano, A. & Torres-Peimbert, S. (1984). *Science,* **224**, 345.

Shaver, P. A. *et al.* (1983). *Mon. Not. Roy. Astron. Soc.* **204**, 53.

Shields, G. A. & Searle, L. (1978). *Astrophys. J.* **222**, 821.

Withbroe, G. L. (1971). *NBS Special Pub.* **353**, 127. Symposium on Solar Physics, Atomic Spectra and Gaseous Nenulae, ed. K. B. Gebbie.

5

The solar chromosphere

ROBERT W. NOYES and EUGENE H. AVRETT

5.1 Introduction: the nature of the solar chromosphere

The solar chromosphere owes its name to the brilliant red emission seen from the region just above the limb at times of total solar eclipse. The red emission is due to the overwhelming contribution of the Hα hydrogen line at the wavelength 6563 Å. This line emission is produced in large part by scattering of photospheric radiation from hydrogen atoms in the chromosphere, and gives little information about the chromospheric temperature. However, other emission lines such as the D_3 line at 5876 Å (the discovery of which gave the name 'helium' to the responsible element) indicate a chromospheric temperature considerably higher than the temperature of the underlying photosphere.

Radiative transfer theory, however, indicates that in an atmosphere in radiative equilibrium the temperature generally decreases with height, reaching a surface value near 4300 K in the case of the Sun. Since the temperature in the chromosphere is substantially higher than this value, there must be a source of non-radiative energy to heat it. The height of the chromosphere seen above the solar limb is many times greater than the density scale height appropriate to chromospheric temperatures; this extension and observed rapid motions both indicate that the chromosphere is in a state of intense dynamic activity. The total energy fed into the chromosphere as heat and kinetic energy is about 4×10^6 erg cm^{-2} s^{-1}, or about 10^{-4} of the solar luminosity. A small fraction of this large flux of energy penetrates even higher and heats the corona.

It is important to understand the physical processes occurring in the solar chromosphere, because these processes transmit energy and momentum into the corona. Therefore they are important in determining both the structure of the corona and its dynamics, including the outward acceleration of the solar wind. The heating of the chromosphere and corona is very sensitive to the presence of magnetic fields, and a study of chromospheric structure and dynamics as they relate to magnetic fields can provide important insights into the interaction between magnetic fields and ionized gases in astrophysics. Also, chromospheres are seen on other stars and are the subject of much current research; we cannot begin to understand observations of stellar chromospheres without first understanding the solar chromosphere.

The chromosphere is generally defined as that part of the solar atmosphere beginning in the temperature minimum region a few hundred kilometers above the visible surface, and extending upward to the altitude where a very steep temperature gradient forms a transition zone separating the chromosphere and the corona. This definition implies that the atmosphere is thermally stratified in horizontal layers. Such a one-dimensional description is of limited validity at best, given the observed highly complex structure of the chromosphere. Yet it does provide a useful framework for understanding many aspects of chromospheric physics, as long as we recognize that in the real dynamic chromosphere, the temperature, density, magnetic field, and velocity all vary in three dimensions and in time – often by amounts too large to be characterized as small perturbations on a mean one-dimensional structure.

A picture of the variation of temperature and density with height in the fictitious one-dimensional chromosphere is given in Fig. 5.1. In this figure, zero height corresponds to unit optical depth at 5000 Å, that is, to the depth of the visible photospheric surface as seen at disk center. The chromosphere, by the definition given above, would then extend from a height of about 500 km to about 2300 km.

The mean structure of the chromosphere has been determined from various studies of the emission in spectral lines and in the continua in different regions of the spectrum. We shall discuss these studies in detail in the next section. Before doing so it is important to relate the fictitious mean structure to the 'real' chromosphere, as illustrated in Figs. 5.2 and 5.3. Fig. 5.2 is a monochromatic image of a magnetic active region of the chromosphere made at the wavelength of the center of the Hα line. The high opacity of Hα allows us to view structures high up in the chromosphere, and reveals that chromospheric structure above an active region is dominated by magnetic fields. This is qualitatively apparent from the elongated and striated structures seen in Hα, and from their relations to magnetic features, such as the sunspot in Fig. 5.2. Time-lapse images show that, although the overall pattern of relationships to magnetic features changes only slowly (the characteristic time is generally of the order of hours or longer), there are large and rapid dynamical fluctuations on a timescale of seconds to minutes.

Views of the 'quiet' chromosphere are given in Fig. 5.3. They show that, in regions of the solar atmosphere far from the closely-packed magnetic structures of active regions, there is still an extreme departure from the idealized one-dimensional chromosphere. More-or-less regular 'hedges' of elongated jets known as spicules are seen in absorption on the disk, and in emission at the limb (Fig. 5.3b). They are concentrated over the boundaries of a large-scale convective flow known as the supergranulation. This flow consists of an upwelling at the centers of convective cells called supergranules, an outflow to their periphery, and a sinking at the boundaries. Typical diameters of supergranulation cells are about 40 000 km, and the speed of the horizontal outflow seen at the surface is about $0.5 \, \text{km s}^{-1}$. The flow imposes structure on the overlying chromosphere, shown schematically in Fig. 5.4. The mechanism by which this is accomplished appears to be that the horizontal outflow entrains concentrations of magnetic field penetrating the photosphere, and sweeps them to the supergranular boundaries, where they collect in a loose *network* of photospheric field. This magnetic network in turn gives rise to enhanced nonthermal heating of the overlying chromosphere. Two important results are (1) the creation of a *chromospheric network* of enhanced emission

Fig. 5.1. Variation with height of the temperature and density in a one-dimensional model of the solar atmosphere. Height $h = 0$ occurs where $\tau = 1$ at 500 nm. From Withbroe & Noyes (1977), *Ann. Rev. Astron. Astrophys.* **15**, 363.

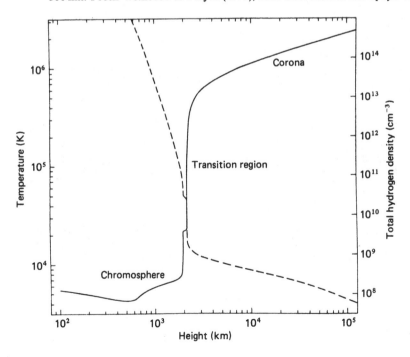

overlying the magnetic network at the supergranular boundaries, and (2) the upward acceleration of chromospheric material in spicules, moving upward along the concentrated magnetic flux tubes there.

The mean stratification of the chromosphere over the supergranular boundaries is thus quite different from that over the center of the supergranules. (These locations are often also termed 'cell boundaries' and 'cell centers' respectively.) The increased chromospheric heating causes both the temperature and density to increase at a given height in the chromosphere. The corresponding emission enhancements are visible in spectral lines and continua formed at temperatures ranging all the way up to several hundred thousand degrees, as indicated in Fig. 5.5. By constructing

Fig. 5.2. Hα filtergram of a solar active region. The width of the field of view is about 88 000 km. (Courtesy Sacramento Peak Observatory, Association of Universities for Research in Astronomy.)

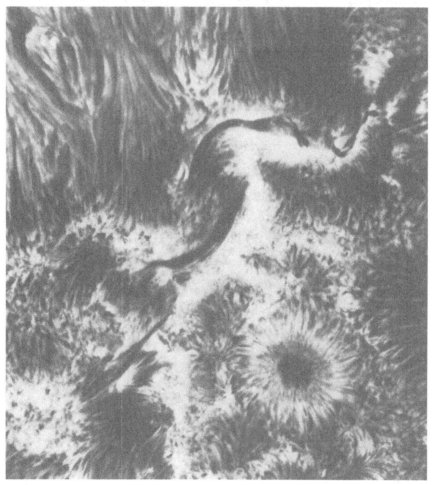

atmospheric models that reproduce the observed emission in the network and quiet Sun, one may determine the increase in energy deposition in the chromospheric network compared with cell centers.

The same physical processes that give rise to enhanced chromospheric heating above the boundaries of supergranules are also at work in the magnetic active regions associated with sunspots (Fig. 5.2), and give rise to other phenomena of magnetic activity, such as flares. Like the photospheric magnetic network, active regions are concentrations of field, but they differ in that the field is concentrated throughout the region and not just at network boundaries. The origins of magnetic active regions can probably be found in the solar magnetic dynamo, whose details are not yet well understood. However, the effect of active region fields in the overlying atmosphere is the same: the chromosphere experiences increased heating just as it does above the magnetic network, and a *chromospheric plage* is formed. The enhancement of density and temperature in the chromospheric

Fig. 5.3. (*a*) Hα + 0.8 Å filtergram of the quiet chromosphere near the limb (at top of figure). Spicules are seen as absorption features projecting upwards at the edge of supergranule cells. (*b*) Hα + 1.0 Å filtergram of the solar limb, showing spicules in emission, projecting several thousand kilometers above the limb. (Courtesy Sacramento Peak Observatory, Association of Universities for Research in Astronomy.)

(*a*)

(*b*)

plages is similar in height-dependence to that in the network, but larger in magnitude.

In the next section we describe in detail the mean (spatially averaged) spectroscopic characteristics of the various chromospheric regions described here, and the use of these characteristics to deduce the mean structure and energy balance of these regions.

5.2   The mean structure of the chromosphere

The only way we can determine parameters such as temperature and density as functions of position in the solar atmosphere is by means of detailed studies of the observed spectrum. The opacity of the solar

Fig. 5.4. Sketch of the granulation–supergranulation– spicule complex, seen in cross-section. *A.* Flow lines of a supergranulation cell. *B.* Photospheric granules. *C.* Wave motions, observable as oscillations of the velocity field in the photosphere and low chromosphere. *D.* The large-scale chromospheric flow field seen in Hα. *E.* Lines of force, pictured as uniform in the upper atmosphere (corona) but concentrated at the boundaries of the supergranules in the photosphere and chromosphere. *F.* The base of a spicule 'bush' or 'rosette', visible as a region of enhanced emission in the Hα- and K-line cores. *G.* Spicules. The separation of the spicule bushes is about 30 000 km.

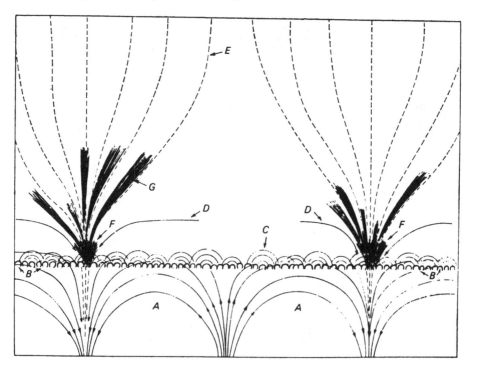

atmosphere varies enormously with wavelength, and these variations give us a way to probe different atmospheric depths.

The spectrum at visible wavelengths consists of a well-defined continuum together with numerous absorption lines. The lines are in absorption because both line and continuum radiation originate in photospheric layers where the temperature decreases in the outward direction, and unit optical depth in the high-opacity lines occurs higher in the atmosphere, at lower temperatures, than unit optical depth in the continuum.

A few spectral lines in the visible part of the spectrum have such high opacity near their central wavelengths that the line core is formed in the chromosphere above the temperature minimum, where the temperature *increases* in the outward direction. These lines show weak emission features near line center. The resonance lines of singly ionized calcium are the best studied lines showing this central emission. Fig. 5.6 shows the observed flux in the Ca II K-line centered at 3933.7 Å. The extended K-line wings exhibit

Fig. 5.5. The chromospheric network near Sun center, as observed from the Harvard College Observatory extreme ultraviolet spectrometer on Skylab in 1973. The field of view of each image is about 210 000 km. In order of increasing temperature of formation, the various spectral features are: H I Lyman continuum 896 Å ($T \approx 10^4$ K); C II 1335 Å ($T \approx 5 \times 10^4$ K); C III 977 Å ($T \approx 10^5$ K); O IV 554 Å ($T + 2 \times 10^5$ K); O VI 1032 Å ($T \approx 3 \times 10^5$ K); and Mg X 625 Å ($T \approx 1.5 \times 10^6$ K). (Courtesy Harvard College Observatory and NASA.)

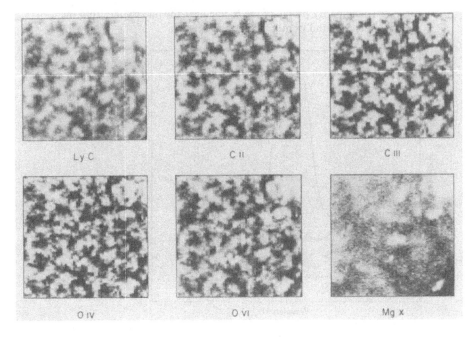

additional absorption due to many other narrower lines. The emission features near the line center, which are barely discernible in Fig. 5.6, are plotted on an enlarged scale in Fig. 5.7. The lower curve in Fig. 5.7 is the quiet-Sun intensity profile from the center of the disk, rather than the flux from the whole disk. The upper curve is the profile from a bright plage region.

The intensity minima located at a wavelength displacement of about 0.3 Å

Fig. 5.6. The flux profile of the Ca II K line, obtained with the Kitt Peak Fourier Transform Spectrograph. The flux, plotted as a function of wavelength in Å, is on a linear scale between zero and a pseudo-continuum level fitted to the observed flux maxima at 3780 and 4020 Å.

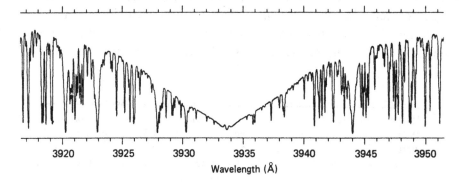

Fig. 5.7. The intensity profile of the K line near the line center, from the average quiet Sun and from a typical plage region.

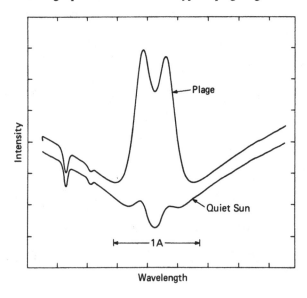

from the line center in both wings are called the K_1 minima. The emission features at a displacement of about 0.15 Å in the wings are called the K_2 maxima or K_2 emission peaks. The line center is called K_3. The K-line is a member of a doublet; the other member is the H-line centered at 3968.5 Å, which differs from the K-line in having opacity just half that of the K-line.

The intensity as a function of $\Delta\lambda$ in the line wing outside the K_2 peak provides direct information about the temperature as a function of optical or geometrical depth. The intensity successively closer to the line center is formed successively higher in the atmosphere. The K_1 minimum corresponds to the temperature minimum, and the intensity rises between K_1 and K_2 because of the temperature rise in the low chromosphere. Why, then, is there a central reversal ($I_{K_3} < I_{K_2}$) if the temperature continues to rise in the chromosphere? The answer is that the central part of the line is formed high in the chromosphere where the electron density is too small to maintain local thermodynamic equilibrium (LTE), and photon scattering causes the line-center excitation temperature to be much smaller than the local kinetic temperature.

Fig. 5.7 clearly shows that at the center of the H- and K-lines, solar plages are brighter than the quiet Sun. As we have noted, this is due to the presence of strong magnetic fields in plages. The spatial correlation between emission in the K-line and underlying photospheric magnetic fields is clearly shown in Fig. 5.8.

The quiet regions themselves show features that are substantially brighter, and others that are substantially fainter, than the average quiet-Sun profile

Fig. 5.8. Three views of a magnetic active region. At left is a white-light image showing a spot group. The magnetogram, center, codes the magnetic fields of opposite magnetic polarity into lighter and darker areas. The right-hand image, in the Ca II K line, shows the chromospheric emission overlying the magnetic areas in the photosphere. (Courtesy Sacramento Peak Observatory, Association of Universities for Research in Astronomy.)

Continuum   Magnetic fields   Calcium II

plotted in Fig. 5.7. This is clearly shown in Fig. 5.5, for ultraviolet emissions; the same holds for the H and K emission cores.

What are the brightness variations within quiet regions? Fig. 5.9 shows the distribution of intensity profiles observed in the Ca II H-line. Ten profiles are plotted; the lowest corresponds to the faintest 10 percent and the highest to the brightest 10 percent of the observed quiet region.

The residual intensities shown on the left in Fig. 5.9 can be converted to brightness temperatures in order to relate these observations to an expected range of atmospheric temperatures. The brightness temperature corresponding to the intensity I at wavelength λ is the temperature of an isothermal atmosphere in LTE, with no scattering, which would emit the intensity $I(\lambda)$. From the brightness temperatures at the H_1 inflections we can infer that the average quiet Sun has a temperature minimum value of about 4400 K and that the minimum temperature from one point to another in quiet regions generally stays within the 4200–4600 K range.

We see evidence for the solar temperature minimum and for the chromospheric temperature rise in the Ca II resonance lines profiles because of the substantial increase of the line opacity with decreasing wavelength displacement $\Delta\lambda$ (measured from line center). The opacity in the continuum also varies with wavelength, but substantial changes occur only with large

Fig. 5.9. Quiet-Sun profiles of the Ca II H line observed by Cram and Dame with high spatial resolution. The 10 profiles, ordered according to brightness, each correspond to $\frac{1}{10}$ of the observed area.

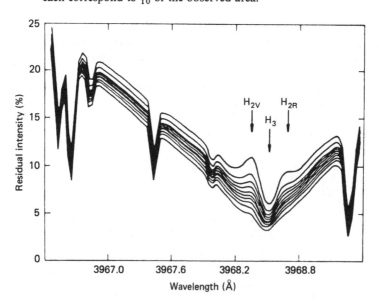

variations in wavelength. In fact, observations are required beyond the visible part of the spectrum and from above the Earth's absorbing atmosphere in order to probe the chromospheric layers sampled within the Ca II resonance line profiles.

The observed brightness temperature of the solar continuum, between spectral lines, has a maximum value near 1.6 μm in the infrared. At this wavelength the opacity (due to H$^-$, the negative hydrogen ion) has a minimum, and we see (with infrared detectors) deeper into the solar atmosphere than at any other wavelength. The 1.6 μm maximum brightness temperature is about 6700 K at disk center and about 6400 K for the averaged disk.

In the far infrared near 150 μm the observed continuum brightness temperature passes through a minimum value of about 4400 K and then becomes larger at larger wavelengths. This observed chromospheric brightness temperature increase between 300 μm and 2 cm (0.3–20 mm) is shown in Fig. 5.10.

Approximately the same minimum value is observed near 1600 Å in the ultraviolet, and at shorter ultraviolet wavelengths the brightness temperature also increases. This increase with decreasing wavelength is shown in the lower part of Fig. 5.11. The upper part of this figure shows the corresponding disk-center intensities. Here we see the broad wings and central emission of the hydrogen Lyman-α line, centered at 1215.7 Å. The

Fig. 5.10. Solar brightness–temperature observations at millimeter wavelengths. (References in the figure are given by Vernazza, Avrett & Loeser (1981).)

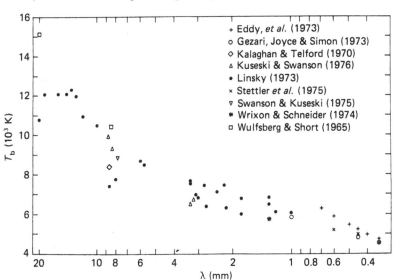

K-line wings in Fig. 5.6 decrease toward line center due to the outwardly decreasing photospheric temperature. The Lyman-α wings increase toward line center due to the outwardly increasing chromospheric temperature.

The three abrupt intensity increases in Fig. 5.11 near 1100, 910, and 500 Å are caused by the onset of photoionization from the lowest energy levels of carbon, hydrogen, and helium, respectively. The opacity is much greater on the shortward side of the threshold wavelength than on the longward side.

Fig. 5.11. Observed values of the disk-center continuum intensity (upper panel) and the corresponding brightness temperatures (lower panel) in the wavelength range 140–400 Å. (References in the figure are given by Vernazza, Avrett & Loeser (1981).)

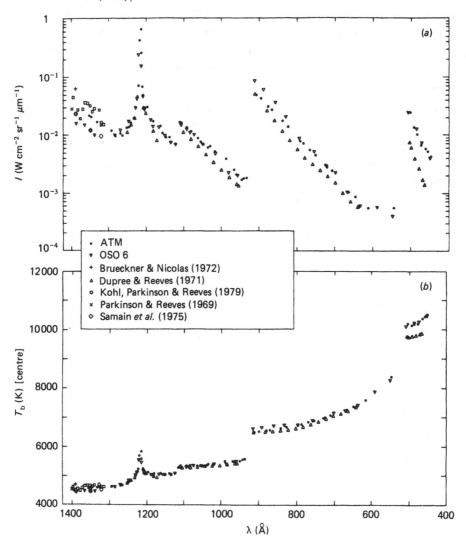

Optical depth unity thus occurs higher in the atmosphere, at a higher chromospheric temperature, on the shortward side.

Apart from the Lyman-α line, the observations in Fig. 5.11 refer only to the continuum values between the many emission lines in the spectrum. The complete spectrum, with the lines labeled, appears in Fig. 5.12. There is evidently a wealth of information in these observations, both about the structure of the solar atmosphere and about the detailed radiative and atomic interactions that take place.

Fig. 5.12 exhibits spectra from average quiet regions, from localized active regions, and from 'coronal holes' where the high-temperature coronal lines are weak or absent. An off-limb spectrum showing strong coronal lines is also included. At 907 Å, near the head of the Lyman continuum of hydrogen, the coronal-hole intensity is about a factor of 2 less than the quiet-Sun intensity, but active regions are 5 or 6 times brighter than quiet regions at this wavelength.

Different components such as quiet and active regions and coronal holes clearly have different intensities. As we have noted, quiet regions also show substantial brightness variations. The pattern of brightness variations in the quiet Sun shown earlier (Fig. 5.5) is a 'network' of bright chromospheric emission enclosing darker areas often called 'cells'. From a histogram of the brightness distribution of this pattern we have selected six components, A through F, ranging from a dark point within a cell (component A, the darkest 8 percent) to a very bright network element (component F, the brightest 4 percent). At 907 Å (cf. Fig. 5.5a) the darkest 8 percent of the observed area has an intensity about 0.4 times the average, while the intensity of the brightest 4 percent is almost 3 times the average.

Such distributions have been obtained at other continuum wavelengths as well, and Fig. 5.13 shows the intensities for components A through F at those wavelengths in the 400–1400 Å range where continuum intensities can be easily identified. The component C intensity as a function of wavelength in Fig. 5.13 is essentially the same as the spatial average intensity distribution in Fig. 5.11.

Now we consider the determination of chromospheric temperatures and densities from these and other observations. The observed brightness temperatures at various wavelengths give us the approximate kinetic temperatures that occur at unit optical depth for these wavelengths, apart from differences between brightness and kinetic temperatures caused by departures from LTE.

The optical depth at any wavelength depends on the wavelength-dependent properties of the absorbing atoms and the number of such atoms

Fig. 5.12. (*a*) Average solar spectra between 1350 and 920 Å from a quiet region, a coronal hole, an active region, and a quiet area off the limb. The individual spectra are displaced from each other for clarity. (*b*) Average solar spectra, as in Fig. 5.11(*a*), between 930 and 490 Å.

(*a*)

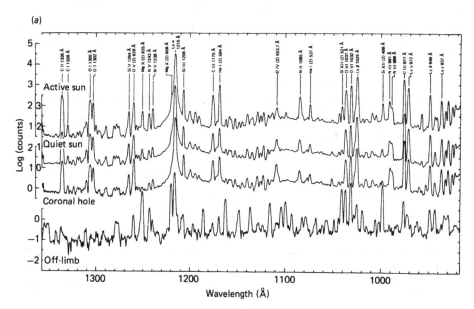

(*b*)

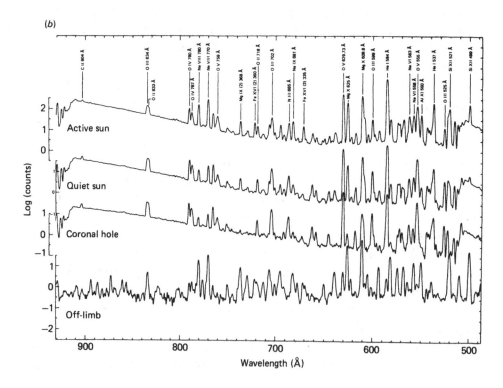

along the line of sight. The atomic number densities depend on a balance between the solar surface gravity, which acts to compress the atmosphere, and the gas pressure (including kinetic and magnetic effects) which acts to expand the atmosphere. In the static case with only gas pressure opposing gravity, the pressure varies with geometrical depth x according to the hydrostatic equilibrium solution

$$p(x) = p(x_0) \exp\left[\int_{x_0}^{x} \frac{dx'}{H(x')}\right]. \tag{5.1}$$

Here H is the (static) pressure scale height given by $H = H_s$, where

$$H_s = \frac{(1 + r_{He} + r_e)kT}{gm_H(1 + 4r_{He})}. \tag{5.2}$$

In this equation r_{He} is the helium fraction n_{He}/n_H (assumed to be 0.1) and $r_e = n_e/n_H$ is the calculated electron number density relative to hydrogen. If there are turbulent motions, the total pressure p is the sum of gas pressure and the

Fig. 5.13. Observed quiet-Sun continuum intensities in the 1350–400 Å wavelength range. The intensity A is from the darkest 8 percent and intensity F is from the brightest 4 percent of the observed quiet area.

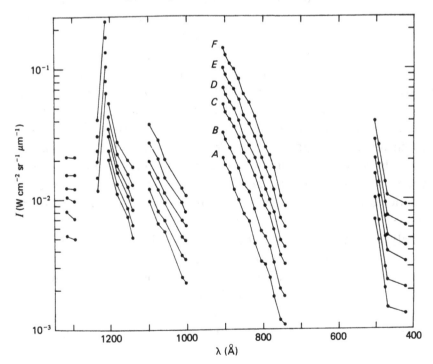

turbulent pressure

$$p_{turb} = \tfrac{1}{2}\rho v_{turb}^2 \tag{5.3}$$

where ρ is the gas density and v_{turb} is an assumed turbulent velocity parameter. Then the pressure scale height is

$$H = H_s + \frac{v_{turb}^2}{2g}. \tag{5.4}$$

At a temperature of 5000 K where hydrogen is neutral and $r_e \ll 1$, the pressure scale height (for solar gravity) is 118 or 147 km s^{-1} for $v_{turb} = 0$ or 4 km s^{-1}, respectively. At a temperature of 10 000 K where hydrogen is ionized and $r_e \approx 1$, H = 452 or 800 km s^{-1} for $v_{turb} = 0$ or 10 km s^{-1}, respectively. The non-zero values of v_{turb} are consistent with the nonthermal line-broadening velocities that must be introduced to account for the observed Doppler widths of lines formed at temperatures of approximately 5000 K and 10 000 K in the solar atmosphere.

Given T, v_{turb}, and r_e as functions of x, we can determine $p(x)$ from (5.1), apart from a normalizing constant. The electron fraction r_e depends on T and p and other calculated parameters, and we normalize the pressure equation so that continuum optical depth unity at 5000 Å occurs at a fixed photospheric reference depth (the zero point of our geometrical height scale). Thus we obtain $p(x)$ given $T(x)$ and $v_{turb}(x)$. Here we consider only the static case, and do not include effects due to magnetic fields.

Given pressure and temperature we can determine atomic number densities, opacities, and monochromatic optical depths. The various atomic number densities are obtained by excitation and ionization calculations that depend not only on pressure and temperature but also on the radiation intensity as a function of depth and wavelength. The radiation intensity depends, in turn, on the atomic number densities. These interactions are determined by solving the equations of radiative transfer and statistical equilibrium. Finally, given $T(x)$ and $v_{turb}(x)$, we can calculate all needed atmospheric parameters and then the emergent intensity as a function of wavelength.

An analysis of the differences between the calculated and observed intensity as a function of wavelength indicates how the atmospheric parameters should be changed in order to get better agreement with observations. In this way we successively adjust $T(x)$ and $v_{turb}(x)$ until we match observations as well as possible. The result is called a semiempirical hydrostatic-equilibrium model of the atmosphere based on the available observations.

The computed spectrum is relatively insensitive to the adopted values of $v_{turb}(x)$, even though the scale height increases substantially as v_{turb} is increased. As noted above, we obtain this turbulent velocity distribution from Doppler line widths that are observed to exceed thermal values. Fig. 5.14 shows our adopted v_{turb} plotted as a function of the height h, in kilometers, above continuum optical depth unity at 5000 Å. For simplicity we use these values for all brightness components A through F. Given this fixed v_{turb} distribution, the atmospheric model then depends only on $T(h)$.

Fig. 5.15 shows the temperature distribution obtained from a variety of quiet-Sun observations. The temperature is shown as a function of height, and as a function of the column mass m (g cm^{-2}) which equals p/g from the hydrostatic equilibrium equation. We also indicate the approximate height regions where the various continua and line components originate.

The ultraviolet continuum near 1600 Å, the far infrared continuum near 150 μm, and the minima in the wings of the Ca II resonance lines and the Mg II resonance lines all are formed in the temperature minimum region near $h = 500$ km. The Ca II and Mg II emission peaks, the hydrogen Hα line

Fig. 5.14. The turbulent velocity distribution used in the model solar atmosphere calculations reported here compared with values inferred from observations. (References in the figure are given by Vernazza, Avrett & Loeser (1981).)

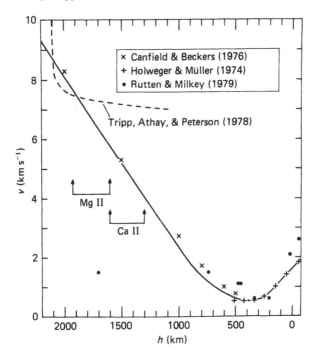

center, the millimeter continuum and carbon 1100 Å continuum, and the wing of Lyman α at 1 Å and 5 Å from line center all are formed in a broad 5500–7000 K temperature plateau which characterizes most of the chromosphere. The Ca II and Mg II resonance line centers, the centimeter continuum, and the hydrogen Lyman continuum ($\lambda < 910$ Å) all are formed at the base of the chromosphere–corona transition region which is characterized by a large temperature gradient. The central portion of the Lyman-α line and the 3-cm continuum are formed at 20 000–30 000 K temperatures.

The plateau near 25 000 K has been introduced to account for (1) the Lyman-α and Lyman-β integrated intensities, (2) the central absorption observed in these lines, and (3) the short-wavelength intensity in the Lyman continuum. We could replace the plateau by a more gradual temperature rise

Fig. 5.15. Model temperature distribution for the average quiet Sun.

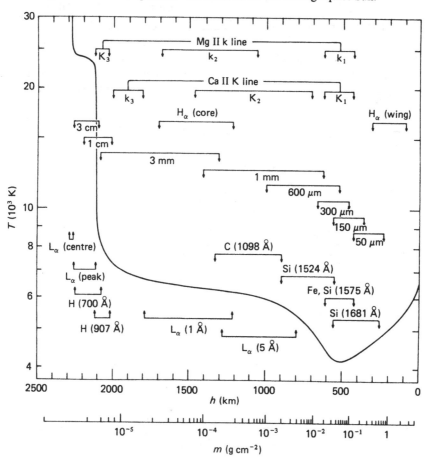

between 10 000 and 30 000 K, adjusted to give the observed values of the integrated Lyman line intensities, but the calculated Lyman-β line would not have the central absorption feature that is observed, and Lyman-α would have less central absorption than observed. No physical mechanism has been identified to explain such a plateau. A one-dimensional model may not properly characterize the real three-dimensional atmosphere in this respect; the plateau may signify only that the line of sight passes through material of the indicated thickness at the indicated temperatures.

Fig. 5.15 also shows the heights at which the hydrogen Hα line core and wing are formed. The reason for the separation of the core and wing heights is the negligible population of the hydrogen $n = 2$ level and the consequent line transparency throughout the temperature minimum region. Spectroheliograms taken in a narrow wavelength band centered on Hα therefore show chromospheric structure and behavior with little photospheric contamination. Displacing the band slightly from the line center enhances the relative brightness of different chromospheric features with different velocities along the line of sight. Spectacular photographs of the inhomogeneous chromosphere such as in Figs. 5.2 and 5.3 are obtained in this way.

The temperature distribution in Fig. 5.15 corresponds to the component C intensities plotted in Fig. 5.13. The temperature distributions found in the same way for components A through F are shown in Fig. 5.16. These results

Fig. 5.16. Temperature as a function of height and of column mass corresponding to the observed intensity components A–F in Fig. 5.12.

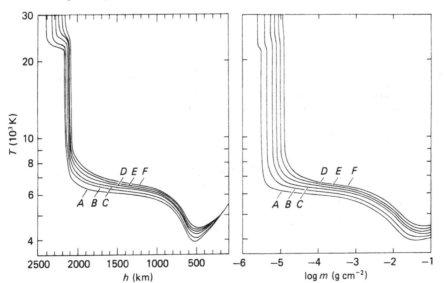

correspond to a range of quiet-Sun intensities, from the faintest 8 percent to the brightest 4 percent of the observed solar surface area. Note that the six curves all resemble each other in showing a gradual increase in the 6000–7000 K range and a very rapid increase above 8000 K. Brighter components have higher temperatures at all depths and have chromosphere–corona transition regions located closer to the photosphere, at higher pressures. Active-region models also have a similar structure, but with a greater chromospheric temperature rise; their temperature enhancements relative to Model *F* are comparable with the temperature differences between Model *F* and Model *A*. Also, active-region models having the same assumed v_{turb} as in Models *A–F* have transition regions even closer to the photosphere, but a choice of larger values of v_{turb} would move the transition region to greater heights.

What can we learn from such models? Given the depth variation of the number densities of the various atomic and molecular energy states and the radiation intensity in the important lines and continua, we can calculate the net radiative cooling rates Φ_{ul} and Φ_m in erg cm^{-3} s^{-1} as functions of depth for lines between upper and lower levels u and l, and for continua associated with levels m. The sum of all of these cooling rates must be the non-radiative energy input at each depth required to maintain the given temperature distribution, assuming a steady state. In radiative equilibrium without any non-radiative energy input, the sum of these cooling rates would have to be zero, but the chromosphere is clearly not in radiative equilibrium.

For continua, the net radiative cooling rate (or radiative flux divergence) in units of erg cm^{-3} s^{-1} is given by

$$\Phi_m = 4\pi \int_m \kappa_\nu (S_\nu - J_\nu) \, d\nu \tag{5.5}$$

where κ_ν is the continuous opacity (in cm^{-1}), S_ν is the source function, and J_ν is the mean intensity of radiation. Both S_ν and J_ν are in units of erg cm^{-2} s^{-1} Hz^{-1}, and ν is frequency in Hz. For a line,

$$\Phi_{ul} = h\nu[n_u(A_{ul} + B_{ul}J) - n_l B_{lu}J] \tag{5.6}$$

where n_u and n_l are the upper and lower level number densities, A_{ul}, B_{ul}, and B_{lu} are the Einstein coefficients for spontaneous emission, stimulated emission, and absorption, and where J is the mean intensity weighted by the line opacity and integrated over the line according to

$$J = \left(\int \kappa_\nu J_\nu \, d\nu \right) \Big/ \left(\int \kappa_\nu \, d\nu \right). \tag{5.7}$$

From the atmospheric model parameters, we have calculated the net radiative cooling rates for (1) the lines and continua of atomic hydrogen, (2) the negative hydrogen ion H$^-$, (3) the resonance and subordinate lines of Ca II, (4) those lines of Mg II, and (5) the lines and continua of a number of other constituents (listed in a subsequent paragraph). The cooling rates we obtain for Model *C* are plotted in Fig. 5.17. This graph shows the five contributions listed above and the sum of these contributions. The upper and lower panels show positive and negative values of Φ, on logarithmic scales, plotted against height in km above continuum optical depth unity at 500 nm.

Three important features of the total rate plotted in Fig. 5.17 are (a) the maximum near 800 km in the low chromosphere and the gradual decreasing

Fig. 5.17. The net radiative cooling rate, in erg cm^{-3} s^{-1}, as a function of height for the model of the average quiet Sun. The total rate is the sum of contributions due to H, H$^-$, Ca II, Mg II, and 'others' as explained in the text.

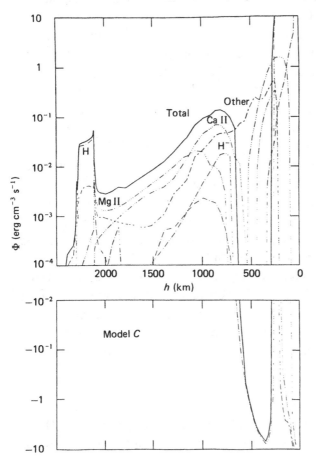

of Φ above this height, (b) the maximum near 2200 km corresponding to the temperature plateau in the chromosphere–corona transition region, and (c) the negative values of Φ near 400 km in the temperature minima region. (The large positive values below the temperature minimum region have less significance; they occur because Φ is proportional to the large number density in this region and because we have not attempted to adjust the photospheric temperature distribution to obtain $\Phi = 0$. In Model C the total hydrogen number density at 100 km is 2500 times larger than the value at 1000 km.)   The maximum of the net radiative cooling rate in the low chromosphere and its gradual decrease with height shows how mechanical energy is dissipated with height to maintain the chromospheric temperature rise. Theories of wave energy dissipation must be consistent with these calculated values. As may be seen from Fig. 5.17, the chromospheric mechanical energy input escapes primarily as Ca II and Mg II line emission.

The maximum value of Φ in the transition region corresponds primarily to emission in the hydrogen Lyman-α line. The area under this maximum is $\approx 3 \times 10^5$ erg cm^{-2} s^{-1}, which agrees with the observed integrated Lyman-α intensity in quiet regions near disk center. Such a localized Lyman-α feature seems required whether there is a plateau in the transition region or a more gradual temperature rise instead. The difficulty is that no localized heating mechanism has been found that can provide the required energy. One proposed explanation is that the energy of 3×10^5 erg cm^{-2} s^{-1} is dissipated in the corona and is brought downward by conduction to the lower transition region to be radiated away by Lyman-α. A difficulty is that the required conductive energy transport can take place only if the temperature gradient is very large, while a much smaller gradient is needed in order to produce the observed emission. This problem remains unsolved.

The negative total cooling rate in the temperature minimum region is perplexing, since negative values imply non-radiative energy extraction, i.e., refrigeration. Possible explanations include (1) the omission of positive contributions that would make the total rate positive, (2) the effect of time-dependent wave motions that, on the average, indeed remove energy from the temperature minimum region, and (3) the conclusion that an average model does not describe the average atmosphere, and that we need to introduce small-scale inhomogeneities which have a more extreme range of variation than components A through F. We consider these three possibilities in more detail below.

(1) Fig. 5.17 includes contributions from H, H^{-}, Ca II, Mg II, and 'others'. This last contribution is due to the calculated lines and continua of He, He II, Ca, Mg, Si, Si II, C, C II, Al, Fe, Na, K, and O. This contribution does not

include carbon monoxide and other molecules, but separate calculations indicate that adding the effects of CO would not greatly affect the results shown in Fig. 5.17. Thus we conclude that Φ calculated from Model C is indeed negative in the temperature minimum region.

(2) If waves travel through the atmosphere and the temperature at each position varies as a function of time, the time-averaged computed spectrum may have characteristics of high and low temperatures that cannot be matched by an average temperature value. Some recent calculations indicate that the negative values of Φ in the temperature minimum region can be explained in this way.

(3) There is also the possibility that neither Model C nor some combination of Models A through F adequately describes the average quiet Sun. Instead, a better description might be achieved by a model combining a background region with temperatures well below that of Model A, with bright points having temperatures much higher than Model F. The bright points would be in approximate radiative equilibrium with $\Phi \approx 0$ in the temperature minimum region because of the higher temperatures. The background areas would have such low temperatures in the minimum region that the additional radiative losses due to carbon monoxide would also give $\Phi \approx 0$. Such an explanation is supported by observations showing the presence of CO lines with brightness temperatures in their cores at about 3800 K, well below the Model A temperature minimum value. There are several difficulties with this explanation, however. In Fig. 5.9 the brightness temperature in the near wing of the H-line is never seen to be as low as such a cool background model would require. Also, the observed quiet-Sun minimum brightness temperature of ≈ 4400 K in the far infrared near 150 μm means that the sum of each component's minimum temperature weighted by the fractional area of each component has to be ≈ 4400 K (since in this wavelength region not only LTE prevails, but also brightness is proportional to the temperature). However, the observed area covered by bright points is not sufficiently large to produce an average brightness temperature of 4400 K, if the background temperature is as low as the observed 3800 K. We discuss this interpretation further in Section 5.4.

The negative total cooling rate in the temperature minimum region thus remains a serious unsolved problem. It aptly illustrates the fact that much further theoretical and observational research is needed before we can claim to have a clear understanding of the structure and behavior of the chromosphere.

We summarize our discussion of semiempirical chromospheric models by listing in Table 5.1 the total radiative losses from various chromospheric layers in quiet and active regions. These losses are derived from models of quiet and active regions that have been adjusted for an optimum fit to the observed spectrum. For comparison, we also include the total energy loss from the corona; this includes not only radiative losses but losses back to the chromosphere by thermal conduction and losses due to the solar wind. This comparison illustrates the fact that it is the chromosphere, rather than the corona, whose existence poses the more serious requirement for mechanical heating.

5.3 The three-dimensional structure of the chromosphere

The observed characteristics of the chromosphere change from cell centers to the chromospheric network and to active regions, but the models described above treat each region as a separate plane-parallel atmosphere in hydrostatic equilibrium. While, as we have already stressed, these assumptions are highly questionable, we will retain them for the purpose of examining the key property that differentiates the grossly different regions of the chromosphere – namely the effect of magnetic fields.

Table 5.1. *Chromospheric and coronal energy losses[a]*

Parameter	Quiet Sun	Coronal hole	Active region
Transition layer pressure (dyn cm^{-2})			
	2×10^{-1}	7×10^{-2}	2
Coronal temperature (K at r≈ 1.1R$_\odot$)			
	1.1 to 1.6×10^6	10^6	2.5×10^6
Coronal energy losses (erg cm^{-2} s^{-1})			
Conduction flux F_c	2×10^5	6×10^4	10^5 to 10^7
Radiative flux F_r	10^5	10^4	5×10^6
Solar wind flux F_w	$\lesssim 5 \times 10^4$	7×10^5	$(< 10^5)$
Total corona loss $F_c + F_r + F_w$	3×10^5	8×10^5	10^7
Chromospheric radiative losses (erg cm^{-2} s^{-1})			
Low chromosphere	2×10^6	2×10^6	$\gtrsim 10^7$
Middle chromosphere	2×10^6	2×10^6	10^7
Upper chromosphere	3×10^5	3×10^5	2×10^6
Total chromospheric loss	4×10^6	4×10^6	2×10^7
Solar wind mass loss (g cm^{-2} s^{-1})			
	$\lesssim 2 \times 10^{-11}$	2×10^{-10}	$(< 4 \times 10^{-11})$

[a] After Withbroe & Noyes (1977). *Ann. Rev. Astron. Astrophys.* **15**, 363–87.

5.3.1 *Plages and network: the role of magnetic fields*

At moderate spatial resolution (averaging over sizes of order 10^3 km) there is a tight spatial correlation between enhanced chromospheric emission in plages or the network and underlying photospheric magnetic fields. This correlation is clearly seen in Fig. 5.8. The magnetic field in active regions is measured from Zeeman splitting of spectral lines to be about 1500 G, but it covers the plage area with a filling factor of only 10 to 20 percent, so that the mean field in the plage is less than about 200 G. The total emission from plages in a 1 Å band and centered on the K-line is enhanced several-fold relative to the quiet Sun (cf. Fig. 5.7). At higher levels in the chromosphere and transition zone the enhancement factor increases to about 5, and increases further with increasing temperature. This can be interpreted in terms of a plane-parallel model for a plage in which (1) the density at and above the base of the chromosphere is enhanced about five-fold, (2) the height above the photosphere of the temperature minimum is smaller than in the quiet Sun, and (3) the minimum temperature value is somewhat higher. The energy requirements to support the roughly five-fold increased radiation from active regions are about five times greater, so that about 2×10^7 erg cm^{-2} s^{-1} of mechanical energy must be deposited in the active region chromosphere above the temperature minimum (cf. Table 5.1).

Outside active regions, emission from network elements is very similar to that of active regions. Just as in the case of chromospheric plages, the elements of the network overlie regions of strong photospheric field, in which the field is concentrated into bundles with field strength of order 1500 G. The surface brightness of chromospheric network elements is somewhat less than that of plages, but this might be explained by the fact that the total magnetic flux per unit area underlying a network element is less than that underlying a similar-sized element in a plage. In other words, the intensities are consistent with the hypothesis that the magnetic field-associated heating mechanism is identical in plages and in the network, but that the fields responsible are simply more closely-packed beneath a plage.

The overall geometry of plages and the network in the chromosphere occurs through enhanced heating by magnetic fields, whose geometry in turn is determined by convective motions and by rotation in the deep interior. The network structure maps the motions of supergranules (Fig. 5.4), because the supergranular flow causes magnetic flux to concentrate at supergranule boundaries.

Why does chromospheric emission map the underlying photospheric magnetic field so faithfully? More specifically, how does the magnetic field enhance the propagation of mechanical energy into the chromosphere? Several possible mechanisms have been advanced.

Let us first consider the case of chromospheric heating in the absence of magnetic fields. M. S. Lighthill first investigated the generation of acoustic energy through turbulent motions in studies of the emission of noise by jet engines. The results of this study are applicable to the generation of acoustic energy in magnetic field-free regions of the solar photosphere by the turbulent motions of granulation. Thus the temperature structure of the centers of supergranules (curve A in Fig. 5.16) may be considered using Lighthill's theory. In this theory the primary emission of acoustic energy is quadrupole emission, giving rise to an emitted acoustic flux per unit area

$$F_{ac} \approx \rho u^3 (u/v_s)^5 \tag{5.8}$$

where ρ and u are the density and characteristic velocity of the turbulent elements and v_s is the sound speed.

In the presence of strong vertical magnetic fields, three additional types of waves are emitted. These are:

(1) Alfven waves, with emitted flux

$$F_A \approx \rho u^3 (u/v_A) \tag{5.9}$$

where v_A is the Alfven speed;

(2) slow-mode waves, with emitted flux

$$F_s \approx \rho u^3 (u/v_s) \tag{5.10}$$

and

(3) fast-mode waves, with emitted flux

$$F_F \approx \rho u^3 (u/v_A)^5. \tag{5.11}$$

The first two of these vary only as the first power of the ratio u/v, where v is the characteristic phase speed of the wave. The last, as well as pure acoustic waves, vary as the fifth power of the ratio u/v, and since that ratio is about 0.5 for the turbulent motions near the top of the convection zone, Alfven or slow-mode waves should emit energy much more effectively than acoustic or fast-mode waves, provided that strong fields are present. In fact, numerical evaluation of wave energy generation for solar conditions (assuming a field strength of 1500 G) yields Alfven and slow-mode fluxes about a factor of 10 greater than acoustic or fast-mode fluxes. The physical explanation is that acoustic or fast-mode waves are radiated isotropically, while slow-mode and Alfven waves are restricted to propagate only along the (vertical) field lines. Therefore it is natural to expect the enhanced chromospheric heating over magnetic regions of the photosphere to be due to slow-mode or Alfven waves.

Slow-mode waves are essentially acoustic waves channelled by the magnetic field, and their dissipation length is similar to that of acoustic waves. By contrast, Alfven waves have a much longer dissipation length. Observations indicate that the actual dissipation length in the solar chromosphere is about the same for cell centers and for bright network elements, since the corresponding empirical models (curves A and F in Fig. 5.16) have similar variations of temperature with height. This is also true for models of plages. These findings then suggest that slow-mode waves are the principal source of heating of the magnetic chromosphere. (This suggestion is reinforced by what is known of the chromospheric heating in other stars, which have different values of surface sound speed and magnetic field strength.) However, Alfven waves probably do play a role in the heating of the solar corona, if not the chromosphere. They can carry a flux comparable to that of slow mode waves, but they dissipate it over much longer distances owing to their greater damping length. The total amount of energy required to maintain the chromosphere and corona in active regions have been given in Table 5.1 as $4 \times 10^6 \, \mathrm{erg \, cm^{-2} \, s^{-1}}$ and $3 \times 10^5 \, \mathrm{erg \, cm^{-2} \, s^{-1}}$ respectively. This is actually less than 1 percent of the energy generated by the Lighthill mechanism or its magnetic variations, according to (5.8)–(5.11), so that the efficiency of upward propagation from the photosphere need not be high.

5.3.2 *Small-scale chromospheric structure*

We have already noted that even when we consider subregions of the chromosphere, such as the various components of the quiet Sun or plages, as individual plane-parallel atmospheres, the best-fit models still do not fit the observed spectrum with satisfactory accuracy. This is scarcely surprising, given the wealth of fine-scale structures visible in high-resolution photographs (cf. Figs. 5.2 and 5.3), and considering the complex dynamics seen in time-lapse sequences or inferred from measured Doppler shifts. Indeed, after examining Fig. 5.2, one is left wondering why plane-parallel models do as well as they seem to do.

One of the most striking disagreements between the chromospheric models discussed in Section 5.2 and the real Sun is provided by observations of the cores of strong lines in the fundamental vibration-rotation band of CO, for which the brightness temperature observed near the limb requires a minimum temperature below 4000 K. As mentioned earlier, a possible resolution of this dilemma is that the region of the temperature minimum consists of cool and hot components which are spatially unresolved in the observational data used to construct plane-parallel temperature models. In the cool component, the CO is such an effective cooling agent that it depresses the minimum temperature to a value well below 4000 K even in the presence of a small amount of mechanical heating. In the hot component the heating is enough to raise the temperature high enough so that CO dissociates and the principal cooling agent disappears. The primary remaining radiative exchange mechanism, due to transitions in H^-, is a source of net radiative heating, rather than cooling, so that the temperature minimum in these components continues to rise, leveling off only at a temperature around 4900 K.

If this picture is correct, what might be the cause of the two different regimes of mechanical heating? One possibility is that the hot components overlie photospheric magnetic flux knots. Such knots are observed as isolated regions, several hundred kilometers across, where the magnetic field is about 1500 G. Between the flux knots are more extended regions where the magnetic field is essentially zero. It is likely that the larger-scale magnetic network and active regions are made up of many closely-spaced flux knots, but flux knots could also exist with much lower density in the quiet Sun away from the network. The flux of mechanical energy within the magnetic knots could greatly exceed that outside, for the same reasons discussed above in the general context of magnetic heating of plages and the network. The regions outside the flux knots may not be heated sufficiently to dissociate CO, and therefore the resulting CO radiation cools them further, at least near the temperature minimum, to temperatures below 4000 K.

In such a picture, the extraordinary limb darkening of radiation in the core of the strong CO vibration–rotation lines is easily explained: At disk center, the observed brightness temperature is an average of that at $\tau_0 = 1$ in the two types of structures. Near the limb, only the cool ($T \approx 3800$ K) structures are seen; they have large opacity in CO and project upward as opaque fingers which occult the hotter-temperature radiation originating deeper within the flux tubes.

This simple picture has some puzzling problems, however. As noted in Section 5.2, one would expect high angular resolution spectra of the Ca II K-line to resolve the cool components, so that somewhere in the line profile (which spans depths of formation all the way from the mid-chromosphere down to the photosphere) radiation would be emerging from the cool components. At that point in the line profile, brightness temperatures below 4000 K should be found. However, this is never observed at any wavelength within the line profile (cf. Fig. 5.9).

A second mechanism noted in Section 5.2 that invokes the thermal bi-stability of CO cooling is a dynamical one. As waves propagate upward, the cooling associated with rarefaction produces localized cooler regions, which are cooled further by CO radiation, so that pockets of cool CO-emitting gas may be produced overlying ordinary warmer chromospheric regions. If these pockets are optically thin in the K-line, the lack of observed brightness temperatures below 4000 K anywhere in the K-line wing becomes understandable. At the same time, the cool pockets could be very opaque in the CO lines, and sufficiently numerous that when viewed toward the limb they block all hotter radiation from below in the CO lines, thus explaining the observed darkening at the extreme limb.

5.3.3 *Structure and dynamics of the high chromosphere*

The energetics of the chromosphere involves much more than the orderly progression and dissipation of acoustic or mhd waves. A large variety of transient dynamic behavior is observed, implying the occurrence of mhd instabilities, perhaps of many sorts. Spicules (Fig. 5.3) are jets of chromospheric material that are ejected upwards along the magnetic field lines of the network at apparent velocities of order 25 km s^{-1}. They rise to heights of about 10 000 km before becoming invisible. The duration of a typical spicule event is about 5 minutes, and at any time there are about 10^5 occurring over the surface of the Sun. The typical diameter of a spicule is about 800 km, so that at any time about 1 percent of the solar surface is covered by spicules.

Whether the spicular matter subsequently falls back or is evaporated into the corona is still unresolved. However, simple considerations of mass balance of the corona show that almost all of the matter ejected upward in spicules must ultimately return to the solar surface, rather than escape from the Sun. The density within spicules is about $n \approx 10^{11}$ cm^{-3} (or $\rho \approx 1.7 \times 10^{-13}$ gm cm^{-3}). This implies an upward mass flux of $\rho v \approx 4 \times 10^{-7}$ gm cm^{-2} within the spicule. With 1 percent coverage of the solar surface by spicules, the spatially-averaged mass outflow in spicules is then about 4×10^{-9} gm cm^{-2}. This is more than 100 times the outward solar wind flux averaged over the solar surface, $M/4\pi R^2 \approx 3 \times 10^{-11}$ gm cm^{-2}, so at most 1 percent of the rising spicular matter could find its way into the solar wind; the remainder must return to the chromosphere.

The upward kinetic energy flux within a spicule is $\rho v^{3/2} \approx 1.3 \times 10^6$ erg cm^{-2} s^{-1}. This is somewhat less than the mean radiative energy loss of 4×10^6 erg cm^{-2} s^{-1} from the chromosphere (Table 5.1), but it is not insignificant. It may be that the spicule energy flux represents the (time-variable) excess energy deposited in the chromosphere, beyond the capabilities of the chromosphere to radiate it. More specifically, it is possible that changes in magnetic field geometry in the network (such as could be produced by the buffeting of photospheric fields by convective motions) would cause the rate of generation or propagation of mhd energy into the chromosphere to increase. If the chromosphere is already radiating at maximum efficiency, increased energy input will cause a rapid non-hydrostatic readjustment, giving rise to the ejection of a spicule. The details of such a mechanism are complex and poorly understood, in part because spicule kinematics are not well determined observationally; the observed spicule motions are consistent either with (a) an initial acceleration followed by coasting to the maximum height under the deceleration of gravity or (b) constant velocity rise to maximum height.

An alternative source of energy to drive spicule motions could be the energy carried downward into the chromosphere by thermal conduction along the very steep temperature gradient of the transition zone. The spatially-averaged conductive flux which penetrates to the bottom of the transition zone is $F_C = \kappa \, dT/dh \approx 2 \times 10^5 \, \text{erg cm}^{-2} \, \text{s}^{-1}$ (Table 5.1). If, as is believed, the magnetic field lines funnel down into the network where the spicules arise (Fig. 5.4), then a significantly greater energy source is available in the chromosphere at the location of the network. However, a serious drawback to thermal conduction as an energy source for spicules is the fact that this energy cannot penetrate to the level where the spicules originate: The thermal conductivity, which is proportional to $T^{5/2}$, becomes much too small to transport energy over significant distances below temperatures of order 10^5 K. While energy might be fed in to low altitudes if the geometry of spicules is complex enough, thermal conduction is an unlikely source of spicule energy if the atmosphere resembles at all the plane-parallel models described earlier. A much more precise description of spicule energetics will require very high angular resolution observations of the upper region of the chromosphere.

Spicules are the most easily seen, and therefore the most thoroughly studied, dynamical events in the chromosphere. With the advent of solar spectroscopy from space, it became possible to study displacements, Doppler shifts, and line broadening of chromospheric features seen in ultraviolet lines formed high in the chromosphere and transition region, such as Lyman-α, C IV, O V, Ne VII, and others. The recent results from the High Resolution Telescope and Spectrograph (HRTS) rocket instrument of the Naval Research Laboratory are particularly interesting in this regard. They show (Fig. 5.18) that, on a scale of about 1 arc sec, the chromospheric motions are much more complex than indicated by Hα spicule data. The C IV emission, which is produced mainly in the network (see the low-resolution image in Fig. 5.5), is resolved by HRTS data into individual structures with dimensions of less than about 2 arc sec. It is likely that the actual size of the structures is less than 100 km (0.14 arc sec) in at least one dimension, based on a comparison of the inferred emission measure $\int n_e^2 \, dV$ and electron density from line ratios. Typical velocities in the high chromosphere and transition zone are of the order of $10 \, \mathrm{km \, s^{-1}}$. Both downflow and upflow velocities of this magnitude are seen in the chromospheric network. Motions are not confined to the network, however. For example, chromospheric 'jets' are seen in ultraviolet lines such as C I 1561 Å or Si II 1533 Å, with upflow or downflow velocities of $10\text{–}20 \, \mathrm{km \, s^{-1}}$.

Fig. 5.18. Far ultraviolet spectrum of the quiet Sun near the limb, showing emission in C I, Si II, and C IV lines (right), along with an Hα filtergram from Sacramento Peak Observatory showing the position of the spectrograph slit. Data are from the Naval Research Laboratory's High Resolution Telescope and Spectrograph (HRTS) experiment. (Courtesy Naval Research Laboratory.)

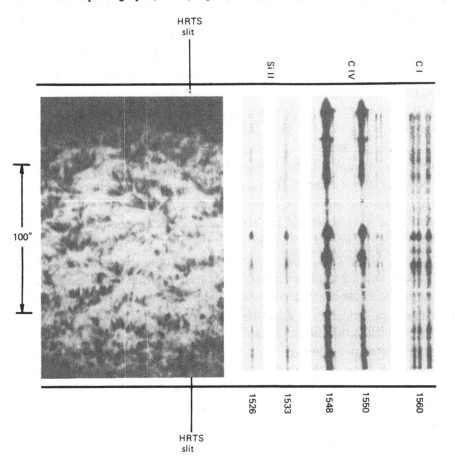

They are accompanied by transient brightenings of chromospheric emission lines, with characteristic lifetimes of a minute or less.

5.3.4   *Magnetic fields in the chromosphere*

It is apparent from the observed shape and velocity patterns of chromospheric features that magnetic fields not only control the heating but also shape the structure and mass flows in the chromosphere. This is not unexpected, since the magnetic pressure $P_B = B^2/8\pi$ above photospheric magnetic field concentrations generally drops off more slowly with height than the gas pressure $P_g = nkT$, and above the point where $P_B > P_g$, magnetic fields will dominate the chromospheric structure.

It is important therefore to measure the magnetic field in the chromosphere. Unfortunately, very few tools exist to explore chromospheric magnetic fields directly. An indirect technique which gives a useful first approximation is the use of potential theory to extrapolate the observed photospheric field upward, that is, to assume that electric currents produce a negligible departure from a potential field configuration. This procedure begs the most interesting questions, which have to do with the departures from a potential field due to dynamic pressure induced by the medium, and due to the associated induced currents. That there are significant departures from a potential configuration is clear from high-resolution Hα observations (cf. Fig. 5.2), which reveal the projected geometrical locus of field lines, if not the strength or sign of the field. The importance of departures from a potential configuration is that the excess energy stored in nonpotential fields is, in principle, available to be deposited in the chromosphere and corona, either continuously as part of the steady-state atmospheric energy balance or perhaps suddenly as a flare energy release.

It would be very desirable if there were a spectroscopic indicator of chromospheric magnetic fields, just as there is for the fields in the photosphere. The problem for chromospheric studies, in contrast to the photosphere, is the paucity of lines that exhibit suitable Zeeman splitting characteristics. These characteristics are: (a) large sensitivity of Zeeman shift $\Delta\lambda_z$ to magnetic field B (i.e., $d\Delta\lambda_z/dB \propto g\lambda^2$ where g is the Lande g-factor and λ the wavelength) and (b) a sharp line profile slope $1/I \, dI/d\lambda$ to give maximum fractional change $\Delta I/I$ for a Zeeman shift $\Delta\lambda_z$. Strong chromospheric lines in the visible like Hα or Ca II K have only modest Lande g-factors and have relatively broad profiles. Far ultraviolet emission lines, like C IV, O VI, etc., even if they have steep profiles, have a wavelength so short that the Zeeman shift, $\Delta\lambda_z$, is relatively small. Some success has been obtained by the Solar Maximum Mission in measuring circular polarization due to the Zeeman effect in the C IV 1550 Å line; this has allowed measurements of the magnetic field in the transition zone above sunspot umbrae, but the sensitivity is too small to permit magnetic field measurements above other solar features.

In the near infrared, the 8542 Å chromospheric line of Ca II exhibits sufficient Zeeman sensitivity, due to its relatively long wavelength, to be a useful indicator of magnetic fields in the middle chromosphere. Simultaneous studies of this line and deeper-lying photospheric lines have provided some of the first direct indications of magnetic structure in the chromosphere. Recently comparative studies have been made of magnetic field morphology in the mid-chromosphere using this Ca II line in comparison with the photospheric lines C I 9111 Å and Fe I 8688 Å. The field in the mid-chromosphere is found to be very diffuse compared with the highly intermittent fields in the photosphere. Furthermore, the field, emerging from the photosphere nearly vertically from concentrated flux knots within plages or the network, appears to bend over to form nearly horizontal 'canopies' at an elevation of only about 500 km above $\tau = 1$. The magnetic field in the canopy seems to have a sharp lower boundary, below which (except over flux knots) the field is essentially zero.

This newly-elucidated canopy structure may be of considerable importance for chromospheric physics. One immediate implication is that magnetic potential theory does not hold in the low chromosphere, and probably should not be considered below a height of order the radius of a supergranule, some 15 000 km. Another important point is that the canopy structure can affect mhd wave propagation, reflection, and chromospheric heating. Thus it must be taken into account in constructing realistic three-dimensional chromospheric models. For example, it is quite possible that in regions above the canopy, where the magnetic field is essentially horizontal, the transition zone extends right to the canopy boundary, that is, to within some 600 km above the photosphere.

It is interesting that the magnetic field in the canopy, as mapped in a chromospheric line like Ca II 8542 Å, is diffuse, while the intensity pattern seen in Hα, originating from the same altitude, is very highly structured. Clearly the Hα structure is not introduced by magnetic field strength variations. The local variations in density and temperature which give rise to the observed Hα fine structure apparently exist within a much more homogeneous field structure. In just the same way the highly structured coronal loops seen in XUV and X-ray images are no doubt embedded within more homogeneous field geometries. It remains a puzzle to explain why some flux tubes or field lines are energized to larger values of density and temperature than their surroundings. Possibly the answer lies in preferential twisting of the photospheric footpoints of these particular flux tubes, which gives rise to wave propagation and energy deposition only along the locus of these particular field lines.

The most magnetically sensitive lines yet discovered in the chromosphere are the Mg I and Al I emission lines near 12.3 μm (Fig. 5.19). For these lines,

Fig. 5.19. Spectrum of two far-infrared Mg I emission lines near 12.3 μm (a) at disk center and at two positions above a sunspot penumbra, showing the very large Zeeman splitting due to predominantly longitudinal (b) and transverse (c) magnetic fields. From Brault & Noyes (1983), *Astrophys. J.* **269**, L61.

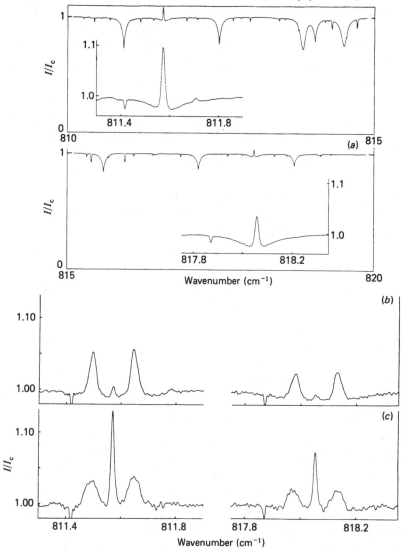

with $g = 1$, the product $g\lambda$ exceeds that of the most sensitive lines in the visible by a factor of 5–10. In addition, the emission is purely chromospheric in origin. This gives very great sensitivity in the chromosphere, partially overcoming the effect of the weaker fields there. These lines could be quite valuable in the future for probing chromospheric magnetic structure. However, significant observational problems must be overcome – especially the present lack of sensitive array detectors for the $12\ \mu m$ region, and the modest angular resolution of conventional solar telescopes at this wavelength. (Even the McMath solar telescope, the world's largest, has a 10-μm diffraction limit of 2 arc sec.)

Recently other chromospheric emission lines have been observed even further in the infrared – namely transitions between high-lying states of hydrogen, such as the $n = 16$ to $n = 15$ transition at $169\ \mu m$. It might be possible in the future to obtain much better magnetic sensitivity from these lines if suitable imaging instruments could be placed above most of the atmosphere, or in space.

5.4   Conclusions

The study of the solar chromosphere is important because it allows us a close-up view of the flow of non-radiative energy in the outer portion of a stellar atmosphere. Through such study we can hope to discover the nature of energy propagation and dissipation, either by direct observation of processes in action (for example, by observing waves or shocks) or indirectly, by studying their effects on the emergent spectrum. The fact that the chromosphere contains both magnetized regions (network and plages) and relatively field-free regions allows us to make comparative studies to isolate the effects of magnetic fields. It is immediately apparent, for example, that magnetic fields play the major role in the heating of the chromosphere in plages and the network, because of the tight correlation between the chromospheric heating rate and photospheric field strength. Furthermore, it appears that slow-mode mhd waves are the most likely mechanism of energy transport, based on the general similarity of the thermal structure in magnetic and nonmagnetic regions. The reason is that slow-mode waves are essentially acoustic waves confined by the magnetic field to propagate within magnetic flux tubes as if inside a piston; thus their upward energy propagation is enhanced over that of isotropically-propagating pure acoustic waves. However, their damping length, determined by the vertical distance for the ratio u/v_s to reach unity, is essentially the same as for acoustic waves, so the relative distribution of energy deposition is expected to be similar in the two cases. To verify this idea, however, very high angular resolution observations will be required, so that the detailed structure, dynamics, and energy balance of individual magnetic flux tubes can be studied. Such observations should be feasible someday from high-resolution solar space telescopes. Similarly, it is reasonable to suppose that the high chromosphere and corona are heated by Alfven waves in magnetic regions, from the fact that, together with slow-mode waves, they are much more efficiently generated than pure acoustic waves in magnetic regions, and that they have a longer dissipation length than slow-mode waves and so should reach greater heights before dissipating. Observations of actual motions transverse to the field lines, characteristic of Alfven waves, are needed to verify this conjecture.

Where do we go from here? Many of the unsolved problems of chromospheric physics would be clarified if we had better observational data. Particularly needed are observations with very high angular resolution (0.1 arc sec, corresponding to 70 km, is a reasonable goal, since this is about the density scale height in the photosphere). Such observations should also cover wavelengths spanning height and temperature ranges from the photosphere through the transition zone. These observations should be attainable with the next generation of solar space telescopes. Refinements to radiative transfer modeling, including the full three-dimensional geometry and the effect of velocity fields, will be needed to interpret such data properly.

It should also be noted that our understanding of solar chromospheric physics is currently improving through spectroscopic study of chromospheres on other stars, which differ from the Sun in their mass, rotation rate, or age. Many of the chromospheric indicators discussed in this chapter are seen and are well studied in visible and ultraviolet stellar spectra. Of particular importance are the Ca II H- and K-lines, which are easily studied from ground-based telescopes. The H and K enhancements in solar active regions (Figs. 5.7 and 5.8) are so strong that even though only a few percent of the Sun is covered by active regions at any one time, the enhancements are detectable in disk-integrated light; that is, in the Sun seen as a star. Similar enhancements are easily detected in other stars having more active chromospheres.

From studies of the H and K emission from other Sun-like stars, as well as ultraviolet and X-ray data, we have learned that chromospheric activity is at a maximum when the star is young, and declines roughly as $(\text{age})^{-1/2}$. The rotational velocity for Sun-like stars also decreases with age in the same way, apparently because of the loss of angular momentum from outflowing stellar winds. Chromospheric activity and rotation are thought to decrease simultaneously because activity depends directly on rotation through the operation of a stellar magnetic dynamo. This idea has considerable support because many slowly-rotating stars like the Sun show clear chromospheric activity cycles very similar to the Sun's activity cycle.

Studies of stellar chromospheres are important for our understanding of the solar chromosphere in at least two ways. First, they allow us to observe the effect on chromospheric structure of different parameters, such as enhanced convection or differing photospheric magnetic field characteristics. This requires the determination of convection and surface field properties on other stars, a problem which presents severe observational and theoretical challenges. Nevertheless, exciting progress is being made in this field at the present time.

Second, the study of stars similar to the Sun provides interesting insights into the past evolution of solar chromospheric emission. From solar studies it appears that when the Sun arrived on the main sequence, it was rotating with a surface rotation rate of several days, and had a very much higher degree of chromospheric activity than at present. During its lifetime, its rotation rate and mean chromospheric emission level have decreased continuously. By an age of order 7×10^8 years (the age of the stars in the Hyades cluster), the Sun's rotation period had lengthened to about 8 days, and its chromospheric emission had decreased to about three times its present value. Its long-term variability was probably much more irregular than at present, so that the pattern of any regular activity cycle would have been hard to discover. Probably only when the rotation rate slowed to about 20 days, at an age of about 2×10^9 years, did a clear quasi-periodic cycle appear. During the entire time of the Sun's main-sequence rotational slow-down, the properties of 'active' and 'quiet' regions of the chromosphere were probably qualitatively similar to those at present, with the greater overall level of activity at a younger age due mainly to a much larger fraction of the solar surface being covered by active regions.

The corresponding changes in solar chromospheric emission, particularly in the ultraviolet, should have affected the Earth's atmosphere. The implications for the evolution of the Earth's atmosphere, and terrestrial biologic evolution, make continued study of solar and stellar chromospheres important, not only for their own sake, but also to quantify the evolution of the solar spectral irradiances at the Earth.

The discussion of chromospheric spectroscopy in this chapter would not be complete without noting the enormous debt that the field of chromospheric physics owes to Leo Goldberg, who has played a key role in the development of many of the concepts discussed here. Without Leo Goldberg's many unique contributions, our physical description of the outer solar atmosphere would be much more primitive than it is today.

References for further reading
General references
Athay, R. G. (1976). *The Solar Chromosphere and Corona: Quiet Sun*. Boston: Reidel.

Jordan, S. D., ed. (1981). *The Sun as a Star*, NASA SP-450. Washington, D.C. NASA Information Branch.

Lites, B. W., ed. (1985). *Chromospheric Modelling and Diagnostics*. Sunspot, NM: National Solar Observatory.

Orral, F. Q., ed. (1981). *Solar Active Regions*. Boulder:Colorado Associated University Press.

Sturrock, P., ed. (1985). *Physics of the Sun*. University of Chicago Press.

White, O. R., ed. (1977). *The Solar Output and Its Variation*. Boulder: Colorado Associated University Press.

Solar flux and intensity
Kurucz, R. L., Furenlid, I., Brault, J. & Testerman, L. (1984). *Solar Flux Atlas from 296 to 1300 nm*. Sunspot, NM: National Solar Observatory.

Neckel, H. & Labs, D. (1984). The solar radiation between 3300 and 12 500 Å. *Solar Phys.* **90**, 202–58.

The calcium H- and K-lines
Cram, L. E. & Dame, L. (1983). High spatial and temporal resolution observations of the solar Ca II H line. *Astrophys. J.* **272**, 355–61.

Linsky, J. L. & Avrett, E. H. (1970). The solar H and K lines. *Pub. Astron. Soc. Pacific*, **82**, 169–248.

Skumanich, A., Lean, J. L., White, O. R. & Livingston, W. C. (1984). The Sun as a star: Three-component analysis of chromospheric variability in the calcium K line. *Astrophys. J.* **282**, 776–83.

White, O. R. & Livingston, W. C. (1981). Solar luminosity variation. III. Calcium K variation from solar minimum to maximum in cycle 21. *Astrophys. J.* **249**, 798–816.

Far-ultraviolet temperature-minimum observations
Bonnet, R. M., Bruner, M., Acton, L. W., Brown, W. A., Decaudin, M. & Foing, B. (1982). Rocket photographs of fine structure and wave patterns in the solar temperature minimum. *Astron. Astrophys.* **111**, 125–29.

Brueckner, G. E. (1980). A high resolution view of the solar chromosphere and corona. *Highlights of Astronomy*, **5**, 557–69.

Cook, J. W., Brueckner, G. E. & Bartoe, J.-D. F. (1983). High resolution telescope and spectrograph observations of solar fine structure in the 1600 Å region. *Astrophys. J.* **270**, L89–93.

Foing, B. & Bonnet, R. M. (1984). On the origin of the discrete character of the solar disk brightness in the 160 nanometer continuum. *Astrophys. J.* **279**, 848–56.

Samain, D. (1980). Solar ultraviolet continuum radiation: The photosphere, the low chromosphere, and the temperature minimum region. *Astrophys. J. Suppl.* **44**, 273–94.

Far-infrared temperature-minimum observations
Cram, L. E. (1984). Interpretation of millimetre and sub-millimetre observations of the solar chromosphere. *Int. J. Infrared Millimeter Waves*, **5**, 1165–77.
Degiacomi, C. G., Kneubühl, F. K. & Huguenin, D. (1985). Far-infrared solar imaging from a balloon-borne platform. *Astrophys. J.* **298**, 918–33.
Lindsey, C., Becklin, E. E., Jefferies, J. T., Orrall, F. Q., Werner, M. W. & Gatley, I. (1984). Observations of the brightness profile of the Sun in the 30–200 micron region. *Astrophys. J.* **281**, 862–69.

Chromospheric models
Ayres, T. R. & Linsky, J. L. (1976). The Mg II h and k lines. II. Comparison with synthesized profiles and Ca II K. *Astrophys. J.* **205**, 874–94.
Avrett, E. H., Kurucz, R. L. & Loeser, R. (1984). New models of the solar temperature minimum region and low chromosphere. *Bull. Amer. Astron. Soc.* **16**, 450.
Basri, G. S., Linsky, J. L., Bartoe, J.-D. F., Brueckner, G. & Van Hoosier, M. E. (1979). Lyman-alpha rocket spectra and models of the quiet and active solar chromosphere based on partial redistribution diagnostics. *Astrophys. J.* **230**, 924–49.
Gingerich, O., Noyes, R. W., Kalkofen, W. & Cuny, Y. (1971). The Harvard–Smithsonian reference atmosphere. *Solar Phys.* **18**, 347–65.
Vernazza, J. E., Avrett, E. H. & Loeser, R. (1981). Structure of the solar chromosphere. III. Models of the EUV brightness components of the quiet Sun. *Astrophys. J. Suppl.* **45**, 635–725.

Energy balance
Avrett, E. H. (1981). Energy balance in solar and stellar chromospheres. In *Solar Phenomena in Stars and Stellar Systems*, ed. R. M. Bonnet & A. K. Dupree, pp. 173–98. Boston: Reidel.
Bonnet, R. M. & Delache, Ph., eds. (1977). *The Energy Balance and Hydrodynamics of the Solar Chromosphere and Corona.* Proc. IAU Colloq. No. 36, p. 504. Clermont-Ferrand: G. de Bussac.
Giovanelli, R. G. (1978). The radiative relaxation time in the chromosphere. *Solar Phys.* **59**, 293–311.
Linsky, J. L. & Ayres, T. R. (1978). Stellar model chromospheres. VI. Empirical estimates of the chromospheric radiative losses of late-type stars. *Astrophys. J.* **220**, 619–28.
Withbroe, G. L. & Noyes, R. W. (1977). Mass and energy flow in the solar chromosphere and corona. *Ann. Rev. Astron. Astrophys.* **15**, 363–87.

Carbon monoxide
Ayres, T. R. (1981). Thermal bifurcation in the solar outer atmosphere. *Astrophys. J.* **244**, 1064–71.

Ayres, T. R. & Testerman, L. (1981). Fourier transform spectrometer observations of solar carbon monoxide. I. The fundamental and first overtone bands in the quiet Sun. *Astrophys. J.* **245**, 1124–40.

Bartoe, J.-D. F., Brueckner, G. E., Sandlin, G. D. & Van Hoosier, M. E. (1978). CO fluorescence in the extreme-ultraviolet solar spectrum. *Astrophys. J.* **223**, L51–3.

Muchmore, D. & Ulmschneider, P. (1984). Effects of CO molecules on the outer solar atmosphere: A time-dependent approach. *Astron. Astrophys.* **142**, 393.

Kneer, F. (1983). A possible explanation of the Wilson–Bappu relation and the chromospheric temperature rise in late-type stars. *Astron. Astrophys.* **128**, 311–17.

Noyes, R. W. & Hall, D. N. B. (1972). Thermal oscillations in the high solar photosphere. *Astrophys. J.* **176**, L89–92.

Wave heating

Kalkofen, W., Ulmschneider, P. & Schmitz, F. (1984). Apparent solar temperature enhancement due to large amplitude waves. *Astrophys. J.* **287**, 952–8.

Schmitz, F., Ulmschneider, P. & Kalkofen, W. (1984). Acoustic waves in the solar atmosphere VII; non-grey non-LTE H$^-$ models. *Astron. Astrophys.* **148**, 217–25.

Stein, R. F. (1981). Stellar chromospheric and coronal heating by magnetohydrodynamic waves. *Astrophys. J.* **246**, 966–71.

Ulmschneider, P. (1981). Theories of heating of solar and stellar chromospheres. In *Solar Phenomena in Stars and Stellar Systems*, ed. R. M. Bonnet & A. K. Dupree, pp. 239–63. Boston: Reidel.

Ulmschneider, P. & Stein, R. F. (1982). Heating of stellar chromospheres when magnetic fields are present. *Astron. Astrophys.* **106**, 9–13.

6

Spectroscopy of the solar corona

JACK B. ZIRKER

6.1   Introduction

The solar corona is the most thoroughly explored example of a hot, dilute, extended stellar atmosphere. Because we are so near the Sun, we can resolve and observe individual structures. Using atomic physics, we can develop and test spectroscopic methods and apply them to construct coronal models that accurately represent the run of temperature, density, and velocity.

The solar corona turns out to be a fascinating astrophysical environment. Beyond its own intrinsic interest, it is also important because it is typical of many other stellar atmospheres. X-ray detectors aboard the HEAO-2 satellite ('Einstein') have revealed soft X-ray emission from stars of every spectral type and luminosity class. Thus, the diagnostic techniques developed with solar observations may help us to understand the nature of stellar coronae in general.

Despite our advantage of being near the Sun, we have as yet very incomplete ideas about some of the basic coronal processes. For example, we are not yet certain how the corona is heated, or how it replenishes the material it loses to the solar wind. Better observations and improved diagnostic techniques will help us to answer these large questions.

In this chapter, I will focus on the spectroscopic diagnostic techniques that have been devised for the study of the solar corona (and which are applicable, in principle, to other coronae) and mention some of the results that have come to light. After briefly discussing monochromatic imaging of the corona, I will consider how coronal temperature density, velocity and magnetic field have been measured by spectroscopic techniques.

First, a bit of history. C. A. Young and W. Harkness independently discovered the first coronal emission line, at 5303 Å (= 530.3 nm) in spectra observed at the total solar eclipse of 1869. More emission lines were discovered at subsequent eclipses until, by 1927, the wavelengths of 16 lines were known. However, no known terrestrial element seemed capable of emitting these lines, and a new element, 'coronium', was postulated. Unfortunately, the periodic table of elements was nearly complete, and allowed no interloper such as coronium. What was the origin of these mysterious lines?

Ira Bowen, a physicist at the California Institute of Technology, faced a similar problem. Gaseous nebulae also emitted unidentified spectrum lines, which were attributed to another mythical element, 'nebulium'. It occurred to Bowen that gaseous nebulae might emit their unidentified lines not because they contain an unidentified chemical element but because they are far more dilute than ordinary laboratory sources, and even than ordinary stellar atmospheres. In a gas of very low density an atom may remain in a metastable excited state for a long time (several seconds or longer) before it finally drops to a state of lower energy, emitting a photon. (In a dense gas, atomic collisions excite or de-excite atoms in metastable states before they have a chance to radiate.) Lines originating from metastable levels are called 'forbidden' lines. Bowen identified the nebular lines with transitions in singly- and doubly-ionized oxygen (O II and O III) and in singly-ionized nitrogen (N II). He also identified lines in the spectrum of a nova as forbidden transitions of multiply-ionized iron (Fe VI and Fe VII). W. Grotrian picked up the idea in 1939. Using B. Edlén's results on the energy levels of highly-ionized atoms, he showed that the coronal red line (at 6374 Å) was emitted by nine-times ionized iron (Fe X). Edlén went on in 1942 to identify most of the remaining observed coronal lines as forbidden transitions in common elements such as iron, nickel and calcium. These pioneering investigations revealed the two most basic properties of the corona: (1) It is highly dilute. Forbidden transitions occur only in a highly dilute gas, where metastable atomic levels become overpopulated because collisional de-excitations are rare. (2) Its temperature is very high – in the range of 1.5–2.5 million degrees. Such temperatures are needed to produce the high degrees of ionization inferred from the coronal spectrum.

Let us turn now to developments of the past decade, beginning with imaging.

6.2   Coronal structures

Fig. 6.1 shows the corona as photographed in white light during a total solar eclipse. Large-scale coronal structures such as streamers and solar plumes are prominent. The visible spectrum of such structures consists of a continuum, on which are superposed bright emission lines. The continuum arises from photospheric light scattered by free coronal electrons in the inner corona and by zodiacal dust particles in the outer corona. The electron-scattered continuum does not show photospheric absorption lines, because the fast random motion of the electrons Doppler-shifts the incident radiation during the scattering process, making the absorption lines very wide and very shallow. On the other hand, the photospheric lines do appear in the dust-scattered continuum. This implies that the dust grains have small random velocities.

Fig. 6.2 shows the corona photographed in soft X-rays by equipment aboard Skylab. At a kinetic temperature of about 2 million degrees, Wien's displacement law (wavelength at peak intensity inversely proportional to temperature) predicts that the corona's *intrinsic* radiation lies predominantly in the soft X-ray region (10–100 Å). The X-ray spectrum

Fig. 6.1. The solar corona, photographed in white light at the total eclipse of 16 February, 1980.

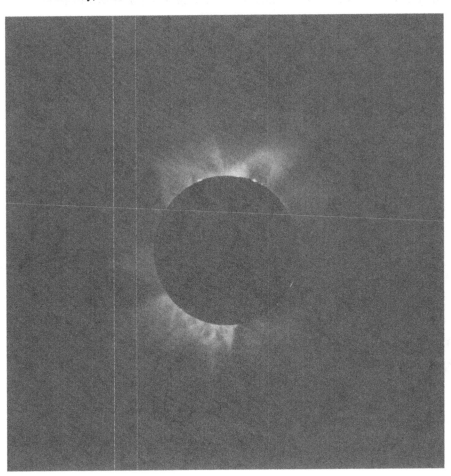

consists of permitted emission lines of highly ionized elements, superposed on a faint continuum. The striking dark lane in Fig. 6.2 is a 'coronal hole'. X-radiation arises from binary collisions and its intensity is therefore proportional to the square of the electron (or ion) density. Since the hole is darker than its surroundings, its electron density must be lower. Maps of the intensity of the extreme ultraviolet lines (especially Mg X 625 Å) obtained from rocket and satellite observations in the late 1960s showed such low-density regions. Photometry and analysis of these maps indicate that the electron density is reduced by a factor of 2 or 3 (Munro & Withbroe, 1972).

The spatial resolution attained in Fig. 6.2 (about 2 arc seconds) reveals that the corona consists mainly of loops or arches, ranging in size from a few thousand to a hundred thousand kilometers. The loops are brighter (e.g. denser) in active regions than in quiet regions.

All this structure arises, we think, from the imbedded coronal magnetic field. The overall structure of the corona changes in phase with the 11-year

Fig. 6.2. The X-ray corona, photographed by equipment aboard Skylab.

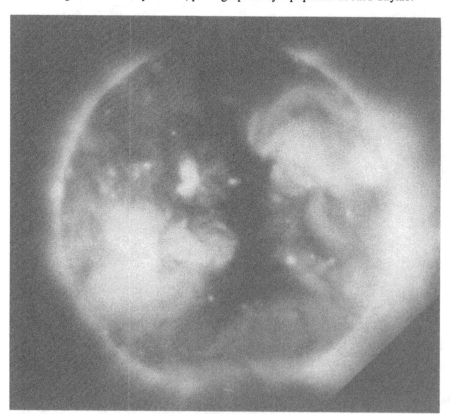

sunspot cycle as the large-scale *fields*, which are associated with spot groups, *evolve*. At sunspot maximum, coronal streamers appear at nearly all latitudes, giving the Sun a spherically symmetric appearance, while at minimum the streamers are confined to low heliographic latitudes.

6.3   Coronal temperature diagnostics

Since the corona is not in thermodynamic equilibrium, it does not have a single well-defined temperature. Several different temperatures can be inferred from spectroscopic data connected with such processes as ionization, excitation, line broadening, etc. Because *collisions* significantly influence all these processes, the primary or fundamental temperature is the *kinetic* temperature of ions or electrons. A theory is needed to relate the kinetic to the other kinds of temperature.

Take ionization, as an example. Atomic theory tells us that coronal ions lose electrons primarily by electron impact, and capture them by radiative and dielectronic recombination. The rates of these processes all depend on the electron kinetic temperature and, of course, the concentrations of ions.

Fig. 6.3 illustrates a typical calculation of the equilibrium of these competing processes. It shows the fraction of iron ions, with prescribed electrical charge, as a function of kinetic electron temperature. A given ion,

Fig. 6.3. The temperature variation of the ionization equilibrium of iron, under coronal conditions. The figures on the curves correspond to the number of missing electrons.

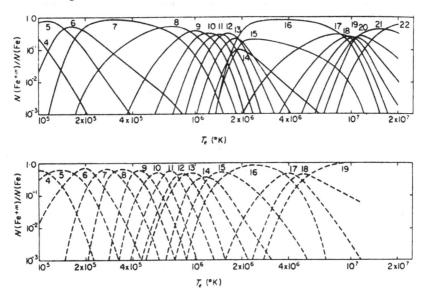

say Fe^{+13}, is abundant only over a relatively narrow temperature range. Thus the *presence* of Fe^{+13} lines in a coronal spectrum immediately signifies a narrow temperature range. This observation suggests a method of quantitative model-building, as follows.

A spectroscopic observation of the corona, properly calibrated, yields the *intensity* of radiation (erg cm^{-2} s^{-1} steradian^{-1}) in a given spectrum line. If we assume the corona is nearly transparent to its own radiation (we can test for self-consistency later on), we may describe the line intensity (I_i) by the following expression:

$$I_i = 1.73 \times 10^{-16} \int_0^\infty A_i f_i Q(T) G(T) \, dT.$$

$Q(T)$, the 'differential emission measure', is proportional to $N_e^2(T)$, the square of the number density of electrons with kinetic temperatures between T and $T + dT$. $G(T)$ is essentially the ionization function shown in Fig. 3. A_i is the element abundance and f_i the line oscillator strength. Because of the peaked character of $G(T)$, the integral may be approximated by

$$I_i = C_i \bar{Q}_i \bar{G}.$$

An observed line intensity thus yields a value of Q, or equivalently N_e^2 at a *mean kinetic temperature*.

Fig. 6.4 illustrates a typical variation of Q with T for the quiet corona at coarse spatial resolution. If we add further information (or asssumptions)

Fig. 6.4. Observations of the differential emission measure, as a function of the coronal temperature.

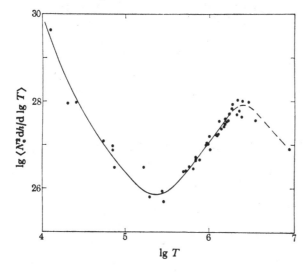

about the *geometry* and *dynamic* state of the coronal gas, we can construct an empirical model of T and N_e. For example, if we assume a spherically symmetric corona, with only radial variations of T and N_e, and assume hydrostatic equilibrium, we find the temperature model shown in Fig. 6.5 from the data shown in Fig. 6.4.

This method has been elaborated extensively and applied to a variety of coronal structures with complicated geometry. For example, H. Mason (1975) derived the T and N_e variations within a coronal condensation (above an active region) that was observed at the 1952 solar eclipse. From a large number of forbidden optical lines, she constructed the model shown in Fig. 6.6. The method has been applied to coronal loops, especially in active regions (Cheng 1980, Pallavicini *et al.* 1981, Levine & Pye 1980, Pye *et al.* 1978, Neupert 1979).

Such models indicate that loops are hotter at their tops than at their footpoints. Kinetic temperatures that lie between 2 and 3 million degrees are common, although some X-ray line data require a small amount of material at up to 5 million degrees. Electron densities of between 10^9 and 10^{10} are common, with the higher values in active regions.

A second method of inferring kinetic temperature (T_K) depends on observations of spectrum line *width*. An optically thin coronal emission line has a Gaussian profile, i.e.

$$I(\Delta\lambda) = I_0 \exp\left[-(\Delta\lambda/\Delta\lambda_D)^2\right],$$

Fig. 6.5. A semi-empirical coronal model of temperature vs. height in a radially-symmetric corona.

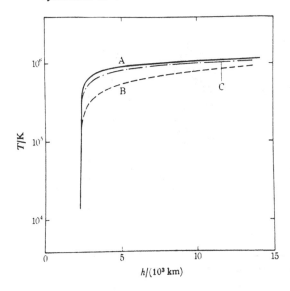

where

$$\Delta\lambda_D = \lambda_0 (2kT_K/m_{ion})^{1/2}/c.$$

Thus, an observation of a line profile yields an estimate of T_K, the ion kinetic temperature. The observations usually indicate that T_K is somewhat larger than the electron kinetic temperature, T_e, i.e., the lines are *non-thermally broadened*. With no better information to use, astronomers have introduced a 'microturbulent' non-thermal motion, specified by an RMS speed v_{turb}, to describe the excess broadening.

Thus, we write $I(\Delta\lambda) = I_0 \exp[-(c\,\Delta\lambda/\lambda_0 v)^2]$, where the broadening speed v is a sum of thermal and non-thermal terms:

$$v^2 = \frac{2kT_e}{m_i} + v_{turb}^2.$$

With two or more line profiles, v_{turb} and T_e can be separated since they influence line widths differently, depending on the ionic mass.

Another method of inferring an ion temperature derives from an extraordinary result of a rocket-borne experiment during the total eclipse of

Fig. 6.6. A semi-empirical model of a coronal condensation.

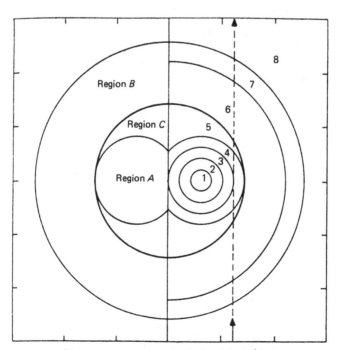

Distance across condensation (units of 5×10^4 km)

March 7, 1970. The rocket carried a slitless spectrograph that produced a spectrum of monochromatic images of the corona, one image for each line in the range 850–2190 Å. This range includes the resonance line of hydrogen, Lyman-α at 1216 Å, and the spectra showed a bright Lyman-α corona, extending to a height of 500 000 km above the limb. A. H. Gabriel (1971) showed that this radiation originates in the chromosphere and is scattered by hydrogen atoms in the corona.

Because of the hydrogen atom's small mass, thermal coronal motions contribute more than nonthermal motions to broadening the scattered Lyman line. Thus the width of the scattered Lyman-α line yields a reliable value for the proton-kinetic temperature. Beckers & Chipman (1974) produced the theory needed to interpret the line shape, shift and intensity.

A group of astronomers (Kohl *et al.* 1980, Withbroe *et al.* 1982) has obtained Lyman-α profiles at several heights in the corona, using a rocket-borne spectrograph. To interpret their data, they adopted a published distribution of electron density and a theory similar to that of Beckers and Chipman. Fig. 6.7 shows the resulting empirical proton temperature variation. A temperature maximum above 1.6×10^6 K appears at a height of $0.5\,R_0$ above the limb.

A quite special (and elegant) method of deriving electron temperatures will be discussed below in connection with solar active regions and flares.

6.4 Electron density

As we have seen above, spectral line fluxes yield emission measures, which are proportional to the square of the electron density averaged along a

Fig. 6.7. The height variation of the coronal proton temperature, as inferred from rocket observations of the Lyman-alpha profile.

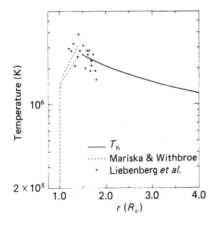

line of sight through the corona. If we have additional information on the distribution of material in space, we may construct a model of electron density (and temperature) as a function of position in the corona.

Free electrons in the corona scatter and linearly polarize photospheric white light, and thus make visible the large-scale structures apparent in a photograph of the corona (e.g., Fig. 6.1). By photographing the corona during total eclipse through a series of suitably chosen and oriented polaroid filters, one can estimate the amount and direction of polarized light. Assuming that the spatial distribution of electrons has axial symmetry, one can then derive the electron density at many positions in the corona. This was first done by Van de Hulst (1950) (see also Saito *et al.* (1977) and Waldmeier (1975)). Fig. 6.8 shows a typical radial variation of electron density above the solar equator during the maximum of the sunspot cycle. The method has been extended to compact, nonsymmetric structures such as a polar coronal hole (Crifo-Magnant & Picat, 1980).

A coronagraph used an occulting disk to produce artificial eclipses. Wlerick & Axtell (1957) used a coronagraph and employed photoelectric recording to measure the brightness and polarization of white light of the corona. Such a device, with various modifications, has been operated by the High Altitude Observatory (HAO) for more than a full solar cycle and has

Fig. 6.8. The radial variation of coronal electron density above the solar equator, near sunspot maximum.

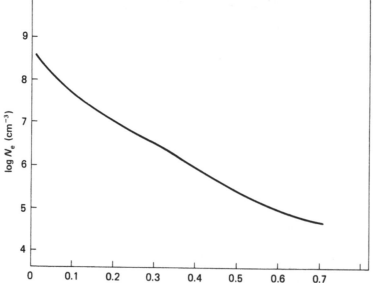

generated a remarkable record of the slow evolution of coronal structures. Eclipse photography retains its usefulness, however. Fig. 6.9 is an enhanced photograph, obtained by the Los Alamos National Laboratory group during the February 16, 1980 eclipse from an aircraft. It shows radial streamer structures out to 16 solar radii – far beyond the normal range (1.5 R_0) of the coronagraph.

A coronagraph with external occulting disks was flown aboard Skylab and produced a valuable 9-month sequence of coronal white-light photographs in three polarizations. Fig. 6.10 is a famous picture of a polar coronal hole, obtained by the HAO instrument aboard Skylab. Munro & Jackson (1977) analyzed a series of similar photographs to obtain the

Fig. 6.9. Coronal streamers photographed from an aircraft during the eclipse of 16 February, 1980.

electron density distribution throughout the hole, assuming cylindrical symmetry. The density along the hole axis is lower by a factor of 2–5 than in the 'quiet' neighborhood at the same height.

Spectral line-flux ratios provide yet another way to derive electron density. This type of diagnostic was developed originally for the interpretation of transition-zone lines, which originate at temperatures between 2×10^4 and 3×10^5 K, and then extended to coronal spectrum lines (Gabriel & Jordan 1972, Munro *et al.* 1971, Nussbaumer 1972, Dupree *et al.* 1976, Vernazza & Mason 1978). Fig. 6.11 illustrates the principle and the result for Mg VIII. At low electron densities, nearly all the Mg^{+7} ions will lie in the ground state ($2p\,^2P$), but at high densities ($> 10^{10}$ cm^{-3}) the

Fig. 6.10. A polar coronal hole, photographed with a coronagraph aboard Skylab, appears at the top (north) limb of the Sun. The dark space on the lower left was caused by the instrument.

populations of the metastable level (^{4}P) and ground state will assume the Boltzmann ratio valid in thermodynamic equilibrium. Thus the population ratio ^{4}P/^{2}P or equivalently the resonance to intersystem line strength ratio (λ430 Å/λ783 Å), will vary with electron density, as shown in Fig. 6.11(*b*). Vernazza and Mason found density-sensitive line ratios in the boron isoelectronic sequence, which includes the coronal ions Mg^{+7} and Si^{+9}. They applied these results to data obtained with the Harvard spectrometer aboard Skylab and determined electron densities in several types of coronal structures. An active region, for example, had an electron density ($> 3 \times 10^9$ cm^{-3}) more than 15 times as high as a coronal hole. The NRL group (Feldman *et al.* 1978a, b) has carried out similar analyses of forbidden lines in the nitrogen isoelectronic sequence (which includes the coronal ion S^{+9}) and of the forbidden lines of Fe IX. They applied their calculations to Skylab observations of active regions and a flare. The Fe IX line intensity ratio (λ241 Å/244 Å) yields a flare electron density of 2×10^{11} cm^{-3}.

Fig. 6.11(*a*). The energy level diagram for Mg VIII, showing the transitions used to determine coronal electron density. Transition wavelengths in Å are indicated.

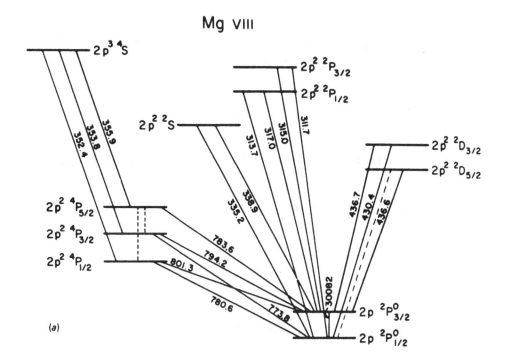

(a)

Fig. 6.11(*b*). The calculated population ratio $^4P/^2P$ in Mg VIII as a function of electron density.

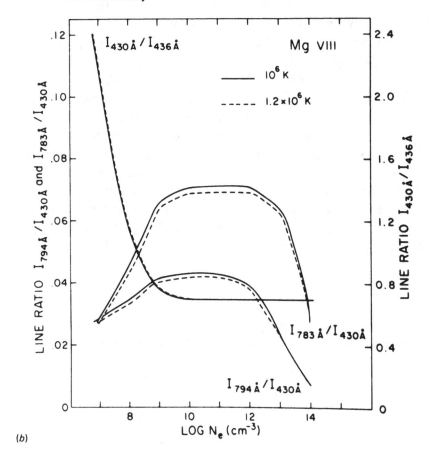

(*b*)

Fig. 6.12. The term diagram for O VII, a helium-like ion that appears in solar flares.

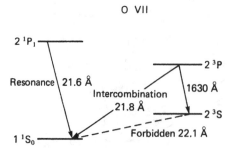

6.5 Electron temperature and density in flares

Each spectroscopic diagnostic is useful over a relatively limited range of T or N_e. Special diagnostics have therefore been developed for the extreme conditions encountered in solar flares, in which the electron temperature may exceed 10^7 K. At such high temperatures, such elements as neon, oxygen, magnesium and calcium become *helium-like*, i.e., stripped down to two electrons. Gabriel & Jordan (1969) first devised a technique, similar to that described above, to determine electron density with line ratios from helium-like ions. Consider the term diagram of O VII (Fig. 6.12), an ion that prevails at $T \approx 2 \times 10^6$ K.

At low electron density ($N_e \approx 10^9$ cm^{-3}), the metastable level, 2^3S, decays radiatively, with the emission of the 'forbidden' line $\lambda 22.1$ Å. The 2^3P level, populated by collisions from the ground state, also decays radiatively in the intercombination line, $\lambda 22.8$ Å. Because of the strong overpopulation of the 2^3P state, the intensity ratio ($\lambda 22.1$ Å/$\lambda 22.8$ Å) exceeds unity. However, at higher electron density ($N \approx 10^{10}$) the 2^3P and 2^3S levels couple by collisional excitations and through the permitted 1630 Å line. As a result, the population of 2^3P increases relative to 2^3S, and the line ratio ($\lambda 22.1$ Å/$\lambda 22.8$ Å) decreases to less than unity. The line ratio is density-sensitive.

Mewe & Schrijver (1977) have carried out extensive calculations, for both stationary and transient plasmas, of the forbidden to intercombination line intensity ratio, for many helium-like ions. The method has been applied to flare spectra obtained from the Solar Maximum Mission by Wolfson & Doyle (1980). Electron densities in the range 10^{11}–10^{12} cm^{-3} have been derived.

In a dense, hot plasma the resonance lines of hydrogen-like and helium-like ions develop 'satellite lines', lying toward longer wavelengths. The interpretation of these lines is discussed below. The intensity ratio of the satellites to the resonance line varies approximately inversely as the electron temperature, almost independently of electron density. Thus, where observable (e.g., in solar flares), the satellite lines provide a reliable indicator of electron temperature.

The development of this diagnostic stems from theoretical work by Gabriel (1972) and Gabriel & Paget (1972). They observed satellite lines in the helium-like spectra O VII and N VI in a Theta Pinch experiment.

Fig. 6.13 illustrates their interpretation of the lines. A lithium-like ion may have several line series, each of which culminates in a different ionization limit. Consider a level (e.g., 1s 2s 2p) that lies above the lowest ionization limit (I_1). The lithium-like ion may *auto-ionize* from this level, liberating the highly-excited electron. The inverse process is the *dielectronic capture* of an electron by a helium-like ion. These two processes balance at a rate determined by the electron temperature. The electron in the highly excited level may occasionally decay radiatively to the lithium-like ground state, producing the satellite line. Thus transitions like 1s n1 → 2p n1 in hydrogen-like or $1s^2$ n1 → 1s 2p n1 for helium-like ions result in satellite lines. Gabriel and Paget showed that the ratio of the intensities of the satellite and resonance lines produced in this way is a pure function of the electron temperature.

Satellite lines may also be produced by direct collisional excitation of the high-lying satellite level from the lithium-like ground state. The rate of this process is proportional to the concentration of lithium ions, and this factor therefore occurs in the intensity ratio of the satellite-to-helium-like resonance lines. Fortunately, this process is negligible in comparison with the previous one, so it does not prevent a clean temperature determination from the satellite-to-resonance line ratio.

Gabriel (1972) calculated the intensities and wavelengths of satellite lines for helium-like ions from Mg through Fe to Cu, and interpreted the available rocket spectra of solar flares. He found electron temperatures (in the range $10–20 \times 10^6$ K) that always lay *below* the temperature at which a particular resonance line has its maximum emission. This result indicated that a rapidly heated flare plasma may not be in ionization equilibrium, i.e., the actual ionization lags that predicted for equilibrium.

Gabriel and his associates have applied this elegant diagnostic to flare observations obtained from the Solar Maximum Mission (1980). The NRL

Fig. 6.13. A term diagram for a lithium-like ion and its associated helium-like ion, illustrating the origin of satellite lines.

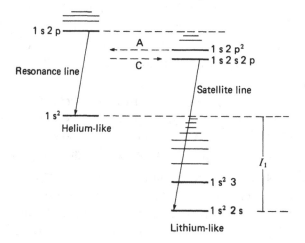

group (Doschek *et al.* 1980) has done similar work, using data from the SOLFLEX satellite. They observed helium-like Fe XXV ($\lambda\lambda 1.82$–1.97 Å) and Ca XIX (3.14–3.24 Å) in several flares. The electron temperature reached 22×10^6 K in the Fe XXV region during the rising phase of the flare, and dropped by 6×10^6 K during the fall.

6.6   Velocity

The speed of coronal material along the line of sight is easily obtained from measurements of the Doppler shifts of spectrum lines. A long history of such measurements exists, but will not be reviewed here. Except for spectacular transients, associated with eruptive prominences or flares, coronal motions are slow – in the range of 1–10 km s^{-1}. The material presumably moves both up and down along coronal arches, but little is known about the source and sinks of the gas. Several experiments have been carried out at recent total solar eclipses (Desai *et al.* 1981, 1982, Singh *et al.* 1982, Livingston & Harvey 1980, 1982: personal communication) to map the Doppler shifts of the forbidden coronal lines Fe XIV 5303 Å and Fe X 6374 Å. Desai and his colleagues obtained extraordinarily high speeds (the mode is 40 km s^{-1} and the maximum is 90 km s^{-1}) from their Fabry–Perot experiment. Other observers agree on smaller speeds.

Solar physicists need measurements of the outward expansion of the corona, into the solar wind, to compare with their calculated models. The most direct route is to measure EUV-line shifts within coronal holes, which are the main sources of the high-speed wind. Cushman & Rense (1976) observed the permitted line Si Xi $\lambda 303$ Å with a rocket-borne spectrograph in a region which later developed into a coronal hole. Its Doppler shift, relative to the surrounding photospheric lines, implies an outflow speed of 3 km s^{-1} at a height where the temperature is about 1.6×10^6 K. Rottman, Orrall & Klimchuk (1981) have repeated this experiment. They observed a shift of the O V $\lambda 629$ Å line by 3 km s^{-1} average and 5 km s^{-1} maximum in the transition zone ($T \approx 2.5 \times 10^5$ K) underlying a hole. In a second experiment with an improved rocket-borne EUV spectrometer (1982), they measured the Doppler shifts of Mg X ($T = 1.4 \times 10^6$ K) and O V ($T = 2.5 \times 10^5$ K) across a coronal hole at disk center. The maximum shifts corresponded to 12 km s^{-1} and 7 km s^{-1}, respectively, implying a height-gradient of outward flow velocity.

Coronal lines are broadened mainly by the thermal motions of radiating ions, e.g. by the Doppler shifts introduced by the Maxwellian distribution of ion speeds along the line of sight. All coronal lines are, however, broader than can be explained in this way. This result implies the existence of small-scale non-thermal turbulent motions in the coronal gas. If these motions have a Gaussian distribution, the velocity dispersion of the distribution can easily be extracted from the line width, after correction for thermal broadening. Values of this 'microturbulence' parameter range from about 20 km s^{-1} in quiet corona to more than 100 km s^{-1} in active regions and flares.

Some astronomers have conjectured that the microturbulence is evidence for small-scale coronal wave motions, which, in turn, might be involved in wave heating of the corona. The NRL group (Feldman *et al.* 1976) has used Skylab EUV lines to derive a general tendency for the microturbulence parameter to correlate with the ionization temperature. This supports the conjecture, but only in a qualitative way.

Tsubaki (1977) and Koutchmy *et al.* (1983) have independently found evidence for oscillatory Doppler shifts of the coronal green line ($\lambda 5303$ Å). The period in each case is near five minutes. This immediately raises a question whether their observations were contaminated with stray photospheric light, since the photosphere oscillates in brightness and velocity with a five-minute period.

6.7   Magnetic field

The most direct way to measure coronal magnetic fields would be to observe the Zeeman splitting of spectrum lines. Unfortunately, the thermal width of the Zeeman components of coronal spectrum lines far exceeds the expected wavelength separation due to a (weak) coronal magnetic field.

Coronal spectrum lines are linearly polarized, however, and this polarization is an indicator of magnetic fields: forbidden lines in the optical spectrum are magnetic dipole transitions between levels of the ground configurations of coronal ions. These transitions are excited by fluorescence: the absorption and reemission of an anisotropic unpolarized photospheric radiation field. During this scattering process, ion motions are briefly aligned along the coronal magnetic field, so that the scattered radiation is partially polarized. Charvin (1965), House (1972, 1977) and Bommier & Sahal-Brechot (1978) have developed a rigorous theory for the polarization (and subsequent partial depolarization) of coronal spectrum lines. The degree of polarization is typically minute (10^{-4}) and its measurement poses severe practical problems.

These problems were overcome by House and by Querfeld (1982). Querfeld built a coronagraph to measure the polarization of the favorable forbidden line Fe XIII 10747 Å. He was able to map the direction and degree of linear polarization in active regions at the coronal limb. The polarization vectors seem to align with coronal structures, and imply magnetic field intensities of order 10 Gauss or less.

6.8 Summary

Spectroscopic diagnostics for the principal thermodynamic, kinematic and magnetic properties of the corona have attained a high degree of sophistication. In the brief account of recent developments given here, the role of Skylab, and its forerunners (OSO 4 and 6) stand out. Leo Goldberg deserves the credit for establishing at Harvard a superb team of solar astrophysicists who, under his guidance, developed advanced instruments for the study of the Sun from such space platforms and who developed many of the diagnostics sketched in this article.

References

Beckers, J. & Chipman, E. (1974). *Solar Physics*, **34**, 151.

Bommier, V. & Sahal-Brechot, S. (1978). *Astron. Astrophys.* **69**, 57.

Charvin, P. (1965). *Ann. Astrophys.* **28**, 877.

Cheng, C. (1980). *Astrophys. J.* **238**, 743.

Crifo-Magnant, F. & Picat, J. (1980). *Astron. Astrophys.* **88**, 97.

Cushman, G. & Rense, W. (1976). *Astrophys. J. Lett.* **207**, L61.

Cushman, G. & Rense, W. (1977). *Astrophys. J. Lett.* **211**, L57.

Desai, J. N., Chandrasekhar, T., Ashok, N. M. & Vaidya, D. B. (1981). *Bull. Astron. Soc. India*, **9**, 68.

Desai, J. N., Chandrasekhar, T., Ashok, N. M. & Vaidya, D. B. (1982). *J. Astron. Astrophys.* **3**, 69.

Doschek, G. A., Feldman, U., Kreplin, R. W. & Cohen, L. (1980). *Astrophys. J.* **239**, 725.

Dupree, A. K., Foukal, P. V. & Jordan, C. (1976). *Astrophys. J.* **209**, 621.

Feldman, U., Doschek, G. A. & Mariska, J. T. (1978). *Astrophys. J.* **226**, 674.

Feldman, U., Doschek, G. A. & Widing, K. G. (1978). *Astrophys. J.* **219**, 304.

Feldman, U., Doschek, G. A., Van Hoosier, M. E. & Purcell, J. D. (1976). *Astrophys. J. Suppl.* **31**, 445.

Gabriel, A. (1971). *Solar Physics*, **21**, 392.

Gabriel, A. (1972). *Mon. Not. Roy. Astron. Soc.* **160**, 99.

Gabriel, A. & Jordan, C. (1972). *Case Studies in Atomic and Collision Physics*, **2**, ed. E. McDaniel & M. R. C. McDowell. N. Holland.

Gabriel, A. & Jordan, C. (1969). *Mon. Not. Roy. Astron. Soc.* **145**, 241.

Gabriel, A. & Paget, T. (1972). *J. Phys.* **5 5**, 673.

House, L. (1972). *Solar Physics*, **23**, 103.

House, L. (1977). *Astrophys. J.* **214**, 632.

Kohl, J. L., Weiser, H., Withbroe, G. L., Noyes, R. W., Parkinson, W. H. & Reeves, E. M. (1980). *Astrophys. J.* **241**, L117.

Koutchmy, S., Zugzda, Y. D. & Locans, V. (1983). *Astron. Astrophys.* **120**, 185.

Levine, R. H. & Pye, J. P. (1980). *Solar Physics*, **66**, 39.

Livingston, W. & Harvey, J. (1980). *Bull. Amer. Astron. Soc.* **12**, 93.

Livingston, W. & Harvey, J. (1980). *Bull. Astron. Soc. India*, **8**, 43.

Mason, H. (1975). *Mon. Not. Roy. Astron. Soc.* **171**, 119.

Mewe, R. & Schrijver, J. (1977). *Astron. Astrophys.* **65**, 99.

Munro, R. H., Dupree, A. K. & Withbroe, G. L. (1971). *Solar Physics*, **19**, 347.

Munro, R. & Jackson, B. V. (1977). *Astrophys. J.* **213**, 874.

Munro, R. & Withbroe, G. (1972). *Astrophys. J.* **176**, 511.

Neupert, W. (1979). *Bull. Amer. Astron. Soc.* **11**, 435.

Nussbaumer, H. (1972). *Astron. Astrophys.* **16**, 77.

Pallavicini, R., Peres, G., Serio, S., Vaiana, G. S., Golub, L. & Rosner, R. (1981). *Astrophys. J.* **247**, 692.

Pye, J. P., Evans, K. D., Hutcheon, R. J., Gerassimonko, M., David, J. M., Krieger, A. S. & Vesecky, J. F. (1978). *Astron. Astrophys.* **65**, 123.

Querfeld, C. (1982). *Astrophys. J.* **255**, 764.

Rottmann, G. J., Orrall, F. Q. & Klimchuk, J. A. (1981). *Astrophys. J.* **247**, L135.

Rottmann, G. J., Orrall, F. Q. & Klimchuk, J. A. (1982). *Astrophys. J.* **260**, 326.

Saito, K., Poland, A. I. & Munro, R. H. (1977). *Solar Physics*, **55**, 121.

Singh, J. N., Chandrasekhar, T. & Angreji, P. D. (1982). *Astron. Astrophys.* **3**, 69.

Tsubaki, T. (1977). *Solar Physics*, **51**, 121.

Van den Hulst, H. (1950). *Bull. Amer. Astron. Soc.* **11**, 135.

Vernazza, J. & Mason, H. (1978). *Astrophys. J.* **226**, 720.

Waldmeir, M. (1975). *Solar Physics*, **40**, 351.

Withbroe, G. L., Kohl, J. L., Weiser, H., Noci, G. & Munro, R. H. (1982). *Astrophys. J.* **254**, 361.

Wlerick, G. & Axtell, J. (1957). *Astrophys. J.* **126**, 253.

Wolfson, C. & Doyle, J. (1980). *Bull. Amer. Astron. Soc.* **12**, 912.

7

Spectroscopy of circumstellar shells

B. ZUCKERMAN

7.1   Overview

A normal main-sequence or red giant star possesses a photosphere, an outer layer of the stellar atmosphere that generates the optical photons that we observe and which is gravitationally bound to the star. Physical conditions in the layer above the photosphere vary greatly among main-sequence stars of different mass and among giant stars, many of which undergo large amplitude photospheric pulsations. Many stars possess chromospheres and, perhaps, also coronae. Mass motions are complicated with some matter falling back onto the photosphere and some being ejected to infinity. It is somewhere in this unsettled region that the outflowing circumstellar shell (CS) begins. For most stars, the circumstellar matter is at least partially transparent and the photospheric spectrum may be observed at wavelengths characteristic of the photospheric temperature. For a few stars with very large mass loss rates, such as IRC + 10216 and CRL 3068, the CS is sufficiently opaque that the photosphere is largely invisible from near infrared to ultraviolet wavelengths. For these stars we see a false photosphere at infrared wavelengths. That is, dust grains that have formed in the outflowing circumstellar gas have sufficient opacity to absorb essentially all of the true photospheric radiation. We then see a cool, roughly black-body, emission spectrum characteristic of the temperature of dust grains in the inner portions of the CS.

There is a very wide range of mass loss rates ($\dot{M}$) among normal stars. At the upper end $\dot{M}$ is perhaps as large as $10^{-4} M_\odot$/year for some infrared giant stars such as IRC + 10216 and CRL 3068 and somewhat less for hot, main-sequence O- and B-type stars. At the low end are stars such as the Sun where $\dot{M} \approx 10^{-14} M_\odot$/year and, presumably, still lower values for many low-mass, main-sequence stars, although, except for the Sun itself, it is not technologically feasible to directly measure such small mass loss rates.

Historically, the term 'circumstellar shell' has usually been applied to stars with very large mass loss rates. (One does not generally refer to the solar wind as the Sun's circumstellar shell.) In this chapter we concentrate on the coolest, 'late-type', stars. Although only a small percentage of main-sequence stars are of class O and B, essentially all main-sequence stars with masses greater than about one solar mass are believed, eventually, to experience an episode of rapid mass loss as red giants. For a giant star with luminosity $L \approx 10^4 L_\odot$, when $\dot{M} > 10^{-7} M_\odot$/year, more matter is lost to interstellar space than is added to the stellar core through nuclear burning processes. So this mass loss is probably critical in determining whether or not the insert core of a red giant star (which begins its life with, say, a few solar masses) ever exceeds the Chandrasekhar limit of $1.4 M_\odot$. That is, if there was not extensive mass loss during the giant phase of stellar evolution, then many more stars than actually do would supernova.

In addition, the combined mass loss from all red giant stars (1) alters the composition of the interstellar medium by enriching it in dust grains and elements such as nitrogen and, probably, carbon and (2) supplies mechanical energy and momentum to the interstellar medium. The O and B stars mentioned above play a much less important roll in (1) but, because of their much higher terminal velocities, 1000 km s^{-1} vs. 10 km s^{-1}, they dominate (2), at least near the galactic plane. We will not consider these two global problems here but will, rather, concentrate on spectroscopic investigation of CS around cool red giant stars. We will interpret 'spectroscopic' rather loosely to include infrared photometry and medium resolution spectroscopy of dust grains as well as high resolution observations of the gas.

Spectroscopic data are available at most wavelengths from ultraviolet to radio. Because of the low photospheric temperature and dusty CS most information is obtained at infrared (IR) and radio wavelengths. Indeed, in the IR the stars that we are considering are the brightest objects in the sky outside of the solar system.

7.2   Observations of gas and dust grains
7.2.1   *Ultraviolet*

Typical late-type giant stars appear faint in the ultraviolet. This is due to their intrinsically cool photospheric temperatures and also because of absorption of photospheric radiation by cool dust and molecules in the CS. For these reasons most ultraviolet spectra obtained with NASA's IUE satellite have been of stars with medium or high surface temperatures. Cool stars, such as α Ori, that are detectable with IUE often apparently possess chromospheres which may be important in the overall mass loss process. Only meager ultraviolet data exist on M-type giants. The strength of the

Mg II lines near 2800 Å is apparently a good measure of the total amount of chromospheric activity in G and K stars and this may also be true for M giants.

7.2.2 *Optical*

As might be expected, considerably more data exist in the visible. Optical data are useful for deducing certain properties of the CS material. Reimers, in the introduction to his 1981 review, remarks that 'optical observations of winds are the only reliable source of mass loss rates'. This statement would have been debatable when it was made in September 1980. Four years later it most certainly is not true: the techniques for analyzing CO microwave emission and broadband far infrared and submillimeter emission from dust grains which are discussed in Section 7.3 are the most reliable methods for obtaining $\dot{M}$ in stars with moderately large ($\gtrsim 10^{-6} M_\odot$/year) mass loss rates.

Optically, the best-studied star is α Ori, which has been investigated for many years, mainly by Goldberg and collaborators, in lines of sodium and potassium. These studies revealed two distinct shells expanding at velocities that differ by $\approx 6\,\mathrm{km\,s^{-1}}$. The KI gas is detectable out to at least 50″ from α Ori. By modelling the KI observations in conjunction with microwave CO observations, it is possible to derive various properties of the gas in the CS.

In addition, there appears to be a variety of evidence for warm chromospheres around α Ori and other M giants. For some stars, these chromospheres may be important in the mass loss process. Assuming that the potassium ionization equilibrium near α Ori is dominated by chromospheric ultraviolet emission, one can construct models for the KI observations.

In addition to these classical spectroscopic investigations, a new technique that might be called 'speckle spectroscopy' has been developed recently. Scattering of starlight by the atmosphere of the Earth (i.e., 'seeing') smears out images that otherwise would be diffraction limited. Thus, for ordinary optical and IR spectroscopy we lose all spatial information in the plane of the sky that is smaller than the seeing disk ($\approx 1''$) and we obtain a one-dimensional picture of CS material along the line of sight between the Earth and a star. With an appropriate model one then attempts to place the various parcels of absorbing CS gas at their proper distances from the star.

It would be much better if we could obtain a diffraction rather than a seeing limited image. The diffraction limit of a 4-m class telescope is reasonably well matched to the photospheric (optical) diameter of the largest stars such as α Ori and Mira, and also to the inner edge of the dust shell that surrounds IR objects such as IRC + 10216. The technique of speckle interferometry introduced by Labeyrie in 1970 is one ground-based method to achieve diffraction limited resolution on medium bright objects. For such objects the spatial resolution of the space telescope will be inferior to that of large ground-based telescopes. A major advantage of the space telescope, however, is that it will be able to achieve diffraction limited images on much fainter objects.

Because of the short time involved in an individual speckle frame and the slow improvement in signal-to-noise with increased total 'integration times', early attempts at speckle interferometry used the largest possible spectral bandwidth that was consistent with coherence length considerations. However, for the brightest visible and IR stars there are sufficient photons to narrow the spectral window, δV, to a few km s^{-1} (i.e., $\delta V < V_{\text{outflow}}$) and still take rapid speckle pictures. In this way it is possible to obtain a two-dimensional picture of the inner regions of the CS. It is still necessary to model the data to take account of smearing along the line of sight of gas elements at different radial distances from the star. But the presence of an extended Hα emitting chromosphere around α Ori is clearly evident.

7.2.3   *Infrared*

A very large body of IR data on evolved stars has been accumulated since 1965. Broadband IR photometry between 1 and 20 μm yields a combination of photospheric and circumstellar emission. The latter, due to dust grains, is often described as an 'IR excess' above that expected from a Rayleigh–Jeans extension of photospheric emission. To specify the dust emission more precisely one requires spectral resolutions $\gtrsim 100$ to resolve structure and to enable one to classify the grains.

The discovery of dust particles in the CS of cool stars was one of the early results of IR astronomy. There is spectroscopic evidence for three distinct types of dust. As expected on the basis of simple thermodynamic equilibrium calculations, the presence of a given species appears to follow the C/O abundance ratio. In stars with C/O < 1, strong 10- and 20-μm silicate bands dominate the IR spectrum. Stars with C/O > 1 display a narrower emission band near 11 μm, usually attributed to silicon carbide (SiC), as well as a 30-μm emission feature carried by an, as yet, unknown grain material. Some S-type stars (these have C/O $\approx$ 1), as well as many carbon stars, also show a featureless continuum indicative of a third species with unknown opacity (apart from lack of IR band structure). The relative strength of the SiC band does not correlate with the degree of emission in the featureless continuum, indicating that two different materials are involved. Both graphite and amorphous carbon have been suggested as the carrier of this continuum. In Section 7.3 we mention how the overall continuum spectrum from ultraviolet to far IR wavelengths can be established and, in conjunction with appropriate laboratory studies, used, perhaps, to deduce the relative importance of various grain materials.

Whatever their composition, dust grains very probably play an important role in the dynamics of the circumstellar gas and its ultimate ejection from the star. We discuss this problem in Section 7.3.

High-resolution ($\lambda/\Delta\lambda \gtrsim 10^4$) infrared spectroscopy has been achieved with a Fourier Transform Spectrometer at wavelengths between 2 and 5 μm and near 10 μm with a heterodyne spectrometer. The molecules that have been identified include CO, HCN, CH_4, SiH_4, C_2H_2, C_2H_4, and NH_3 around carbon-rich stars and CO, NH_3, and possibly H_2O in oxygen-rich envelopes. These measurements of vibration–rotation transitions are important in establishing relative chemical and isotopic abundances in the inner CS. They may also be used to determine density, temperature, and velocity as a function of radial distance above the stellar photosphere. Sometimes multiple absorption features, for example in the first overtone band of CO, are clearly resolved. These are suggestive of intermittent ejection of material from the stellar photosphere rather than a smooth continuous flow. It is possible to follow secular changes in the line profiles and characterize major motions in the photosphere. It has been suggested that the observed motions in χ Cyg, a bright S-type Mira variable, are consistent with a shockwave–pulsational model and that, in addition, there is a quasi-stationary layer a few stellar radii above the photosphere. If this layer exists around most Mira variables, then it might be the place where dust grains form.

As we mentioned in Section 7.2, speckle spectroscopy is a recently implemented technique whereby high spectral and spatial resolution are achieved simultaneously. In the infrared, the first successful application of this technique was to the fundamental 4.7-μm band of CO toward IRC + 10216 and Mira. This technique holds great promise for the future, as discussed in Section 7.4.

7.2.4 *Microwave*

The most extensive spectroscopic observations of circumstellar molecules have been in the microwave band. More than a dozen have been identified in the carbon-star IRC + 10216 and four (OH, H_2O, SiO, and H_2S) in oxygen-rich stars. OH and H_2O are in maser emission and SiO is in both thermal and maser emission. All observed microwave lines are pure rotational transitions except for inversion transitions in NH_3 and Λ-doublet transitions in OH masers. SiO masers, although rotational transitions, originate in excited vibrational states, $v = 1$, 2, and 3.

Except in a very few cases such as IRC + 10216, the beamwidth of an individual radio telescope is larger than the measurable size of the CS. Therefore, usually, the line profiles are assumed to be generated in a spherically symmetric envelope that is flowing outward at constant velocity. VLBI and/or VLA maps have been obtained for OH, H_2O, and SiO masers. These indicate that the H_2O and SiO masers are formed within a few stellar radii but the OH masers originate much further out in the envelope. Recently, millimeter wavelength interferometers at Hat Creek and the Owens Valley have begun to synthesize HCN and CO emission from the strongest carbon-rich emitters.

7.3   Analysis and interpretation
7.3.1   *Physical conditions*

The mass outflow affects the CS line profiles in essential ways. Most of the acceleration of the material in the envelope takes place in a relatively small region near the star. Therefore, over the vast majority of the volume of the CS the outflow velocity is equal to the terminal velocity, v_∞, which is much larger than velocities associated with local thermal and turbulent motions. Because of the gradient in radial velocity due to the divergent outflow, each parcel of gas is effectively decoupled from the rest of the envelope except in directions that are nearly parallel to the line of sight from the parcel to the star. This greatly simplifies the radiative transfer. In addition, the physical and chemical conditions in the CS are determined by the mass outflow so that one has considerable motivation to understand the underlying physical mechanisms that drive it.

Radiation pressure on dust grains has been the most favored mechanism for supplying the momentum in the outflowing wind. Among the reasons that one might advance in support of this belief is the IR excess which is believed to be due to conversion of UV, optical, and near IR photospheric radiation to longer IR wavelengths by absorption and reemission by cool dust in the CS. The outward momentum in this absorbed radiation goes into the dust and, as we discuss below, is then transferred to the gas. If there is sufficient dust the matter will be accelerated outward and one would predict that the maximum momentum in the outflowing gas would be $\approx L_*/c$ and, debatably, for the dustiest stars it is. In addition, v_∞ calculated in this model is of the same order as the observed velocities.

Much effort has been devoted to calculations of the dynamics and related phenomena in a dust-driven outflow. The most ambitious model of an oxygen-rich star couples the gas temperature to the variation of velocity in the accelerated inner portions of the flow. The coupling is present because: (1) the escape of photons produced by rotationally excited H_2O (and, less importantly, other molecules) will cool the gas and escape of these photons depends on the local velocity gradient, and (2) heating is predominantly supplied by the viscous interaction between the gas and grains, which in turn depends on the difference between their outflow velocities. The coupled equations are complicated and must be solved numerically.

The radiation pressure acts on the dust which collides with and drags along the gas. That is, locally the outward dust velocity is larger than the outward gas velocity – the grains drift supersonically through the gas. Collisions occur only infrequently in that the mean free path of a gas atom between collisions with the dust grains is large compared with the radius of the envelope. There are, however, sufficient collisions between a given dust grain and various gas molecules to transfer to the latter essentially all of the momentum that the grain gains from the radiation field. Molecule–molecule collisions then redistribute this momentum among all of the gas particles. The gas and dust are therefore momentum coupled.

A proper treatment of the dynamics then would consist of modeling a two-component (dust and gas) fluid, each component having its own density, temperature, and velocity as a function of distance from the central star. The components are coupled through gas–dust collisions and one must explicitly consider the drag force between dust and gas in the momentum equation. Here we simplify matters by assuming that the motion of the gas is strictly coupled to that of the dust; that is, collisions are frequent, the momentum carried by the radiation field is quickly shared between the dust and the gas, and their relative drift velocity is small with respect to v_∞. This assumption is more nearly correct for stars with large mass loss rates $\dot{M}$ (see (7.9)).

Conservation of mass in the envelope may be written as:

$$\dot{M} = 4\pi r^2 \rho v \tag{7.1}$$

where r is the distance from the star and ρ and v are the gas density and the outflow velocity at that distance. For steady outflow the momentum equation for a unit mass of gas may be written:

$$v\frac{dv}{dr} + \frac{1}{\rho}\frac{dP}{dr} = \frac{\int K_v L_v \, dv}{4\pi r^2 c} - \frac{GM_*}{r^2} \tag{7.2}$$

where P is the gas pressure, $L_v/4\pi r^2$ is the radiant flux at frequency v incident on a grain located at r and c, G, and M_* are the speed of light, gravitational constant and stellar mass, respectively. K_v is the opacity (cross section per gram of material) at frequency v.

For radiation pressure on dust, $(K_v)_{gr}$ is often written as:

$$(K_v)_{gr} = \frac{3}{4} \frac{Q_v}{a\rho_{gr}} \qquad (7.3)$$

where a and ρ_{gr} are the radius and density of a 'typical' grain. Q_v is a dimensionless emission efficiency. For $\lambda \gg a$, so that extinction is due mainly to true absorption (not scattering), $Q_v = f_v \cdot (2\pi a v/c)$. So, in this limit, $(K_v)_{gr}$ is independent of grain size; f_v is a function of specific grain material.

We must distinguish between $(K_v)_{gr}$ and the K_v that appears in (7.2) because, in our approximation, the radiation pressure force acts only on the dust, but it must accelerate the entire mass of gas as well. Therefore, the cross section per gram of circumstellar material is smaller than $(K_v)_{gr}$ by a factor of the dust-to-gas ratio by mass.

To have outflow at all, the radiation pressure term in (7.2) must be larger than the force of gravitational attraction or:

$$\frac{\int K_v L_v \, dv}{4\pi c G M_*} > 1. \qquad (7.4)$$

That is, $L_*(=\int L_v \, dv)$ must be greater than the so-called 'Eddington luminosity' if material is to be blown away from the star. Inequality (7.4) will generally be true if the majority of the available silicon or carbon is incorporated into the grains in oxygen- and carbon-rich stars, respectively.

Data for three of the most famous giant stars in the sky, IRC + 10216, Mira and α Ori, suggest that inequality (7.4) is probably satisfied for the first two, but the situation is unclear for α Ori. It seems that, usually, for stars with infrared excesses, L_{IR}, greater than a few percent of L_*, the dust, once it forms, has sufficient opacity to accelerate the gas to the observed terminal velocity (see below). However, within a few stellar radii temperatures are too high to permit more than a few percent of the mass of silicon or carbon to condense into dust grains. In this case, K_v is very very much smaller than $(K_v)_{gr}$ and inequality (7.4) is not likely to be true. Therefore, within a few stellar radii, CS that have substantial outflow velocities are likely to be driven by a mechanism other than radiation pressure on dust grains.

One such potentially promising mechanism involves an outward-directed flux of Alfvén waves in a stellar atmosphere. However, it is not clear that wave-driven winds can produce simultaneously the large mass loss rates but small asymptotic velocities that are often measured for red giant stars. ('Small' in this context implies that $v_\infty \ll v_{esc}$, where v_{esc} is the escape velocity from the stellar photosphere.) This difficulty for the Alfvén wave mechanism is common to other mechanisms for initiating massive winds and remains one of the outstanding problems in this field.

The pressure gradient term in (7.2) must be retained if one explicitly considers transition from subsonic to supersonic flow. Here we consider only the latter after the CS material has attained a good fraction of v_∞. In this case, the pressure term may be neglected.

Neglecting the pressure and gravity terms, the momentum equation simplifies to:

$$v\frac{dv}{dr} = \frac{\bar{K}L_*}{4\pi r^2 c} \tag{7.5}$$

where $\bar{K} = \int K_v L_v \, dv / \int L_v \, dv$. If $\bar{\tau}$ is the mean luminosity weighted dust optical depth, then $d\bar{\tau}/dr = K\rho$ and, using (7.1), one obtains:

$$\int dv = \int \frac{L_*}{\dot{M}c} \, d\bar{\tau}.$$

The integrals are taken between the inner edge of the dust shell and infinity. If we assume that the outflow velocity at the inner edge is negligibly small with respect to v_∞, then:

$$\dot{M}v_\infty = \bar{\tau}L_*/c. \tag{7.6}$$

So the mass loss rate is determined by the fraction ($\bar{\tau}$) of the stellar luminosity that is absorbed and scattered by the dust. The absorption component can be estimated from L_{IR} and the scattered component from optical images. In principle, $\bar{\tau}$ can be greater than unity if the average stellar photon is scattered and/or absorbed more than once before it escapes from the CS. However, we do not expect that $\bar{\tau}$ will be very large because absorption will probably dominate scattering near the peak of L_v and because absorption and reemission rapidly converts photons to longer wavelengths where $(K_v)_{gr}$ is small. Observationally, this appears to be the case, although the exact value of $\bar{\tau}$ for the reddest stars is still somewhat uncertain.

Whatever the exact value of $\bar{\tau}$, the fact that for the reddest giant stars $\dot{M}v_\infty$ is of order L_*/c is reasonably suggestive of a wind driven by radiation pressure. (If pulsations or some other physical mechanism were the primary driving force, then one might expect $\dot{M}v_\infty$ to bear little or no relationship to either L_*/c or L_{IR}/c.) The derived terminal velocities also agree with observations which indicate that, typically, $v_\infty \approx 15 \text{ km s}^{-1}$. We can see that this is the case by integrating (7.5) from the inner boundary (i) where the dust forms (and where $v \ll v_\infty$) to infinity. Then:

$$v_\infty = \left(\frac{\bar{K} L_*}{2\pi c r_i} \right)^{1/2}. \tag{7.7}$$

The inner boundary, r_i, of the dust envelope is likely to be found not far from that distance from the star, r_c, where the equilibrium dust temperature equals the condensation temperature ($\gtrsim 1000 \text{ K}$) for silicate or carbonaceous grains. Observations suggest that the actual condensation point might be farther from the star than r_c. At any rate, one may use (7.7) to estimate outflow velocities that are not too different from observed terminal velocities. Note that, according to this simple theory, for stars that we are considering (i.e., $L_* \approx 10^4 L_\odot$), v_∞ should not vary greatly from one star to the next unless $\bar{K}$ is very small (in which case inequality (7.4) probably does not hold anyway). Therefore, according to (7.6), the main effect of increasing the optical thickness of the envelope will be to increase $\dot{M}$.

In the preceding discussion we treated the gas and dust as a single fluid and derived a few useful expressions from the momentum equation. However, to calculate the thermal structure of the gas as a function of r, it is necessary to treat the dust and gas separately. The major heating source for the gas is collisions with the dust grains as they drift supersonically through the gas. The grains are heated directly by the stellar flux (which may be modified by absorption and reemission by dust grains close to the star). Their internal vibration temperature is not really a kinetic temperature since they collide infrequently with each other. This internal temperature is quite a bit higher than the gas kinetic temperature. For example, a detailed model suggests that at $r \approx 10^{17}$ cm from IRC + 10216, $T_{\text{dust}} \approx 100$ K and $T_{\text{gas}} \approx 15$ K. None the less, because of the supersonic drift velocities, the energy gained by the gas due to the bulk motion of the dust through the gas is larger than the energy gain due to the gas–dust temperature difference.

One can estimate the drift velocity, v_{dr}, by equating the radiation force on a grain to the drag force from the gas:

$$\frac{\pi a^2 L_* Q}{4\pi r^2 c} = \rho \pi a^2 v_{\text{dr}}^2. \tag{7.8}$$

Using (7.1) one obtains:

$$v_{\text{dr}} = \left(\frac{L_* v_\infty Q}{\dot{M} c} \right)^{1/2}. \tag{7.9}$$

Since v_∞ can be measured easily, $L_* \approx 10^4 \, L_\odot$ (from kinematically estimated distances and theoretical models of stellar evolution), and $\dot{M}$ can be estimated in a variety of ways (see below), the largest uncertainty in calculating v_{dr} probably comes from Q which is known only poorly. For a plausible model of IRC + 10216, one calculates $v_{dr} \approx$ few km s^{-1} compared with $v_\infty = 15$ km s^{-1}. Note that v_{dr} will be larger for stars with smaller $\dot{M}$. As mentioned above, when $\lambda \gg a$, $Q \propto a$, so larger grains have larger drift velocities and the grains can collide, albeit infrequently, with each other.

The heating rate per unit volume, $(n_{gr}\pi a^2 v_{dr}) \cdot \frac{1}{2}\rho v_{dr}^2$ is a very strong function of v_{dr}. The rate of change of gas temperature with radius r is calculated by balancing this heating rate against adiabatic cooling and molecular line emission which, for the carbon star IRC + 10216, is often attributed primarily to CO molecules. Actually, in the inner portions of the CS the observed energy emitted by $J = 1 \rightarrow 0$ HCN plus H^{13}CN is comparable with that emitted by $J = 1 \rightarrow 0$ CO plus ^{13}CO. It remains to be shown, observationally, that CO dominates at the higher J levels. For oxygen-rich stars, H$_2$O is likely to be the dominant molecular coolant. There is also some heating due to radiative excitation followed by collisional deexcitation of vibrational levels of the most abundant molecules. Because both grain–gas collisional heating and molecular cooling (per hydrogen molecule) decrease with increasing $\dot{M}$, the derived gas temperature distribution is insensitive to $\dot{M}$.

In addition to the two temperatures discussed above, there is another temperature of interest: the rotational excitation temperature of abundant molecules such as CO, HCN and SiO. Models of the expected intensity of rotational line emission as a function of r may be compared with the observations to derive $\dot{M}$ and relative molecular and isotopic abundances. These model calculations couple the equation of transfer with the equations of statistical equilibrium for the rotational level populations and are quite elaborate. Various simplifying approximations are made including, for example, spherical symmetry, excitation of the rotational levels by either collisions or the near infrared radiation field, and local line widths which are small with respect to v_∞.

The assumption of spherical symmetry may be quite reasonable for most Mira variables, at least those that are oxygen-rich. However, there are many 'bipolar' nebulae with large $\dot{M}$ that clearly are not spherically symmetrical. Models for such objects that include both dust shielding and molecular self-shielding indicate that the maximum spatial extent of the undissociated molecules is not much affected by the anisotropy of the mass loss. The computed intensity of the total CO emission in an anisotropic outflow suggests that Ms inferred from models that assume spherical symmetry are not overestimated by more than a factor of 2 unless the flattening, the ratio of density at the equator to that at the pole, is more than a factor of 100.

The rotational levels may be excited primarily either directly by collision or by radiative absorption in vibrational lines followed by cascade back to the ground vibrational state. The choice depends on $\dot{M}$, the abundance relative to H_2 of the molecule in question, the ratio of local line width, Δv, to v_∞, the gradient with radius of the expansion velocity, the cross sections for collisional excitation of the rotational levels by H_2 and the intensity of the radiation field at the wavelengths of the fundamental, $v = 0 \rightarrow 1$, vibration–rotation transitions. The latter intensity, at some r_0, is determined by the direct stellar flux modified by line absorption due to molecules located at $r < r_0$ and by energy reradiated by warm dust grains.

Most molecules have fairly small abundances relative to H_2 and their excitation is dominated by the IR radiation field for likely values of $\Delta v/v_\infty$ ($\gtrsim 0.05$). For CO, which is very abundant ($[CO]/[H_2] \approx 10^{-3}$), collisional excitation dominates at large $\dot{M}$ ($\approx 10^{-5}\ M_\odot$/year) and radiative excitation dominates at small $\dot{M}$ ($\approx 10^{-7}\ M_\odot$/year). The situation is still unclear for HCN at large $\dot{M}$.

The assumption that Δv, which is due to local thermal and turbulent motions, is small with respect to v_∞ enables one to treat the radiative transfer at a given point independently of most of the rest of the envelope. The assumption that $\Delta v/v_\infty$ is small is supported by infrared and radio observations of IRC + 10216 and of SiO, OH, and H_2O masers in oxygen-rich stars. In this case, the radiative transfer can be solved by using Sobolev's local escape probability formalism. This method is appropriate because the outflow velocity gives rise to a large velocity gradient transverse to the radial direction so that line photons produced at a given point interact, except along a narrow, roughly conical, region in the radial direction, only with local molecules. Castor has shown that when $v \propto r$, where v and r represent velocity and distance measured with respect to a point in the envelope, then $\beta(\tau)$, the probability that a photon emitted at that point will escape from the cloud without further interaction, is given by $(1 - e^{-\tau})/\tau$. Here:

$$\tau = \frac{\lambda^3 g_u A_{ul}}{8\pi\, dv/dr} \left(\frac{n_l}{g_l} - \frac{n_u}{g_u} \right) \tag{7.10}$$

represents the line optical depth in a given direction between infinity and the point in question. In this case, except for nearly radial directions, which must be treated more carefully, the mean intensity in a rotational transition when integrated over the line profile thus depends only upon the local value of the source function and upon β.

In the outer parts of the envelope where H_2 densities are too low to thermalize the rotational level populations, optical depths for abundant molecules such as CO and HCN may still be large. Trapping of these photons will enhance the rotational excitation temperatures above the values appropriate for optically thin conditions. In addition, and more generally, optically thick and thin transitions produce millimeter wavelength lines with parabolic and flat-topped shapes, respectively, for the simple case of an envelope that is unresolved by the telescope beam. The thin and thick cases generally correspond to the radiative and collisionally-dominated regimes discussed above. In the case where the envelope is partially resolved, the line profile becomes doubly peaked because gas at the largest angular distances from the star is no longer sampled by the telescope beam when pointed directly at the star and it is this gas that contributes to the center of the line profile (see Figs. 7.1 and 7.2).

Fig. 7.1. Theoretical millimeter wavelength line profiles. The profile shape is determined primarily by two parameters: the optical depth of the line and the ratio of the angular size of the emitting region to the beam size of the telescope. (Figure courtesy of Dr. H. Olofsson.)

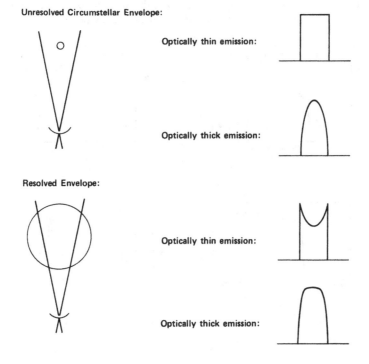

Models may be used to estimate the shapes and the relative and absolute intensities of emergent CO lines as measured with a telescope that, typically, partially resolves the expanding envelope. These quantities may be compared directly with 'on-source' observations made with appropriate telescopes to derive CO loss rates. The on-source profiles alone do not enable one to determine $[CO]/[H_2]$ and, therefore, $\dot{M}$. The models are not applicable to profiles obtained in situations where the envelope is resolved and the telescope is not pointed directly at the star (i.e., the CO emission has been 'mapped'). For the nearby, spatially well-resolved envelope of IRC + 10216, one may utilize mapping data to calculate values for $\dot{M}$ and $[CO]/[H_2]$ that are consistent with the view, based on chemical equilibrium calculations, that most of the oxygen in the envelope is in CO.

In addition to these rather involved techniques for obtaining CO loss rates and $\dot{M}$, a number of other simpler methods have also been suggested. One, that was discussed above, is to assume that radiation pressure on grains is responsible for the outflow momentum at infinity, $\dot{M}v_\infty$. If this is the case, then one may set:

$$\dot{M}v_\infty = L_{IR}/c \tag{7.11}$$

Fig. 7.2. Actual millimeter wavelength spectra showing the dependence of line shape on optical depth (τ) and on the ratio of the angular size of the emitting region (R) to the telescope beam size (B). (Figure courtesy of Dr. H. Olofsson.)

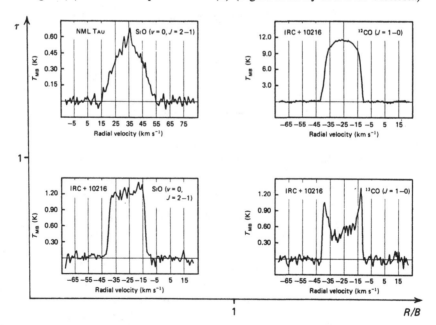

where L_{IR} is the infrared excess defined in Section 7.2.3. For a few stars, M has been derived independently by techniques such as those described above and below and, within the sometimes considerable uncertainties, is in reasonable agreement with the value obtained from (7.11).

Simple models for estimating $\dot{M}$ that assume collisional excitation of CO have been suggested. In the outer portions of envelopes with large $\dot{M}$ there are three regions of interest for CO excitation. An inner one where $n_{H_2} \cdot \langle \sigma_v \rangle > A_{ul}$, CO rotational depths are large, and $T_{excitation} = T_{kinetic} = T_{brightness}$. Here $\langle \sigma_v \rangle$ is a sum of cross sections for excitation to all levels with J equal to or greater than the upper state J in the transition of interest. Surrounding this region may be one where $n_{H_2} \cdot \langle \sigma_v \rangle \lesssim A_{ul}$ but τ (see (7.10)) is large and photon trapping enhances $T_{excitation} (= T_{brightness})$ possibly to a value as large as $T_{kinetic}$. Finally, at large r, when $\tau < 1$ and $n_{H_2} \cdot \langle \sigma_v \rangle < A_{ul}$, then $T_{brightness}$ declines rapidly with r as indicated in (7.13) below.

In this outer region every collision from $J = 0$, where most of the population resides, into level J'' produces one line photon for every transition $J' \rightarrow J' - 1$ for which $J' \leqslant J''$. So one chooses a $J' \rightarrow J' - 1$ transition that has been mapped, counts collisions to all levels $J \geqslant J'$ and assumes that this number equals the number of emergent photons. Observations along a given line of sight (y-axis) to the star yield:

$$E_{J' \rightarrow J' - 1} = \int n(H_2) n(CO) \langle \sigma_v \rangle (h v / 4\pi) \, dy \qquad (7.12)$$

where E is the specific intensity integrated over the spectral line. E may also be written as $2kT_B \Delta v / \lambda^2$ where $\Delta v \ (= v_\infty v/c)$ is the line width. From arguments given in Section 7.3.2 we expect that CO and H_2 are not affected by photochemical processes in the region of the envelope that we are considering. As a result, assuming that $\dot{M}$ is constant with time, we set $n(H_2) = \dot{N}/4\pi r^2 v_\infty$ and $n(CO) = q n(H_2)$. Consequently:

$$T_B = \frac{q \dot{N}^2 \pi}{4} \left(\frac{c}{4\pi v_\infty} \right)^3 \frac{\langle \sigma_v \rangle h}{d^3 k v^2} \qquad (7.13)$$

where $d \ (= \theta D)$ is the impact parameter of the line of sight with respect to the star, θ is the angular offset, and D is the distance to the star. With $q = 10^{-3}$, all the other quantities in (7.13) may be evaluated or measured and $\dot{N} \propto \theta^{3/2} D^{3/2}$. Estimating the appropriate value of θ from a mediocre map can introduce significant uncertainty into $\dot{N}$.

All methods for estimating $\dot{M}$ are sensitive functions of D which is usually not well determined. For all the methods that we have mentioned, excepting the above, where $\dot{M} \propto D^{3/2}$, $\dot{M} \propto D^2$. A technique that combines interferometric and monitoring measurements of time-varying stellar OH masers may someday yield more reliable distances to these stars.

7.3.2   *Chemical composition*

The spectroscopic characteristics of a circumstellar envelope are determined by whether carbon or oxygen is the more abundant element. Calculations of chemical equilibrium under photospheric-like conditions suggest that essentially all of the element of lesser abundance will be tied up in CO. The preponderance of most other molecules will therefore depend on whether there is an excess of carbon, so that, for example, C_2H_2 and HCN are abundant, or of oxygen in which case H_2O and SiO are plentiful.

Various important processes determine the chemical mix in the envelope. The first is chemical 'freeze out' which should generally be a good approximation for abundant and stable molecules such as CO, HCN and C_2H_2. That is, suppose that some molecule A can react with a variety of other molecules and atoms B, C, D, ... under circumstellar conditions. If the fastest of these reactions proceeds at a rate (s^{-1}) $k_x n_x$ where n_x is the number density of the reactant x, then $k_x n_x t_{exp} \propto (rv^2)^{-1}$. Here $t_{exp} (= r/v)$ is a characteristic expansion time at r. So, at some r_f, the time for reaction A + X to proceed becomes long compared with t_{exp} and the abundance of A at r_f is frozen into the envelope for $r_f < r < r_{ph}$ where r_{ph} is the distance from the star at which photodissociation becomes important. (Below, r_{ph} is denoted $(r_{ph})_0$ for the case where there is no shielding by dust or gas.) For highly reactive species such as free radicals and some atoms the rate constants are sufficiently large that reactions may occur far out in the envelope and freeze-out never occurs.

If the sticking probability is not too small, grains will play a role as sinks for some molecules and for most elements, especially refractory ones such as Si, Fe, and Al. The collision rate and hence an upper limit to the reaction rate with grains is $(t_{gr}) = \sigma_{gr} v_{dr} n_{gr}$. Outside of $r \approx 10^{16}$ cm, where grain growth has stopped, t_{gr}/t_{exp} increases with r and the importance of grain–gas reactions is diminished.

In the envelope, then, gas phase reactions and collisions with grains are expected to be important only for certain molecules and probably not for the most abundant such as H_2O, CO, HCN, and C_2H_2. However, a process that at large r must become important for all molecules is photodissociation by the ambient interstellar UV radiation field. There is observational evidence that CN is produced from photodissociation of HCN in the outer envelope of IRC + 10216 and OH from H_2O in oxygen-rich stars. Probably the C_2H observed in IRC + 10216 is also produced this way, although the data are not yet sufficiently complete to differentiate between constant relative abundance or photoproduction from C_2H_2 in the outer envelope.

Outflowing molecules may be shielded against photodissociation by the interstellar UV field by dust grains and, for the most abundant species such as H_2 and CO, by self-shielding. In envelopes around stars with small $\dot{M}$, in the very outer parts of envelopes with large $\dot{M}$, and in envelopes with dust-to-gas ratios much smaller than interstellar, there is little or no shielding. In this case, ignoring the molecule formation processes mentioned above, one may write:

$$d\dot{N}_M = -4\pi r^2 n_M P\, dr$$

as the rate at which molecules are photodissociated in a shell of thickness dr, at a distance, r, from the central star. Here P is the rate of photodissociation (s^{-1}) and:

$$\dot{N}_M = 4\pi r^2 n_M v_\infty \tag{7.14}$$

is the molecule loss rate in the absence of sources and sinks. If there is no shielding, we set $P = P_0$ where P_0 is the dissociation rate in the ambient interstellar radiation field. Then:

$$\frac{1}{n_M r^2}\frac{d}{dr}(n_M r^2) = -\frac{P_0}{v_\infty}. \tag{7.15}$$

The solution to (7.15) with the inner boundary condition given by (7.14) is:

$$n_M = \frac{\dot{N}_M}{4\pi r^2 v_\infty}\exp(-P_0 r/v_\infty). \tag{7.16}$$

Here we have assumed that r is much larger than the radius at the inner boundary (r_I) where $v_{\text{outflow}} = v_\infty$. The characteristic distance where dissociation occurs in the unshielded outer envelope is given by $(r_{ph})_0 = v_\infty/P_0$. However, the interstellar UV photons can penetrate inside r_I in these unshielded envelopes. Therefore, these photons as well as those from a chromosphere or an interior H II region such as exists around CRL 618 may play a significant role in the chemistry in the region where the outflow velocity is small (subsonic).

Equation (7.16) is valid as long as dust attenuation is small at $(r_{ph})_0$. At this location the dust grain column density (cm^{-2}) to infinity is equal to $\dot{N}_{gr}P_0/4\pi v_\infty^2$. If m_{gr} is the mass of an individual grain and $K_{gr}m_{gr}$ is the UV extinction cross section of grains, then (7.16) is valid only when:

$$K_{gr}m_{gr}\dot{N}_{gr}P_0/4\pi v_\infty^2 \ll 1.$$

This may not obtain for stars with large $\dot{M}$ that are sufficiently cool so that $n_{gr}/n(H_2)$ is not very much smaller than the interstellar value. In this case, one must estimate the penetration of light into a spherical dusty region with density $\propto r^{-2}$. Because $n_{gr}/n(H_2)$ and the scattering properties of the grains (the albedo and phase function) are uncertain, one cannot hope to solve this problem exactly at this time. In the approximate solutions that have been obtained, the $P_0 r/v_\infty$ in the exponential in (7.16) is multiplied by an exponential integral which itself is a function of $P_0 r/v_\infty$ and the dust optical depth at $(r_{ph})_0$. Calculations with reasonable parameters for the dust and gas indicate that the radial extent of most molecules that are currently observable is determined by photodissociation and not, for example, by inadequate excitation or low optical depth due to decreasing column density with increasing r.

A very interesting example is the photodestruction of H_2O into OH which then emits maser radiation in the outer ($r \approx 3 \times 10^{16}$ cm) portion of the envelope. There is an inner envelope where $(H_2O)/(OH) \gg 1$ and where the H_2O is protected from the external UV radiation field by a combination of dust shielding, continuum self-shielding (see below) and, for $\lambda \lesssim 1100$ Å, by the Lyman and Werner bands of H_2. At some point in the envelope, which depends on the unshielded photodissociation rate, the amount of shielding, $\dot{M}$, and v_∞, the H_2O is photodissociated into OH and H. The OH is, in turn, photodissociated so that the photoproduction model leads to the formation of a well-defined OH shell. The radius of the shell increases with increasing $\dot{M}$ and decreases with increasing v_∞. The peak OH densities and column densities through the shell are, respectively, slowly decreasing and increasing functions of $\dot{M}$. The column density is very important for maser radiation and this calculation implies that stars with small $\dot{M}$ will have weaker OH masers than those with large $\dot{M}$ but not so much weaker as to be unobservable. The predictions of the OH photoproduction model agree reasonably well with existing observations. When the OH itself is photodissociated, the resulting neutral oxygen atoms will radiate 6300 Å and 63-μm photons that may be detectable in stars with very large mass loss rates. The latter line, as well as the 157-μm line from ionized carbon, have been detected toward NGC 7027 but this is probably the result of molecular photodissociation due to the hot central star rather than the ambient interstellar radiation field.

In addition to the dust shielding discussed above, self-shielding can be very important for molecules such as H_2 that are photodissociated mainly by line rather than by continuum radiation. Searches for 21-cm radiation from atomic hydrogen in IRC + 10216, IRC + 10011 and various other red giant stars have been unsuccessful. According to chemical equilibrium models, for stellar effective temperatures $\gtrsim 2600$ K, most of the hydrogen will be in atomic form in regions near the photosphere. Conversion to H_2 in the outflowing envelope seems unlikely, so if a fairly nearby star with $T_{eff} > 2600$ K and a substantial $\dot{M}$ could be found, then 21-cm radiation should be detectable. Unfortunately, α Ori, which is the best such case by far, lies in front of an interstellar cloud complex that produces appreciable 21-cm emission. So the signal, if any, from α Ori is hopelessly lost in the confusion.

For cooler stars such as IRC + 10216 and IRC + 10011, chemical equilibrium in the photosphere implies that essentially all the hydrogen that flows outward into the envelope will be in the form of H_2. There are a number of mechanisms that have been suggested for producing at least some atomic hydrogen. Of these, a simple photoproduction model ($H_2 + h\nu \rightarrow H + H$) yields predicted 21-cm line intensities that lie somewhat below the observational upper limits. Because these limits were established with the Arecibo antenna and the VLA, it will, unfortunately, be very difficult to substantially improve upon them.

Another molecule that is probably self-shielding is ^{12}CO. From laboratory data, it has not been entirely obvious whether the photodissociation of this molecule by radiation with $\lambda > 912$ Å is dominated by the continuum or by discrete lines. The astrophysical data suggest that line dissociation probably dominates. Unless ^{12}CO also shields ^{13}CO (which might be the case), in stars with reasonably large $[^{12}C]/[^{13}C]$, ^{13}CO will not be self-shielding. If so, then one must be careful when deducing $^{12}C/^{13}C$ from observations of CO. Analysis of molecules such as HCN, which are not self-shielding, should imply smaller values for $^{12}C/^{13}C$ than does CO if the latter molecule is self-shielded.

Because CO is apparently self-shielding around IRC + 10216, the decline of surface brightness with radial distance is less rapid than would be predicted by a photodestruction model. Rather, the $\dot{r}$ dependence of the cutoff seems to be consistent with declining collisional excitation. It will be interesting to see if this is also the case for ^{13}CO or if it cuts off more sharply than does ^{12}CO.

For a few very abundant molecules such as H_2O and C_2H_2, continuum self-shielding could be comparable with dust shielding. To estimate their relative importance, one may use a simple model for self-shielding. That is, the critical radius r_c, at which molecules are protected in the case of self-shielding, may be roughly estimated by balancing the flux of incoming dissociating photons, F, with the flux of outflowing molecules:

$$F = n_M \cdot v_\infty = \frac{\dot{N}_M}{4\pi r_c^2}. \tag{7.17}$$

For dust shielding we assume that the molecules are shielded when the grain optical depth at the relevant UV wavelengths from a critical radius r_{gr} outward to infinity is equal to unity or:

$$\tau_{gr} = 1 = \frac{K_{gr} m_{gr} \dot{N}_{gr}}{4\pi r_{gr} v_\infty}. \tag{7.18}$$

Grain opacity dominates self-shielding when $r_{gr} > r_c$, that is:

$$\frac{K_{gr} m_{gr}}{v_\infty} \left(\frac{F}{4\pi}\right)^{1/2} \frac{\dot{N}_{gr}}{\dot{N}_M^{1/2}} > 1.$$

So grains will tend to dominate for stars with large mass loss rates and/or small outflow velocities. In general, only the most abundant molecules are self-shielding at any value of $\dot{N}$.

The total grain mass, M_{gr}, in a circumstellar envelope is difficult to measure accurately. The problem is that we really do not know what the grains are composed of. The IR observations mentioned in Section 7.2.3 yield some clues but none of the observed spectral features enable us to specify the grain material uniquely. These features are broad and are influenced not only by chemical composition but also by crystalline structure, impurities, and grain size. Therefore, $(K_v)_{gr}$ is uncertain by at least an order of magnitude and this translates into a concomitant uncertainty in M_{gr}. By combining considerations of photodissociation, as discussed above, with far infrared or submillimeter photometry, it is possible to specify the ratio of $(K_v)_{gr}$ at UV and long wavelengths or, equivalently, the index p if we write $(K_v)_{gr} = k_0 v^p$. This is an observational constraint in addition to the broadband photometric spectrum that may help to pin down the nature of the dominant grain material. Of course, if there are two distinct grain types for which p differs greatly, it could be that we are measuring one at UV wavelengths and the other at long wavelengths and fooling ourselves greatly.

Uncertainties in K_{gr} notwithstanding, M_{gr} is probably measured best at far infrared and submillimeter wavelengths since the dust is certainly optically thin and, for the most part, sufficiently warm to be in the Rayleigh–Jeans portion of the spectrum so that the emission is only linearly proportional to the uncertain dust temperature (T_{gr}). Sopka *et al.* (1985) give a prescription for determining T_{gr} that probably reduces uncertainties in T_{gr} to negligible proportion (compared with uncertainties in K_{gr}) in their estimate of M_{gr}.

One may express the submillimeter flux density F_v measured at the Earth as:

$$F_v = \frac{\pi a^2}{D^2} B_v Q_v n_{gr} V \qquad (7.19)$$

where D is the distance to the star and V is the volume of the circumstellar envelope subtended by the telescope beam. One may relate the number density of grains, n_{gr}, to the mass loss rate in dust through (7.1) (but applied to the grains rather than the gas). Although it might appear at first glance that the received flux is proportional to the grain radius (a), the dust mass that one derives from (7.19) is dependent only on the quantity $Q_v/a\rho_{gr}$ which, from the Kramers–Kronig relations, is independent of 'a' provided that $\lambda \gg a$. Also, the shape of the grains will have little influence on the determination of dust mass unless the bulk of the dust is made up of very thin needles or flakes which is not likely if the dust is amorphous as argued below.

Sopka *et al.* (1985) derive values for the ratio of mass loss rate in gas to that in dust for a total of about a dozen infrared giant stars and young planetary nebulae. For a few of the objects, $\dot{M}_{gas}/\dot{M}_{dust}$ would be much smaller than the interstellar gas-to-dust ratio unless a large $Q(400\,\mu m)$ is used in (7.19). 'Large', in this context, implies amorphous rather than crystalline grains in the outflowing envelopes around both carbon-rich and oxygen-rich stars. Such choice of the absolute value of Q_v at 400 μm is also consistent with the value of the index 'p' (≈ 1) which is believed to be appropriate for amorphous grain structures and which matches the photometric spectra.

Another method for obtaining mass loss rates in dust is to model the near infrared photometric spectrum (2–10 μm). This technique also suffers from uncertainties in Q_v and, in addition, is sensitive to the fact that the dust emission is a product of the uncertain run of both dust temperature and density with distance from the star. That is, in addition to standard constant outflow velocity models with $n_{gr} \propto r^{-2}$ one must also consider flatter radial distributions which are possible (indeed likely) if the dust is forming within 10^{16} cm of the central star. It is just this region that is responsible for the emission observed between 2 and 10 μm wavelength. So, although one can measure, as a function of λ, an inner dust shell radius with heterodyne, Michelson, and speckle interferometry, it is not yet possible, given the above uncertainties, to uniquely invert the visibilities.

7.4   Some near future developments

The broadband photometric spectrum of dusty envelopes can be easily measured from the ground out to 20 μm. IRAS has extended this to 100 μm. New submillimeter telescopes with diameters between 10 and 30 m

Table 7.1. *Famous evolved stars*

Object	D (pc)	L_* ($10^4 L_\odot$)	Sp. class	v_{lsr} (km s^{-1})	v_∞ (km s^{-1})	$\dot{M}_{gas}$ ($M_\odot$/yr)	$\dfrac{\dot{M}_{gas}}{\dot{M}_{dust}}$	θ_{IR} (arc sec)	θ_{OH} (arc sec)	Notes
Mira	77–114	0.4–1	M6e	47	6	5–10×10^{-7}	>120	see note 15		1, 2, 3, 4
IRC+10216	150	1.5	N	−26	17	2×10^{-5}	170	0.43	—	2, 4, 5, 19
CIT 6	190	0.23	N	−2	17	4×10^{-6}	>110	0.1	—	2, 4, 5
α Ori	200	11	M2	3	14	4×10^{-6}	>280	see note 15		2, 6
R Aqr	260	0.27	M7e	−28	30	10^{-7}		see note 15		7, 8
NML Tau	270–400	1–2	M6–10e	34	22	4–8×10^{-6}	>96	0.05	6	1, 2, 3, 4, 17
AFCRL 915	280	0.07	A0					0.13		9
IRC+10011	510	0.63	M	9	23	1.4×10^{-5}	200	see note 15	8.8	2, 4, 10
AFCRL 3068	1000	1	N	−31	15	2×10^{-5}	230		—	2, 4
AFCRL 2688	1000	2.5	F5	−36	20	7×10^{-5}	31		—	2, 4, 5
NGC 7027	1090	0.6	see note 11	25	18	8×10^{-5}	61		—	2, 4, 5
OH231.8+4.2	1300	0.4	M6	28	39	1×10^{-4}	13	0.5	10	2, 16, 20
VY CMa	1500	35	M5	19	36	2×10^{-4}	300	0.12	4.2	2, 10
AFCRL 618	1700	2.8	B0	−21	22	1×10^{-4}	120	0.13	—	2, 4, 5
OH26.5+0.6	1800	2.5	M	27	15	7×10^{-5}		0.086	7	10, 12
NML Cyg	2000	46	M6	0	29	$>6 \times 10^{-5}$	>95	0.14	6	2, 10
M1-92	3000	1.6	B1	2	23			0.06		14, 18
IRC+10420	3400	30	F8	74	43	2×10^{-4}		0.14	6	10, 13

Table column headings:

D = distance to the star in parsecs.

L_* = bolometric luminosity in units of $10^4 L_\odot$.

Sp. class = spectral class, where N indicates a spectrum that is similar to but redder than a typical N-type carbon star.

v_{lsr} = stellar radial velocity with respect to the local standard of rest.

v_∞ = outflow velocity at large distance from the star.

$\dot{M}_{gas}$ = mass loss rate in gas.

θ_{IR} = 'uncorrected' infrared angular diameter as defined and measured by Dyck *et al.* (1984). All measurements refer to a wavelength of 3.8 μm, except for IRC+10216 for which θ_{IR} corresponds to 2.2 μm.

θ_{OH} = angular diameter of 18-cm OH maser emission region. A '—' in this column indicates a carbon-rich object for which no OH maser emission would be expected (and indeed none has been detected).

on Mauna Kea and other dry sites will enable the determination of 400-μm, 1-mm, and perhaps 2-mm fluxes for the reddest objects. Thus the complete photometric spectrum should be measured for a large sample of giant stars. An orbiting 10-m-class telescope could measure the 100-μm size of the envelopes around numerous stars which would be useful in determining the frequency dependence of Q_v. Unfortunately, measurements that enable us uniquely to specify composition and structure of circumstellar grains may still be far in the future.

Studies of the gas will benefit greatly from the construction of large optical and millimeter-wave telescopes and interferometers. The CO $J = 3 \rightarrow 2$ and/or $J = 2 \rightarrow 1$ rotational transitions near 1-mm wavelength should be detectable in many objects that have smaller mass loss rates and/or are further away than those that are listed in Table 7.1. Additional molecules may also be found. The mm-λ interferometers will map various molecules in the stronger sources and the results may be compared with the photochemical models described in Section 7.3.2.

In the optical and infrared, speckle interferometry at high spectral resolution (speckle spectroscopy) should enable the measurement of the density, temperature, and velocity structure of the circumstellar gas within a few stellar radii of the central star. With multiple spectral elements it will be possible to measure the structure of the envelope in a spectral line simultaneously with the adjacent continuum. The latter serves as a (pseudo)

Notes
 (1) The range given for L_* and $\dot{M}_{gas}$ corresponds to the range of D. For Mira, luminosity corresponds to maximum light.
 (2) Sopka *et al.* (1985).
 (3) Cahn & Wyatt (1978). *Astrophys. J.* **221**, 163.
 (4) Knapp & Morris (1985). *Astrophys. J.* **292**, 640.
 (5) Jura (1983).
 (6) Huggins (1985). In *Mass Loss From Red Giants*.
 (7) Zuckerman (1979). *Astrophys. J.* **230**, 442.
 (8) Spergel *et al.* (1983). *Astrophys. J.* **275**, 330.
 (9) Schmidt *et al.* (1980). *Astrophys. J.* **239**, L133.
(10) Bowers *et al.* (1983). *Astrophys. J.* **274**, 733.
(11) The nebula has C/O $\gtrsim 1$ (Olofsson *et al.* (1982). *Bull. Amer. Astron. Soc.* **14**, 895). The underlying star is very hot but unseen.
(12) Werner *et al.* (1980).
(13) Mutel *et al.* (1979). *Astrophys. J.* **228**, 771.
(14) Cohen & Kuhi (1977). *Astrophys. J.* **213**, 79.
(15) Star is unresolved at 3.8 μm with 3-m-class telescope. Mira and α Ori are partially resolved at longer wavelengths (see papers by McCarthy *et al.* and Sutten *et al.* referenced in Zuckerman (1980)).
(16) Bowers & Morris (1984). *Astrophys. J.* **276**, 646.
(17) Baud & Habing (1983). *Astron. Astrophys.* **127**, 73.
(18) Davis *et al.* (1979). *Astrophys. J.* **230**, 434.
(19) Zuckerman *et al.* (1986). *Astrophys. J.* **304**, 401.
(20) Jewell (1985), in preparation.

point source phase reference which enables one to eliminate seeing noise and reconstruct complete images. These techniques should be especially powerful on proposed telescopes in the 7- to 15-m class. Even smaller angular scales may be studied with multiple telescope interferometers.

I am grateful to my UCLA colleagues, Drs. M. Jura and M. Morris, for reading and commenting on the manuscript and especially to Dr. Jura for helpful comments on life in general. This work was partially supported by the National Science Foundation through grant AST 82-08793 to the University of Hawaii.

References

A comprehensive source of information on circumstellar envelopes around red giant stars is:

Morris, M. & Zuckerman, B. (eds.) (1985). *Mass Loss From Red Giants*. Reidel Publishing Co.

Other relevant reviews and basic research papers include:

Section 7.2

Dupree, A. K. (1981). Mass loss from cool stars. In I.A.U. Colloquium No. 59, *Effects of Mass Loss on Stellar Evolution*, ed. C. Chiosi & R. Stalio, p. 87. Reidel Publishing Co.

Goldberg, L. (1979). Some problems connected with mass loss in late-type stars. *Quart. J. Roy. Astron. Soc.* **20**, 361.

Goldberg, L. (1981). Acceleration of mass flow in the chromosphere of α Orionis. In *Physical Processes in Red Giants*, ed. I. Iben, Jr. & A. Renzini, p. 301. Reidel Publishing Co.

Hinkle, K. H., Hall, D. N. B. & Ridgway, S. T. (1982). Time series infrared spectroscopy of the Mira variable χ Cygni. *Astrophys. J.* **252**, 697.

Labeyrie, A. (1970). Attainment of diffraction limited resolution in large telescopes by Fourier analyzing speckle patterns in star images. *Astron. Astrophys.* **6**, 85.

Linsky, J. L. (1981). Outer atmospheres of late-type stars. In *Physical Processes in Red Giants*, ed. I. Iben, Jr. & A. Renzini, p. 247. Reidel Publishing Co.

Reimers, D. (1981). Winds in red giants. In *Physical Processes in Red Giants*, ed. I. Iben, Jr. & A. Renzini, p. 269. Reidel Publishing Co.

Zuckerman, B. (1980). Envelopes around late-type giant stars. *Ann. Rev. Astron. Astrophys.* **18**, 263.

Zuckerman, B. (1981). Circumstellar envelopes. In *Scientific Importance of High Angular Resolution at Infrared and Optical Wavelengths*, ed. M. H. Ulrich & K. Kjar, p. 379. European Southern Observatory.

Section 7.3.1

Bowers, P. F., Johnston, K. J. & Spencer, J. H. (1983). Circumstellar envelope structure of late-type stars. *Astrophys. J.* **274**, 733.

Castor, J. I. (1970). Spectral line formation in Wolf–Rayet envelopes. *Mon. Not. Roy. Astron. Soc.* **149**, 111.

Gehrz, R. D. & Woolf, N. J. (1971). Mass loss from M stars. *Astrophys. J.* **165**, 285.

Goldreich, P. & Scoville, N. (1976). OH-IR stars. I. Physical properties of circumstellar envelopes. *Astrophys. J.* **205**, 144.

Hildebrand, R. H. (1983). The determination of cloud masses and dust characteristics from submillimeter thermal emission. *Quart. J. Roy. Astron Soc.* **24**, 267.

Jura, M. (1983). Mass loss rates and anisotropies in the outflows from late-type stars. *Astrophys. J.* **275**, 683.

Kwan, J. & Linke, R. A. (1982). Circumstellar molecular emission of evolved stars and mass loss: IRC + 10216. *Astrophys. J.* **254**, 587.

Kwok, S. (1975). Radiation pressure on grains as a mechanism for mass loss in red giants. *Astrophys. J.* **198**, 583.

Morris, M. (1975). The IRC + 10216 molecular envelope. *Astrophys. J.* **197**, 603.

Morris, M. (1980). Molecular emission from expanding envelopes around evolved stars. III. Thermal and maser CO emission. *Astrophys. J.* **236**, 823.

Sobolev, V. V. (1960). *Moving Envelopes of Stars.* Harvard University Press.

Tielens, A. G. G. (1983). Stationary flows in the envelopes of M giants. *Astrophys. J.* **271**, 702.

Section 7.3.2

Draine, B. T. (1981). Infrared emission from dust in shocked gas. *Astrophys. J.* **245**, 880.

Dyck, H. M., Zuckerman, B., Leinert, Ch. & Beckwith, S. (1984). Near-infrared speckle interferometry of evolved stars and bipolar nebulae. *Astrophys. J.* **287**, 801.

Lafont, S., Lucas, R. & Omont, A. (1982). Molecular abundances in IRC + 10216. *Astron. Astrophys.* **106**, 201.

Rowan-Robinson, M. & Harris, S. (1982). Radiative transfer in dust clouds. II. Circumstellar dust shells around early M giants and supergiants. *Mon. Not. Roy. Astron. Soc.* **200**, 197.

Rowan-Robinson, M. & Harris, S. (1983). Radiative transfer in dust clouds. III. Circumstellar dust shells around late M Giants and supergiants and IV. Circumstellar dust shells around carbon stars. *Mon. Not. Roy. Astron. Soc.* **202**, 767, 797.

Sopka, R. J., Hildebrand, R. H., Jaffe, D., Gatley, I., Roellig, T., Werner, M., Jura, M. & Zuckerman, B. (1985). Submillimeter observations of evolved stars. *Astrophys. J.* **294**, 242.

Werner, M. W., Beckwith, S., Gatley, I., Sellgren, K., Berriman, G. & Whiting, D. L. (1980). Simultaneous far-infrared, near-infrared, and radio observations of OH/IR stars. *Astrophys. J.* **239**, 540.

8

The gaseous galactic halo

BLAIR D. SAVAGE

8.1   Introduction

The neutral interstellar gas in the Milky Way is largely confined to the galactic plane with a density stratification away from the plane that is approximately given by $n(\mathrm{H\,I}) \approx n(\mathrm{H\,I})_0\, e^{-|z|/h}$, with $n(\mathrm{H\,I})_0 \approx 1.2$ atoms cm^{-3} and $h \approx 0.12$ kpc. However, this simple exponential distribution does not adequately describe a very extended and highly ionized component of the interstellar gas that is commonly referred to as galactic halo gas or galactic corona gas. The density stratification of the halo gas is very uncertain but may have a scale height that exceeds by a factor of 30 that of the neutral disk gas.

The observational study of galactic halo gas is a very young field – the youngest of those subjects included in this volume. That a hot (10^4–10^6 K) and extended ($z \approx 10$–30 kpc) gaseous halo surrounds the Milky Way was suggested by Shklovsky (1952) based on measurements of non-thermal radio emission from the galaxy and by Spitzer (1956) based on the apparent stability of high-latitude interstellar clouds. Spitzer noted that stars at z distances exceeding 0.5 kpc more frequently exhibit high velocity interstellar Ca II optical absorption lines than do stars at smaller z. He concluded that an appreciable fraction of the high-velocity clouds producing these absorption features must exist more than 0.5 kpc from the plane. The basic problem associated with a cloud at $z > 0.5$ kpc is its instability to outward expansion unless the cloud is in near pressure equilibrium with an external medium. The external medium was postulated to be a high-temperature low-density gas – the galactic corona.

High-velocity clouds detected in optical Ca II absorption line studies provided indirect evidence for the existence of galactic halo gas. Complementary results have come from 21-cm emission line studies of high-velocity clouds. However, the direct detection of the hotter and more highly ionized phases of gas in the galactic halo has come from ultraviolet satellite telescopes. In particular the ultraviolet absorption line studies with the *International Ultraviolet Explorer* (*IUE*) satellite have provided a wealth of new information about halo gas. The ultraviolet studies with the *IUE* will be emphasized in this chapter.

8.2 Ultraviolet observations of halo gas

Most atoms have their resonance absorption lines (transitions out of the ground state) at wavelengths shortward of the atmospheric cut-off which is near 3100 Å. Thus, the first direct detection of the highly ionized component of galactic halo gas had to wait for the development of sensitive ultraviolet space observatories. The *Copernicus* ultraviolet satellite which was launched in 1972 and operated until 1980 provided important new information about the interstellar medium in the immediate vicinity of the Sun but lacked the sensitivity to observe stars with distances away from the plane exceeding ≈ 2 kpc. The instrumental breakthrough for the direct study of halo gas came with the launch of the *IUE* satellite in January 1978. *IUE* contains a 0.4-m telescope and two echelle spectrographs capable of producing ultraviolet spectra between 1150 and 3200 Å with a velocity resolution of approximately $25 \, \mathrm{km \, s^{-1}}$. The observatory is in a geosynchronous orbit and spends long periods of time far from the highly populated regions of the Earth's radiation belts. With the low radiation background, very long integration times are possible and the *IUE* has obtained high dispersion spectra of hot stars in the Large and Small Magellanic Clouds. In addition, *IUE* has been directed at distant high-latitude O and B stars, at globular cluster stars, and at various extragalactic objects. These observations have provided unique information about the nature of galactic halo gas (for reviews see York 1982 and de Boer 1985).

Fig. 8.1 illustrates the Milky Way, Large Magellanic Cloud, and Small Magellanic Cloud geometry. The LMC lies at galactic longitude $l \approx 270°$ and galactic latitude $b \approx -33°$ and has a radial velocity with respect to the local standard of rest of $V_{\mathrm{LSR}} \approx 270 \, \mathrm{km \, s^{-1}}$. The SMC lies at $l \approx 300$, $b \approx -45$ and has $V_{\mathrm{LSR}} \approx 150 \, \mathrm{km \, s^{-1}}$. The 55–70-kpc path to each galaxy samples an extensive region of our galactic halo. Each galaxy has a radial velocity large enough to Doppler shift local Milky Way absorption away from Magellanic Cloud absorption.

As an introduction to the absorption features we will soon consider, the absorption line profile in Fig. 8.1 has been drawn to illustrate the absorption one might expect to see in the direction of the LMC. Material in the immediate vicinity of the Sun will absorb near $V_{\mathrm{LSR}} \approx 0 \, \mathrm{km \, s^{-1}}$ while material associated with the LMC will absorb near $270 \, \mathrm{km \, s^{-1}}$. If there are clouds in the halo and if they participate in galactic rotation like the material in the disk, they will absorb at intermediate velocities as shown in the schematic profile. The particular wavelength or velocity of the absorption will depend on the rotation curve and the distance to the cloud. If corotation of halo gas is a valid assumption, the absorption velocity can be used to estimate a cloud's distance provided the rotation curve is known.

Fig. 8.2 shows an *IUE* spectrum of the region 1510–1580 Å for the star HD 5980 (WN3 + 0B) in the SMC. Superposed on the broad stellar wind P Cygni profile of C IV near 1550 Å are narrow, interstellar components of the C IV doublet at $\lambda = 1548.20$ and 1550.77 Å. Each line of the doublet has components near $V_{LSR} \approx 0$, 150, and 300 km s^{-1}.

A representative set of ultraviolet line profiles for many ions for a LMC star (HD 36402) and a SMC star (HD 5980) is shown in Fig. 8.3. In preparing

Fig. 8.1. The most detailed ultraviolet measurements of Milky Way halo gas are for lines of sight extending through the halo to bright O and B stars in the LMC and SMC. These two galaxies have radial velocities of $V_{LSR} \approx 270$ km s^{-1} and 150 km s^{-1}, respectively. Therefore Milky Way absorption features are displaced from absorption occurring in the vicinity of the two galaxies. If gas in the Milky Way halo above or below the disk follows the rotation of the gaseous disk, the motion is referred to as corotation. Magnetic fields attached to disk material and extending into the halo may force corotation. In this situation, halo clouds at various distances along the line of sight will produce absorption at a velocity that depends on the rotation curve and the distance to the cloud as illustrated in the schematic absorption line profile shown for the LMC line of sight. If a rotation curve is known or assumed, the velocity of absorption in an observed profile can be used to estimate the distance to the absorbing cloud. The assumption of corotation is probably valid in the lower halo ($|z| < 1$–2 kpc) but may become invalid at larger $|z|$.

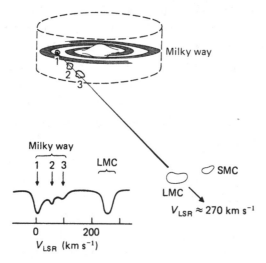

this figure, data similar to those shown in Fig. 8.2 were transformed into plots of intensity versus velocity. The data for the LMC star should be compared with the schematic profile shown in Fig. 8.1. For both lines of sight the strong absorption near $V_{LSR} \approx 0$ km s^{-1} is certainly associated with the Milky Way. For the HD 36402 measurements, the feature near 270 km s^{-1} is likely produced near the LMC while for the HD 5980 measurements the SMC absorbs near 150 km s^{-1}. The origin of the 70 km s^{-1} and 120 km s^{-1} absorption features toward HD 36402 is more uncertain. If the simple model presented in Fig. 8.1 applies, then, these features may be due to clouds at very large distances from the Milky Way plane. However, other possibilities exist. The Magellanic Clouds are known to be associated with a stream of gas – the Magellanic Stream. If this gas stretches in front of the LMC stars, it might explain some of the ultraviolet absorption at intermediate velocities.

The absorption extending from 0 km s^{-1} along each line of sight is detected in a wide range of ionization stages. For example, C I, II, and IV are

Fig. 8.2. A high dispersion *IUE* ultraviolet spectrum of the SMC star HD 5980 (WN3 + OB) between 1510 and 1580 Å. The broad absorption and emission is the stellar P Cygni profile of C IV formed in the rapidly expanding wind of HD 5980. Superposed on the broad stellar feature are narrow interstellar absorption lines of Si II $\lambda1526.72$ and C IV $\lambda\lambda148.20$ and 1550.77, and C I $\lambda1560.31$. For each Si II and C IV transition three interstellar components are visible. Each component is spaced by ≈ 1 Å or 150 km s^{-1}. The shortest wavelength components absorb near $V_{LSR} \approx 0$ km s^{-1} and are produced in the Milky Way. The intermediate velocity features absorb near $V_{LSR} \approx 150$ km s^{-1} which implies an origin in the SMC. The weakest component which absorbs near $V_{LSR} \approx 300$ km s^{-1} is of unknown origin. Bright spots on the detector are identified with a 'B'. Detector registration marks or reseaux are identified with an 'R'.

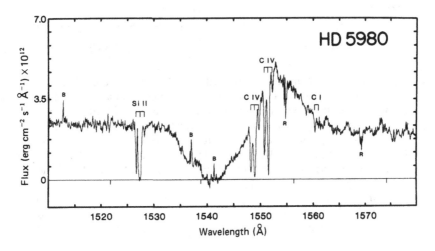

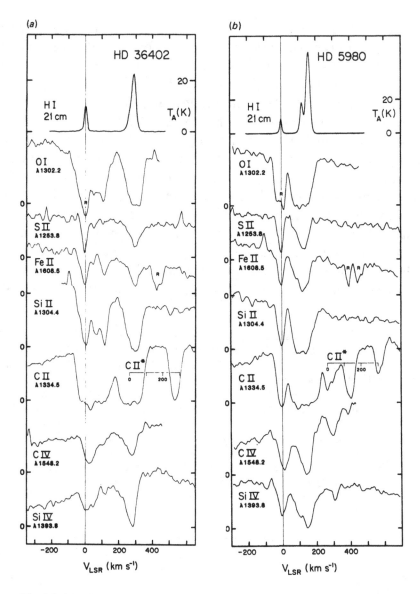

Fig. 8.3. Sample UV line profiles toward (*a*) HD 36402 in the LMC and
(*b*) HD 5980 in the SMC. The profiles plotted represent an average of five
individual spectra. The profiles are plotted as a function of LSR velocity. The
intensity zero points for each profile are indicated with the tick marks. The
absorption lines plotted from top to bottom include O I λ1302.17, S II λ1253.81,
Fe II λ1608.46, Si II λ1304.37, C II λ1334.53, C IV λ1548.20, and Si IV λ1393.76.
Features extending from $V_{LSR} = 0$ are formed in the Milky Way. LMC and
SMC absorption occurs near $V_{LSR} = 270$ and 150 km s^{-1} respectively. The
pronounced difference in the Milky Way absorption profiles toward HD 36402
in the LMC ($l \approx 280°$ and $b \approx -33°$) vs. HD 5980 in the SMC ($l \approx 330°$ and
$b \approx -45°$) is possibly due to differential galactic rotation. At the top of each

detected along with Si II, III, and IV. C III is probably present but does not have a resonance line in the region accessible to IUE. Most noteworthy is the presence of the highly ionized gas as revealed by the Si IV and C IV lines. For the HD 5980 line of sight, N V at $\lambda = 1238.20$ and 1248.20 is also marginally detected. These features provide information about the highly ionized gas of the galactic halo.

8.3   Interpretations of the observations
8.3.1   *Column densities*

Absorption line data of the type illustrated in Fig. 8.3 can be used to estimate the line of sight column densities of various species by employing the 'curve of growth'. This is illustrated in Fig. 8.4 for the important lines of Si IV and C IV near 0 km s^{-1} in the spectrum of HD 5980. The curve of growth is a logarithmic plot of the line strength, W_λ/λ against the product $Nf\lambda$. Here, W_λ is the line equivalent width, λ is the wavelength, N is the column density, and f is the line oscillator strength or f value. The observed values of W_λ/λ are shifted horizontally until they line up with a theoretical curve of growth as shown in the figure. Theoretical curves of growth represent the expected W_λ/λ vs. $Nf\lambda$ relation for various line broadening assumptions. It is common to make the simplifying assumption that the line broadening is produced by a Maxwellian distribution of velocities described by the velocity spread parameter, b (km s^{-1}). If the Maxwellian distribution is thermal in origin then $b = (2kT/M)^{1/2}$ where T is the kinetic temperature and M is the mass of the absorbing ion. In Fig. 8.4, theoretical curves are shown for $b = 1, 2, 5, 10, 20, 40,$ and 80 km s^{-1}. The measured data points have been shifted horizontally to provide a fit to one of the theoretical curves. For the lines of both Si IV and C IV a reasonable fit is achieved for a theoretical curve near $b = 25 \text{ km s}^{-1}$. From this fit, the product $Nf\lambda$ can be determined for each absorption line for each species. Knowing f and λ then yields N. The resultant column densities are $\log N(\text{Si IV}) \approx 13.6$ and

(Fig. 8.3 continued)

figure the 21-cm emission profiles in the direction of each star are illustrated. The very different velocity extent of the H I emission and UV absorption toward HD 36402 illustrates the extreme sensitivity of the strong UV lines to small amounts of gas. The Milky Way halo cloud component structure is best revealed in the lines of Si II ($\lambda 1304.37$) and Fe II ($\lambda 1608.46$) toward HD 36402. Along this particular line of sight, halo components occur near 70 and 120 km s^{-1}, while the component near 0 km s^{-1} is mostly from disk gas. The instrumental velocity resolution for the UV data is 25 km s^{-1}. Many additional sample profiles can be found in Savage & de Boer (1979, 1981), de Boer & Savage (1980), and Fitzpatrick & Savage (1983).

log N(C IV) $\approx$ 14.4. These values of N are quite uncertain since the actual velocity dispersion for the absorption is certainly more complex than that represented by a single cloud with one value of b.

For measuring column densities it is desirable to measure weak lines for which the column density is linearly related to the equivalent width and independent of the line broadening. Unfortunately the photometric precision and resolution of the *IUE* makes such weak line measurements difficult. Thus the accuracy of column densities based on the *IUE* data is generally poor.

8.3.2   *The z extent of galactic halo gas*

If the gas in the halo participates in galactic rotation like the gas in the disk, then the observed line profiles can be analyzed to yield the run of density with distance away from the galactic plane. The procedure is similar to that employed by radio astronomers to map the distribution of H I in the galaxy through the 21-cm line. The only difference is we are dealing with

Fig. 8.4. Measures of absorption line equivalent widths for various ions can be converted to estimates of column density via the 'curve of growth'. For the line of sight to HD 5980, the Milky Way absorption measurements for the Si IV and C IV doublets are plotted as the open and solid circles, respectively. The measured line strengths are reasonably well described by a simple one component curve of growth with $b \approx 25 \, \mathrm{km \, s^{-1}}$. The resulting column densities are log N(Si IV) $\approx$ 13.6 and log N(C IV) $\approx$ 14.4.

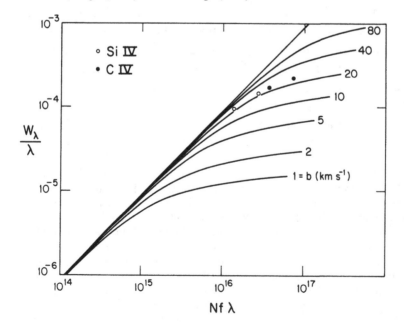

absorption rather than emission lines and we know very little about the actual motions of halo gas. Fig. 8.5 shows the result of an analysis of the asymmetric Si IV and C IV profiles toward HD 36402 in the LMC. The analysis assumes the rotation of halo gas follows that of disk gas with a constant rotational velocity of 220 km s^{-1} for galactocenter distances greater than 8 kpc. The fall off of density with distance away from the galactic plane is approximately described by an exponential with a scale height of 4 kpc. A change in the assumed character of the rotation curve will of course modify the derived C IV and Si IV density distribution. One might suspect that corotation of halo gas is a reasonable assumption for small z and invalid at large z. The density distribution shown in Fig. 8.5 should be considered very uncertain.

Fig. 8.5. C IV and Si IV densities (atom cm^{-3}) toward HD 36402 as a function of z (distance below the galactic plane). The derived curves assume that the observed velocities in the line profiles represent the radial component of corotating gas in the galactic halo. Due to velocity blending and saturation near $V = 0$ km s^{-1}, the densities between 0 and -3 kpc are only approximate. At large z the corotation assumption is probably invalid. The presence of gas flows along this particular line of sight would of course invalidate the result.

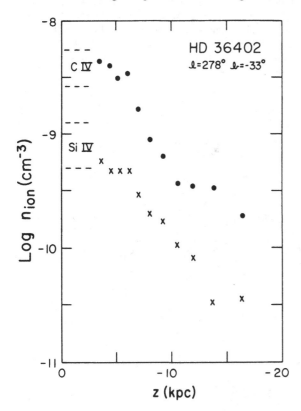

An independent way of obtaining information about the z distribution of the gas consists of measuring column densities toward galactic stars at different z distances and plotting the column density projected in the z direction, $N|\mathrm{Sin}\, b|$, versus the absolute value of the z distance to each star. Here b is the galactic latitude of each star. An extensive amount of this data is shown in Fig. 8.6 for C IV. The data are roughly approximated by an exponential stratification with a scale height of 4 kpc and a midplane density of $n(\mathrm{C\,IV})\, 8 \times 10^{-9}$ atoms cm^{-3} (see the solid curves in Fig. 8.6). This result is compatible with the derived C IV density distribution shown in Fig. 8.5.

8.3.3   *Ionization and temperature*

The rich range of ionization existing in galactic halo gas is clearly revealed in the line profiles for $V < 100$ km s^{-1} recorded toward HD 36402 in

Fig. 8.6. Logarithmic plot of the C IV column density perpendicular to the galactic plane versus z distance from the plane for various galactic and extragalactic stars. The data are mostly from Savage & de Boer (1981), Pettini & West (1982), and Savage & Massa (1985). The solid curves illustrate the $N(\mathrm{C\,IV})|\mathrm{Sin}\, b|$ versus $|z|$ relation for a simple exponential stratification, $n(\mathrm{C\,IV}) = n(\mathrm{C\,IV})_0 \exp(-|z|/h)$, with $n(\mathrm{C\,IV})_0 = 8 \times 10^{-9}$ cm^{-3} and $h = 0.3, 1.0,$ and 3.0 and 10 kpc. The measurements are compatible with h of about 4 kpc. However, the large scatter in the plot suggests a very patchy C IV distribution. There is also evidence for an enhancement in $N(\mathrm{C\,IV})|\mathrm{Sin}\, b|$ for $z \gtrsim 1$ kpc. This enhancement might be produced by UV photoionization of high z gas by the extragalactic EUV background radiation.

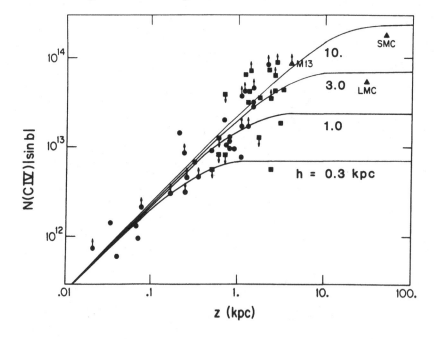

the LMC. The simultaneous occurrence of absorption by C II and C IV along with Si II, Si III, and Si IV with dissimilar line profile structure implies a very inhomogeneous absorbing region with possibly a wide range of temperatures. The low ionization stages of Si II, Mg II, C II, Mn II, Fe II, etc., are probably associated with relatively dense halo matter in the form of condensations while the more highly ionized absorption of Si IV, C IV and in some cases N V may represent the detection of hotter halo gas. However, it is not simple to make inferences about the temperature of the absorbing gas since the origin of the ionization is not well understood.

The highly ionized halo gas might arise via a number of different processes. For example, XUV or X-ray photoionization by an extragalactic background is one possibility. Collisional ionization in a hot gas is another. If these ions are produced by collisional ionization balanced by recombination – often called 'coronal ionization' – then Fig. 8.7 can be used to estimate the gas temperature. The various curves show the expected ionic

Fig. 8.7. The fractional abundance relative to ionized hydrogen of various ionized species as a function of temperature for steady state 'coronal' or collisional ionization. Solar abundances are assumed. The curves for Si are from Baliunas & Butler (1980) and those for C, N, and O are from Shapiro & Moore (1976). If the assumption of equilibrium coronal ionization is valid, observed ion ratios can be used to provide crude estimates of the temperature of the absorbing plasma. For example, the C IV to Si IV column density ratio toward HD 5980 is ≈ 6.3 (see Fig. 8.4). The temperature at which coronal ionization predicts this ratio is 75 000 K.

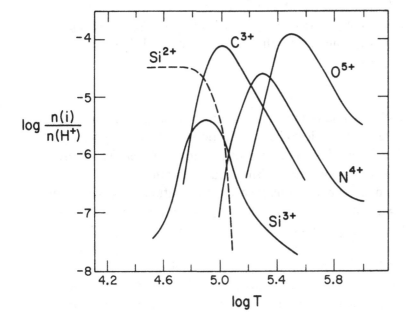

abundances with respect to the ionized hydrogen abundance for a range of temperatures. The various curves assume solar abundance ratios apply. As can be seen, the species Si^{3+}, C^{3+}, N^{4+}, and O^{5+} peak in abundance at temperatures of approximately 80 000, 100 000, 200 000, and 300 000 K, respectively. Thus far, Si^{3+}, C^{3+}, and N^{4+} have been detected in halo gas implying a range of temperatures from $\approx 80\,000$ to $\approx 200\,000$ K, respectively. Unfortunately, *IUE* is not capable of recording spectra near 1038 and 1032 Å, the wavelengths of the O VI doublet. Therefore, there is very little information about the possible existence of O VI in galactic halo gas. Jenkins (1978) on the basis of *Copernicus* satellite data has claimed an O VI exponential scale height of $0.3^{+0.2}_{-0.15}$ kpc. However, only three stars were observed with $|z| > 1$ kpc and this result should be considered very uncertain.

If photoionization dominates over collisional ionization, the various high ionization lines detected by the *IUE* could be produced in relatively cool gas, e.g., $T \approx 10\,000$ K. A more direct measure of the gas temperature might be provided by the measurement of thermal Doppler widths of spectral lines. The b values obtained for Si IV and C IV via the curve of growth analysis of Section 8.3.1 was 25 km s^{-1}. If this b value has a thermal origin with $b = (2kT/M)^{1/2}$, it would imply temperatures of 0.5×10^6 K and 10^6 K for C IV and Si IV respectively. However, the observed line profiles themselves imply that multiple absorption components are contributing to the large b derived via the curve of growth analysis. Furthermore, at $T > 0.5 \times 10^6$ K very little C IV and Si IV should exist (see Fig. 8.7). A direct measure of halo gas temperatures from a line profile analysis will probably require the higher spectral resolution capabilities of the Hubble Space Telescope high-resolution spectrograph.

8.3.4   *Densities*

Crude density estimates for halo gas can be obtained for particular ions via the profile analysis described in Section 8.3.2 (see Fig. 8.5). To convert these estimates into information about the total gas density requires a knowledge of the ionization mechanisms and the abundances.

If C^{+3} is produced under conditions of coronal ionization in a gas of solar abundances at $T \approx 10^5$ K, then the curves of Fig. 8.7 imply the H^+ density will be $\approx 2 \times 10^4$ times larger than the C^{+3} density. At $z \approx 2$–5 kpc, Fig. 8.5 shows $n(C^{3+}) \approx 5 \times 10^{-9}$ atoms cm^{-3}. From this it follows that $n(H^+) \approx 10^{-4}$ atoms cm^{-3} in that gas producing the C^{3+} absorption. So many assumptions were required to produce this estimate it should be considered extremely uncertain. In fact, the resultant density is so low the assumption of time equilibrium should be questioned since the recombination times are on the order of 10^9 years.

8.3.5 *Abundances*

Unfortunately the *IUE* studies of extragalactic stars have only provided very crude abundance information about galactic halo gas. However, the gas certainly contains processed elements and the abundances are within an order of magnitude of being solar.

More accurate information on halo gas abundances is being obtained through high-resolution high-precision ground-based studies. In particular, the studies of Albert (1983) have demonstrated that the abundance of Ti II relative to hydrogen increases with z distances and with LSR velocity by factors exceeding 20. In the most extreme cases, the Ti abundance approaches the solar value. The low Ti II to H I ratio for disk gas is very likely due to depletion of gaseous Ti into interstellar dust grains. The more normal Ti II to H I ratio for halo gas probably means halo gas contains very little dust. Similar results were found for Fe II in the *Copernicus* satellite survey of Savage & Bohlin (1979).

Improved information on the abundances of halo gas may provide clues about its origin. However, the possible existence of dust in the halo will always tend to confuse the interpretation of gas phase abundance studies. The work of Jenkins (1983) suggests that Si II and Al II behave differently from Fe II or Ti II. Perhaps some halo clouds contain dust grains.

8.4 Origin of galactic halo gas

A possible model of the halo gas is the galactic fountain model in which hydrodynamical flow supplies gas to the galactic halo (Shapiro & Field 1976).

Briefly, the basic ideas are: Hot gas ($T \approx 0.3$–1×10^6 K) is known to exist in the galactic plane from measures of soft diffuse X-rays and from interstellar O VI absorption line studies toward O and B stars located in the disk. The hot disk gas is buoyant, since gas at $T \approx 10^6$ K has a thermal scale height in the galactic gravitational field of about 7 kpc. The hot gas will therefore attempt to flow outward away from the galactic plane. If the outflowing gas cools, clouds may form which will fall back to the galactic plane. Calculations of the resulting flow are found in Bregman (1980) and Habe & Ikeuchi (1980). The rate of cooling and heating of the outflowing gas will determine whether or not a fountain occurs as described above or whether the flow continues outward as a galactic wind. The currently favored view is the fountain – at least for the local region of the galaxy. Near the galactic center, where the rate of heating may be enhanced, the flow might be in the form of a wind.

As a result of the flow associated with the galactic fountain, a highly inhomogeneous region in density and temperature develops. Therefore, it is perhaps not too surprising to observe an extremely complex interstellar absorption line spectrum when viewing background sources through this material. For the galactic fountain to work, an appreciable fraction of the disk volume must be occupied by hot gas. At present, the filling factor of hot gas in the disk is only poorly known.

In the galactic fountain model, the large z extent of galactic halo gas is explained by thermal pressure support. Furthermore, the large range of temperatures in the ascending and descending collisionally ionized gas provides a simple explanation for some of the high and low ionization stage lines seen by the IUE. However, there are other support possibilities for halo gas. In particular the support might be provided by the galactic magnetic field inflated by the pressure of cosmic rays. In this case, the halo gas might exist at a lower temperature and be ionized by galactic and extragalactic EUV and X-radiation. Detailed ionization equilibrium calculations (Chevalier & Fransson 1984; Hartquist, Pettini & Tallant 1984) are able to explain the IUE C IV and Si IV measurements. However, the predicted N V column density through the halo is $10\times$ smaller than the observed value.

In summary, it appears likely that collisional ionization in halo gas is necessary to explain the observed N V but that photoionization and collisional ionization may produce the observed Si IV and C IV. A galactic fountain operating in the presence of known galactic and extragalactic sources of ionizing radiation may explain the basic observations.

8.5   Implications for extragalactic astronomy
8.5.1   *The evolution of interstellar elemental abundances*

A key to the ultimate understanding of galactic evolution is through an understanding of the evolution of elemental abundances in galaxies. The evolution of elemental abundances in the interstellar medium depends on the various input, loss, and mixing processes occurring within a galaxy. The various input processes operating include gradual and catastrophic mass loss of processed elements from stars of all types and the infall of unprocessed matter from intergalactic space. In the case of the Milky Way, there is ample evidence for stellar mass loss. The high velocity clouds seen in the H I 21-cm line may contain evidence for the infall of intergalactic matter (Mirabel & Morras 1984).

Loss processes include matter accumulating into stars and loss via a galactic wind. Stellar formation is observed to occur in dense molecular clouds. It is not known if the Milky Way experiences mass loss via a galactic wind. However, Garmire & Nugent (1981) have interpreted their X-ray observations of the Milky Way center as suggestive of a wind. In the solar neighborhood the existence of cool clouds in the halo possibly implies a bound circulation in the form of a fountain rather than an unbound wind (Chevalier & Oegerle 1979).

Processes mixing material on a galactic scale might include the dynamic phenomena leading to spiral arm formation and destruction and the mixing that might be produced by the flow associated with a galactic fountain. These mixing processes will tend to smooth out abundance inhomogeneities that result from stellar evolution.

Computing realistic models for interstellar abundance evolution is a complex task involving many assumptions about the production and loss processes discussed above. To these one now needs to add the effects of mixing via fountain flows and/or loss via galactic winds. However, these latter two processes are linked to the element production process because the interstellar heating from supernova explosions and stellar mass loss are important in driving a fountain or a wind.

8.5.2 *Quasar absorption lines*

The origin of quasar absorption line systems probably involves a number of different physical phenomena. However, many of the narrow line absorption systems may arise in the gaseous halos of intervening galaxies (Bahcall & Spitzer 1969). For a number of quasar absorption line systems for which the most detailed data exists, the species observed and the approximate line strengths are roughly comparable with those seen in the Milky Way halo when viewing extragalactic sources (Savage & Jeske 1981).

The confirmation via the *IUE* satellite that our own Galaxy possesses a gaseous halo has lent support to the intervening halo hypothesis for the origin of some of the quasar absorption line systems. However, a major complication with this idea is the high frequency of occurrence of the absorption line systems. If most of these systems are produced in gaseous halos, the halos must be exceedingly large. For flattened halos, Young, Sargent & Boksenberg (1982) estimate that halo radii must approximate $3 R_{HO}$ while with somewhat different assumptions Weymann *et al.* (1979) estimate $6 R_{HO}$. R_{HO}, the Holmberg radius, is a measure of the extent of a galaxy at visual wavelengths at a brightness level of 26.5 mag arc s^{-2}. Although these size requirements are large it is important to note that ultraviolet absorption lines are very sensitive probes of small amounts of gas.

An observational example of this sensitivity is provided in Fig. 8.3. Toward HD 36402 in the LMC the Milky Way C II line exhibits nearly total absorption between -40 and $+140$ km s^{-1} while the Milky Way H I 21-cm emission line in the same direction shows appreciable emission only over the range -40 to $+40$ km s^{-1}.

The sizes of spiral galaxies in H I 21-cm emission have been studied by Briggs (1982). At a detection limit of $N(\text{H I}) \approx 10^{18}$ atoms cm^{-2}, spirals are typically not more extended than $2\,R_{\text{HO}}$. If this result is generally true and if the statistical results of Young, Sargent & Boksenberg (1982) and Weymann *et al.* (1979) are correct, then halo absorption must typically occur in regions with $10^{17} < N(\text{H I}) \leqslant 10^{18}$ atoms cm^{-2} at $R > 2\,R_{\text{HO}}$. This would only be possible if the outer fringes of galaxies contain matter with nearly solar abundances.

8.6   Future prospects

The *IUE* has provided us with just a glimpse of the wealth of information contained in high-resolution ultraviolet spectra of targets in and beyond the galactic halo. The *IUE* is a small observatory and has a resolution that is less than ideal for investigating narrow interstellar absorption lines. With the high-resolution spectrograph (HRS) on the Hubble Space Telescope, it will be possible to expand greatly on the *IUE* studies of the Milky Way halo. In its high-resolution mode, the HRS has a resolution $\lambda/\Delta\lambda \approx 90\,000$ or 3.3 km s^{-1} compared with 25 km s^{-1} for the *IUE*. The HRS detector is a photon counting digicon which is capable of producing spectra with signal-to-noise ratios exceeding 100. With these features and the 2.4-m aperture of the Hubble Space Telescope many important studies of halo gas will be possible. Some of these include accurate abundance determinations and direct measures of line Doppler widths and hence temperature. Unfortunately the space telescope HRS will not be capable of studying O^{+5} through its resonance lines near 1030 Å. Studies at wavelengths below 1150 Å will have to wait for the Lyman Observatory. Lyman will be specifically designed to have high efficiency in the 900–1200 Å region. With this instrument, measurements of O^{+5} and S^{+5} toward galactic and extragalactic stars should be easy to obtain.

The Space Telescope and Lyman spectrographs will have the sensitivity to probe extragalactic halos by viewing Seyfert galaxy nuclei and quasars situated off the edges of foreground galaxies. With this type of measurement we will for the first time have the ability to interrelate the properties of nearby halos with those presumably detected from the ground in the spectra of high redshift quasars. The future of halo gas studies looks very promising.

References

Albert, C. E. (1983). *Astrophys. J.* **272**, 509.

Bahcall, J. L Spitzer, L. (1969). *Astrophys. J. Lett.* **156**, L63.

Baliunas, S. L. & Butler, S. E. (1980). *Astrophys. J. Lett.* **235**, L45.

Bregman, J. N. (1980). *Astrophys. J.* **236**, 577.

Briggs, F. H. (1982). In *The Comparative H I Content of Normal Galaxies*, ed. M. Haynes & R. Giovanelli, p. 50. Green Bank: NRAO Publication Division.

Chevalier, R. A. & Fransson (1984). *Astrophys. J. Lett.* **279**, L43.

Chevalier, R. A. & Oegerle, W. R. (1979). *Astrophys. J.* **227**, 398.

de Boer, K. S. (1985). *Mitt. der. Astr. Gesell.* **63**, 21.

de Boer, K. S. & Savage, B. D. (1980). *Astrophys. J.* **238**, 86.

Fitzpatrick, E. L. & Savage, B. D. (1983). *Astrophys. J.* **267**, 93.

Garmire, G. P. & Nugent, J. J. (1981). *Bull. Aner. Astron. Soc.* **13**, 786.

Habe, A. & Ikeuchi, S. (1980). *Prog. Theo. Phys.* **64**, 1995.

Hartquist, T. W., Pettini, M. & Tallant, A. (1984). *Astrophys. J.* **276**, 519.

Jenkins, E. B. (1978). *Astrophys. J.* **219**, 845.

Jenkins, E. B. (1983). In *Kinematics, Dynamics and Structures of the Milky Way*, ed. W. L. H. Shuter, p. 21. Holland: Reidel.

Mirabel, I. F. & Morras, R. (1984). *Astrophys. J.* **279**, 86.

Pettini, M. & West, K. A. (1982). *Astrophys. J.* **260**, 561.

Savage, B. D. & Bohlin, R. C. (1979). *Astrophys. J.* **229**, 136.

Savage, B. D. & de Boer, K. S. (1979). *Astrophys. J. Lett.* **230**, L77.

Savage, B. D. & de Boer, K. S. (1981). *Astrophys. J.* **243**, 460.

Savage, B. D. & Massa, D. (1985). *Astrophys. J. Lett.* **295**, L9.

Savage, B. D. & Jeske, N. A. (1981). *Astrophys. J.* **244**, 768.

Shapiro, P. R. & Field, G. B. (1976). *Astrophys. J.* **205**, 762.

Shapiro, P. R. & Moore, R. J. (1976). *Astrophys. J.* **207**, 460.

Shklovsky, I. S. (1952). *Astr. J. U.S.S.R.* **29**, 418.

Spitzer, L. (1956). *Astrophys. J.* **124**, 20.

York, D. G. (1982). *Ann. Rev. Astron. Astrophys.* **20**, 221.

Young, P., Sargent, W. L. W. & Boksenberg, A. (1982). *Astrophys. J. Suppl.* **48**, 455.

Weymann, R. J., Williams, R. E., Peterson, B. M. & Turnsek, D. A. (1979). *Astrophys. J.* **234**, 33.

Note added in proof: There have been numerous new results on the galactic halo since the writing of this chapter. Many of them are summarized in the *Proceedings of the NRAO Workshop on Gaseous Galactic Halos*, eds. J. Bregman and F. Lockman, Greenbank: NRAO Publication Division, 1986.

9

Astrophysical shocks in diffuse gas

CHRISTOPHER F. McKEE

9.1 The nature of astrophysical shocks

Energetic phenomena are common in the Universe: Stars inject matter into the interstellar medium via high-velocity winds at their birth and during their lifetime, and massive stars explode as supernovae when they die. Observations of active galaxies reveal rapidly expanding radio sources and jets which appear to be blasting into the ambient medium. On yet larger scales, over-dense regions of the Universe draw matter in at high velocity by gravitational attraction, whereas under-dense regions can expand at high velocity into their surroundings. In many of these cases, the sound speed of the ambient medium is less than the velocity of the gas expanding into it, especially if radiative cooling is efficient. In this case there is no way for the ambient medium to respond to the energy injection in a smooth way. Instead a near discontinuity is produced, a shock, which suddenly accelerates, heats, and compresses the ambient gas. Shocks often govern the dynamics of astrophysical plasmas: they transmit energy from stars to the interstellar medium, they can compress gas past the point of gravitational instability so that stars form, they may terminate the growth of protostars by driving the ambient gas away, they efficiently destroy dust grains and thereby determine interstellar gas phase abundances, and they are efficient accelerators of high-energy particles. Shocks are particularly important in astronomy because they heat gas and cause it to radiate, providing valuable diagnostics for the underlying energetic phenomena and for the ambient medium.

Shocks can generally be divided into four regions, which are not always distinct (Zel'dovich and Raizer 1966): (1) the *precursor*, in which radiation and/or fast particles heat and possibly ionize the gas ahead of the shock; (2) the *shock front*, in which the gas is accelerated, heated, and compressed through collisional or collisionless processes; (3) the *post-shock relaxation layer*, in which collisional processes such as collisional excitation, ionization, recombination, dissociation, and molecule formation alter the 'chemical' state of the gas and cause the gas to radiate; and, if the column density is large enough, (4) the *thermalization layer*, in which radiation from the shocked gas is absorbed downstream and reradiated as quasi-black-body radiation.

The sizes of these various regions are conveniently characterized by the column density $N = n_0 v_s t$ of the material which has passed through the shock front; here n_0 is the pre-shock number density of hydrogen nuclei, v_s is the shock velocity, and t is the time since the gas entered the shock front. The critical column density characterizing the post-shock relaxation layer is N_{cool}, in which half the relative energy of the shocked and unshocked gas is radiated away. Shocks with $N > N_{cool}$ are termed *radiative*; for example, supernova remnants typically become radiative when the blast wave velocity falls below 200 km s^{-1}, corresponding to $N_{cool} \approx 2 \times 10^{18}$ cm^{-2}. Radiative shocks can be either steady or unsteady, depending on whether N_{cool} is smaller or larger than the column density over which n_0 or v_s change. Shocks with $N < N_{cool}$, such as occur in young SNRs, are *non-radiative* in the sense that radiative losses are not strong enough to affect the dynamics. Non-radiative shocks generally have steady shock fronts because the fronts are so thin; there is no well-defined post-shock relaxation layer. The column density N_{th} required to actually thermalize the radiation is generally much larger than N_{cool}; allowing for interstellar dust, Hollenbach & McKee (1979; hereafter HM) found $N_{th} > 10^{23}$ cm^{-2} for shocks in diffuse gas.

Astrophysical shocks occur under a wide variety of conditions, ranging from the dense inner regions of evolved stars to the rarefied gas of the intergalactic medium. Shocks in stars generally have $N \gg N_{th}$, whereas those in diffuse gas in interstellar or intergalactic space generally have $N \ll N_{th}$. This chapter will focus on shocks in diffuse gas, so that thermalization can be ignored. Even in this case, emission lines from the shock are subject to absorption by dust if N is large enough ($N \gtrsim 10^{21}$ cm^{-2} for optical lines from an interstellar shock with $v_s \lesssim 100$ km s^{-1}, so that a significant amount of dust survives). The emphasis will be on steady, radiative shocks, since radiative shocks are more readily observable, and since in many cases steady shocks can be characterized in terms of just three parameters, n_0, v_s, and B_0, the ambient magnetic field strength.

For radiative shocks in ionized or neutral atomic gas, or in molecular gas with a substantial degree of ionization ($x_e = n_e/n \gtrsim 10^{-5}$, where n_e is the electron density), there is a clear separation between the shock front, in which the gas is impulsively heated and accelerated, and the post-shock relaxation layer, in which the radiation is emitted. Such shocks are termed *J-shocks* because the hydrodynamic variables suffer a jump across the shock front in a distance short compared with the radiation length. On the other hand, shocks in a magnetized molecular gas which is weakly ionized have a completely different structure: the dominant, neutral component of the gas is accelerated and heated by collisions with the charged particles, which are driven through the neutrals by the compressed magnetic field. The acceleration and neutral heating occur on a length scale comparable with the radiation length so that the hydrodynamic quantities vary continuously through the shock; such shocks are termed *C-shocks* (Draine 1980). In radiative C-shocks there is no distinction between the shock front and the post-shock relaxation layer. Non-radiative molecular C-shocks can also occur, but only if the shock is weak. Shocks can also be of mixed type, with J-shocks having C-type precursors or C-shocks having embedded jumps. Observationally, radiative C- and J-shocks are quite different: the high temperature behind a J-shock leads to atomic line emission, primarily in the uv and optical, whereas C-shocks are cooler and radiate primarily infrared molecular lines.

C-shocks can occur whenever the dissipation in the shock is effected by a trace constituent over a significant length scale. It is now generally believed that cosmic rays are efficiently accelerated in shocks, with the conversion of the flow energy to cosmic-ray energy occurring in a precursor ahead of the jump (Axford 1981). Such a C-type precursor is completely different from the radiative C-shocks in weakly ionized molecular gas discussed above since it is collisionless and can occur only in highly ionized gas. The theory of cosmic-ray mediated shocks is incomplete at present, so we shall discuss them only briefly at the end of Section 9.2 on the shock jump conditions. Thereafter we shall describe the structure and emission from radiative J-shocks in atomic gas (with no cosmic rays) and radiative C-shocks in molecular gas. More detailed discussions of shocks (primarily J-shocks) can be found in the monographs by Zel'dovich and Raizer (1966) and Tidman and Krall (1971), and in the reviews by Chu and Gross (1969) and McKee and Hollenbach (1980).

9.2   Shock jump conditions

In contrast to most nonlinear fluid mechanics problems, which are analytically intractable, a number of the important features of shock waves can be deduced by application of conservation laws across the shock front. The resulting Rankine–Hugoniot relations, or jump conditions, give the density, temperature, magnetic field, etc., behind the shock in terms of the values ahead of the shock and the shock velocity. Most of the jump conditions remain valid in the post-shock relaxation layer as well, and can be used in determining its structure.

9.2.1 *Derivation*

The basic assumption required to derive the jump conditions is that the shock front is thin: in a frame comoving with the shock, the flow is steady and one-dimensional since the thickness of the shock front is small compared with the length scales over which the ambient medium varies. The first conservation law we consider is conservation of mass, which applies in all nonrelativistic shocks. It is expressed by the equation of continuity

$$\frac{\partial \rho}{\partial t} + \mathbf{\nabla} \cdot \rho \mathbf{v} = 0, \tag{9.1}$$

which reduces to

$$\frac{\mathrm{d}}{\mathrm{d}z} \rho v = 0 \tag{9.2}$$

in the shock frame, which has been chosen so that the shock velocity $\mathbf{v}_s$ is normal to the plane of the shock front and lies in the z-direction; here ρ is the density and v is the fluid velocity in the shock frame. Integration of (9.2) shows that the mass flux ρv is constant in the shock frame. This gives the 'jump condition'

$$[\rho v] = 0, \tag{9.3}$$

where the quantity in square brackets is to be evaluated behind and ahead of the shock front and the difference taken. The evaluation can be made at arbitrary points ahead of and behind the shock front, so long as the planar, steady flow approximation remains valid.

Next consider momentum conservation. We focus on the region outside the shock front, so that viscosity and relative streaming between different species can be neglected. To simplify the discussion, we assume that the magnetic field B is perpendicular to the shock velocity $\mathbf{v}_s$, which is a reasonable approximation for strong shocks (HM). The steady, planar equation of motion is

$$\rho v \frac{\mathrm{d}v}{\mathrm{d}z} + \frac{\mathrm{d}p}{\mathrm{d}z} + \frac{1}{8\pi} \frac{\mathrm{d}B^2}{\mathrm{d}z} = 0, \tag{9.4}$$

where p is the gas pressure. Since ρv is constant, this becomes

$$[\rho v^2 + p + B^2/8\pi] = 0. \tag{9.5}$$

This equation as written applies only outside the shock front. For J-shocks this is not a significant restriction since the shock front has a negligible thickness, and this equation is useful in determining the structure of the post-shock relaxation layer. For radiative C-shocks, however, the shock front and the post-shock relaxation are coincident, so that this equation is only approximate unless p and ρv^2 are generalized to include the relative streaming between the neutral and ionized components of the plasma.

Before considering the energy jump condition, let us determine the behavior of the magnetic field in the shock. In diffuse astrophysical plasmas the magnetic field is 'frozen' to the charged component of the plasma (Spitzer 1978; McKee, Chernoff & Hollenbach 1984) so that

$$E + \frac{1}{c}(\mathbf{v}_c \times \mathbf{B}) = 0, \tag{9.6}$$

where $\mathbf{v}_c$ is the velocity of the charged component of the plasma. For steady flow, Faraday's law reduces to

$$\frac{d}{dz}\mathbf{E}_\perp = 0, \tag{9.7}$$

where $\mathbf{E}_\perp$ is perpendicular to $\mathbf{v}_s$. Together with (9.6), this gives the jump condition for B,

$$[v_c B] = 0. \tag{9.8a}$$

In the absence of relative streaming between the charged and neutral components of the plasma (i.e., outside the shock front), v_c reduces to v, and the mass jump condition (9.3) gives

$$[B/\rho] = 0. \tag{9.8b}$$

The jump conditions (9.3), (9.5), and (9.8) apply to any steady, planar, nonviscous flow with $\mathbf{B} \cdot \mathbf{v} = 0$, and hence they are quite general. The solution of these equations has been studied in connection with the theory of ionization fronts (see Mathews & O'Dell 1969 for a review, and Lasker 1966 for the effects of magnetic fields). The fronts are classified as R-type if they are supersonic relative to the upstream gas and D-type if they are not. R fronts compress the gas, D fronts expand it. The flow behind the front can be either subsonic or supersonic. R fronts with subsonic flow behind them are termed 'strong' because the jump in the density is greater than when the flow behind the front is supersonic. Shocks correspond to strong R fronts.

The final shock jump condition is provided by energy conservation. It is complicated by the emission and absorption of radiation in the shock, and generally requires numerical evaluation when radiation is important. For steady flow, energy conservation gives

$$\frac{d}{dz}\left(\frac{1}{2}\rho v^3 + (u+p)v + \frac{1}{4\pi}B^2 v\right) = 4\pi\kappa J - n^2\Lambda, \tag{9.9}$$

where u is the internal energy per unit volume, κ is the absorption coefficient, J is the mean intensity of the radiation, and Λ is the cooling function (see HM for a more detailed discussion). The right-hand side of this equation is just the divergence of the radiative flux F, so integration yields

$$\left[\frac{1}{2}\rho v^3 + \frac{\gamma}{\gamma-1}pv + \frac{1}{4\pi}B^2 v + F\right] = 0, \tag{9.10}$$

where γ is the effective ratio of specific heats, defined such that $p=(\gamma-1)u$. We now consider the implications of these jump conditions for shock structure.

9.2.2 Non-radiative jump conditions (J-shocks)

In a J-shock, radiative losses are negligible in the shock front itself, so that the conditions just behind the shock front are governed by the jump conditions with $[F]=0$. These conditions become particularly simple in the absence of a magnetic field:

$$[\rho v] = 0$$

$$[p + \rho v^2] = 0 \quad , \quad \text{(J-shock, } B=0). \tag{9.11}$$

$$\left[\frac{1}{2}v^2 + \frac{\gamma}{\gamma-1}\frac{p}{\rho}\right] = 0$$

There is only one nontrivial solution to these equations: a shock, corresponding to a particular case of a strong R front (supersonic flow upstream, subsonic downstream, and compression across the front). The strength of a shock is characterized by the Mach number M, the ratio of the shock velocity to the upstream signal velocity. For strong shocks ($M \gg 1$), these equations have the solutions

$$\rho_s = \frac{\gamma+1}{\gamma-1}\rho_0 \quad \rightarrow \quad 4\rho_0$$

$$v_{ps} = \frac{\gamma-1}{\gamma+1}v_s \quad \rightarrow \quad \frac{1}{4}v_s \quad , \tag{9.12}$$

$$p_s = \frac{2\rho_0 v_s^2}{\gamma+1} \quad \rightarrow \quad \frac{3}{4}\rho_0 v_s^2$$

where ρ_s, p_s, and v_{ps} (the post-shock flow velocity in the shock frame) are evaluated just behind the shock front; the numerical evaluation is for the case $\gamma = \frac{5}{3}$. Note that ρ_s/ρ_0 is bounded as v increases, whereas p grows without limit. In the frame of the unshocked gas (the 'laboratory' frame), the velocity of the gas just behind a strong shock front is

$$v_1 \equiv v_s - v_{ps} = \frac{2}{\gamma + 1} v_s \quad \rightarrow \quad \frac{3}{4} v_s. \tag{9.13}$$

The spectrum of the radiation emitted by a strong J-shock depends on the temperature behind the shock front, which follows from (9.12)

$$kT_s = \frac{\mu p_s}{\rho_s} = \frac{2(\gamma - 1)}{(\gamma + 1)^2} \mu v_s^2 \quad \rightarrow \quad \frac{3}{16} \mu v_s^2, \tag{9.14}$$

where μ is the mean mass per particle. Alternatively, T_s can be derived directly from energy conservation in the frame of the shocked gas by equating the upstream energy per particle $\mu v_1^2/2$ to the downstream value $3kT_s/2$. For a gas of cosmic abundances with 10% helium by number the post-shock temperature is

$$T_s = \begin{cases} 1.38 \times 10^5 \; v_{s7}^2 \, \text{K} & \text{(fully ionized)} \\ 2.9 \times 10^5 \; v_{s7}^2 \, \text{K} & \text{(neutral atomic)} \\ 5.3 \times 10^5 \; v_{s7}^2 \, \text{K} & \text{(molecular)} \end{cases} \tag{9.15}$$

where $v_{s7} = v_s/(10^7 \, \text{cm s}^{-1})$. For a fully ionized gas, the quoted temperature is the mean of the electron and ion temperatures (see Section 9.3.1). The entry for molecular shocks is somewhat academic since if the shocked gas is molecular the shock is generally C-type; indeed, for strong C-shocks one has $T \ll T_s$ throughout the shock.

The presence of a magnetic field does not significantly alter these results for strong shocks. Since the magnetic field is frozen to the plasma, the magnetic pressure $B^2/(8\pi)$ is amplified by at most a factor $(\rho_s/\rho_0)^2$, whereas the gas pressure increases without limit as the shock velocity increases. Equation (9.14) for T_s is accurate to within 30% for the Alfvén Mach number $M_A \equiv v_s/v_{A0} \equiv (4\pi\rho_0 v_s^2/B_0^2)^{1/2}$ exceeding 6 and the thermal Mach number $M \gg 1$ (see HM). Note that inside the shock front the steady state approximation may break down and B may reach values in excess of (ρ_s/ρ_0); observations of the Earth's bow shock show oscillating magnetic fields which overshoot the final downstream value by a factor of order 2 on a length scale of order the ion gyroradius (Russell & Greenstadt 1979).

An essential attribute of shocks is that they are irreversible: dissipation in the shock front results in an increase in the entropy. The jump in the entropy across the shock is determined by the jump conditions. Write the entropy per unit mass as $(k/\mu) \ln s^*$, where $s^* = T^{3/2}/n$ for a gas with $\gamma = \frac{5}{3}$. Behind a strong shock, s^* has the value

$$s_s^* = \frac{1}{4n_0} \left(\frac{3}{16} \frac{\mu v_s^2}{k} \right)^{3/2} = 1.29 \times 10^7 \, v_{s7}^3/n_0 \, \mathrm{K}^{3/2} \, \mathrm{cm}^3, \tag{9.16}$$

where the numerical evaluation is for an ionized gas with cosmic abundances. The jump in the entropy across the shock is then proportional to $\ln (s_s^*/s_0^*)$, where

$$\frac{s_s^*}{s_0^*} = \frac{1}{4} \left(\frac{3}{16} \frac{\mu v_s^2}{k T_0} \right)^{3/2} = \left(\frac{M}{3.67} \right)^3 \tag{9.17}$$

for strong shocks (isothermal Mach number $M \equiv v_s (k T_0/\mu)^{-1/2} \gg 4$).

The quantity s^* is particularly useful in characterizing astrophysical shocks because of its simple relation to the cooling time t_c of the shocked gas (Kahn 1976, McKee 1982). Shocks are non-radiative for $t \ll t_c$. In general, the cooling time $t_c \propto nkT/n^2\Lambda$. Over the temperature range $10^{7.5}\,\mathrm{K} \gtrsim T \gtrsim 10^5\,\mathrm{K}$, the cooling curve of Raymond, Cox & Smith (1976) can be approximated as $\Lambda \approx 1.6 \times 10^{-19} T^{-1/2} \, \mathrm{erg} \, \mathrm{cm}^3 \, \mathrm{s}^{-1}$, so that

$$t_c = 2.0 \times 10^3 s^* \beta^{-1} \, \mathrm{s}, \tag{9.18}$$

where β is the ratio of the actual cooling rate to the equilibrium value calculated by Raymond *et al.* Non-equilibrium ionization effects will generally make $\beta \gtrsim 1$. Since the cooling time is directly related to the entropy, (9.18) remains valid even if the shocked gas is compressed or expanded, provided only that the changes are adiabatic. We turn now to a consideration of what happens for $t \gtrsim t_c$, after radiative losses set in.

9.2.3 *Radiative shocks (J- and C-shocks)*

When radiative losses become important, the gas in the shock cools and compresses, maintaining approximate pressure balance. Because the compression in a radiative shock can be much greater than in a non-radiative shock, the magnetic pressure may become dominant; if so, the post-shock region is said to be magnetically supported. On the other hand, if thermal pressure is dominant, the region is said to be thermally supported. Strong radiative interstellar shocks generally become magnetically supported. (The possibility that shocks become temporarily supported by the pressure of shock-accelerated cosmic rays is discussed in Section 9.3 below). For J-shocks radiative losses become important at column densities

N behind the shock front comparable with $N_c = n_0 v_s t_c$. In radiative C-shocks radiative losses are important throughout the entire shock. Since radiation pressure is negligible for the conditions under consideration, these losses do not affect the mass and momentum jump conditions (9.3) and (9.5).

In the magnetically supported case, it is convenient to introduce a fiducial density n_m and field strength B_m such that the magnetic pressure $B_m^2/(8\pi)$ balances the dynamic pressure $\rho_0 v_s^2$ of the unshocked gas (HM). The mass density ρ and the density of hydrogen nuclei n are related by $\rho = n\mu_H$, where the mass per hydrogen nucleus μ_H is 2.3×10^{-24} g for cosmic abundances. Outside the shock front itself, (9.8b) implies $B \propto \rho \propto n$, so that

$$\frac{B_m^2}{8\pi} = \left(\frac{n_m}{n_0}\right)^2 \frac{B_0^2}{8\pi} = n_0 \mu_H v_s^2. \tag{9.19}$$

Numerically, this is

$$n_m = 76.7(n_0^{3/2} v_{s7}/B_{0-6}) \quad \text{cm}^{-3}, \tag{9.20}$$

where $B_{0-6} = B_0/(10^{-6}\,\text{G})$. To define a fiducial temperature T_m corresponding to n_m, write the pressure as $p = x_t nkT$, where $x_t = \mu_H/\mu$ is the number of particles per hydrogen nucleus; for example, $x_t = 2.3$ for a fully ionized cosmic gas. Then T_m is defined by

$$x_t n_m k T_m = n_0 \mu_H v_s^2, \tag{9.21}$$

or

$$T_m = 2.2 \times 10^4 \frac{B_{0-6} v_{s7}}{x_t n_0^{1/2}} \quad \text{K}. \tag{9.22}$$

The cooling shocked gas is thermally supported for $T \gg T_m$, magnetically supported for $T \lesssim T_m$. If we ignore any possible variation in x_t for simplicity, the momentum jump condition (9.5) has the approximate solution

$$\frac{n}{n_0} \approx \frac{16}{3} \frac{T_s}{T + T_m} \quad \text{(J-shocks)}. \tag{9.23}$$

This relation does not apply inside C-shocks because of the relative streaming between the neutrals and magnetic field; it is valid behind the shock front, where the gas is nearly in its final state. For J-shocks it is least accurate just behind the shock front, where it gives $n = \frac{16}{3} n_0$ rather than $4n_0$ for $T = T_s \gg T_m$. As the gas cools below T_s, it becomes increasingly accurate, and is exact in the limit $T \to 0$, when $n \to n_m$.

The final compression in the shock varies linearly with the Alfvén Mach number M_A in the magnetically supported case, but quadratically with M in the thermally supported case. Equation (9.19) gives

$$M_A \equiv \left(\frac{4\pi \rho_0 v_s^2}{B_0^2}\right)^{1/2} = \frac{1}{\sqrt{2}}\left(\frac{n_m}{n_0}\right) \tag{9.24}$$

in the magnetically supported case, whereas equations (9.14) and (9.23) give

$$M^2 \equiv \frac{\rho_0 v_s^2}{p_0} = \frac{16}{3}\frac{T_s}{T_0} = \frac{n}{n_0}\bigg|_{T=T_0} \tag{9.25}$$

in the thermally supported case ($T \gg T_m$) when T drops to T_0.

It is sometimes convenient to characterize the ambient magnetic field by a quantity proportional to the Alfvén velocity v_A; in particular, Mouschovias (1976) has shown that for self-gravitating clouds the typical field strength varies as n_0^κ, with $\frac{1}{3} \lesssim \kappa \lesssim \frac{1}{2}$, so that v_A is a weak function of the ambient density n_0. Define b by

$$B_0 \equiv 10^{-6} b n_0^{1/2} \quad \text{G}, \tag{9.26}$$

so that

$$b = v_A/(1.84 \text{ km s}^{-1}). \tag{9.27}$$

Typically b is of order unity in atomic and molecular clouds in the interstellar medium. In terms of b, the parameters governing shock compression are then

$$M_A = 54.2\frac{v_{s7}}{b}, \quad \frac{n_m}{n_0} = 76.7\frac{v_{s7}}{b}, \quad T_m = 2.2 \times 10^4 \frac{v_{s7} b}{x_t}. \tag{9.28}$$

The total amount of radiation emitted by a radiative shock is determined by the energy jump condition (9.10). The upstream and downstream radiative fluxes F are equal in magnitude and opposite in sign by symmetry and, since the incoming kinetic energy flux is almost completely converted to radiation in a radiative shock, we have

$$F = \frac{1}{4}n_0 \mu_H v_s^3 = 5.8 \times 10^{-4} n_0 v_{s7}^3 \quad \text{erg cm}^{-2} \text{ s}^{-1}. \tag{9.29}$$

In J-shocks, much of this is emitted in the uv, whereas in C-shocks it is primarily in the infrared. The effective temperature of this radiation – i.e., the temperature of a black body with this flux – is

$$T_{\text{eff}} = 1.79(n_0 v_{s7}^3)^{1/4} \quad \text{K}. \tag{9.30}$$

In the interstellar medium this is generally negligible; furthermore, the column density in the shock is small compared with the value N_{th} required to thermalize the shock emission. On the other hand, at the much higher densities encountered in protostellar accretion shocks, the radiation is thermalized and T_{eff} becomes large enough (≈ 2000 K) to destroy dust grains (Stahler, Shu & Taam 1981).

9.2.4   *Effects of cosmic rays on J-shocks*

It is now believed that shocks can efficiently accelerate relativistic particles, and that shocks associated with supernova remnants (SNRs) are responsible for the acceleration of cosmic rays in the Galaxy (Axford, Leer & Skadron 1977; Bell 1978a, b; Blandford & Ostriker 1978, 1980; Krimsky 1977). A shock forces the convergence of the upstream and downstream gas; cosmic rays are coupled to the gas by scattering in the turbulent magnetic fields in this gas and undergo first-order Fermi acceleration in the converging flow. This process is efficient only in ionized gas because the hydromagnetic waves which scatter the cosmic rays are damped by charge exchange if neutrals are present. This theory of particle acceleration is under active investigation, and has recently been reviewed by Axford (1981).

One of the central questions in the theory of shock acceleration, as yet unresolved, is the efficiency of the acceleration process: what fraction of the post-shock pressure is in relativistic particles? For strong shocks, much of the cosmic-ray energy is in the highest energy particles, where time-dependent effects are important, so there may be no simple answer to this question (Lagage & Cesarsky 1983). Cosmic rays are accelerated over a length scale much larger than the thickness of the shock front in the thermal gas, forming a C-type precursor. If the acceleration is sufficiently efficient, the J-shock in the gas could disappear entirely, resulting in a cosmic-ray mediated C-shock (Drury & Volk 1981); observation of sharp edges in the X-ray emission of young SNRs suggests that this extreme case does not occur in nature, however.

Consider a shock with $v_s > 100$ km s^{-1}, so that the gas is fully ionized (see below) and ion–neutral damping is negligible. For a strong shock the relative streaming between the cosmic rays and the gas at the Alfvén velocity is small. Then the effect of the cosmic rays on the shock jump conditions is determined by

$$w = p_{rs}/p_s, \tag{9.31}$$

the fraction of the post-shock pressure in relativistic particles (Chevalier 1983). Since roughly 10 percent of the energy injected by supernovae is needed to power the cosmic rays in the Galaxy, we expect $w \gtrsim 0.1$. The cosmic rays have negligible mass, so the mass jump condition (9.3) is unaffected. The momentum jump condition (9.5) is also unchanged if we interpret the pressure p as the total particle pressure, including that of the cosmic rays. The enthalpy appearing in the energy jump condition (9.10) is the sum of the gas enthalpy $\frac{5}{2}p_g$ and the cosmic-ray enthalpy; the cosmic rays, being relativistic, have $\gamma = \frac{4}{3}$ and an enthalpy $4p_r$. The effective specific heat ratio for the shock, denoted γ_s, is then

$$\frac{\gamma_s}{\gamma_s - 1} p_s = \frac{5}{2} p_{gs} + 4 p_{rs}, \tag{9.32}$$

so that in terms of w,

$$\gamma_s = \frac{5 + 3w}{3(1 + w)} \tag{9.33}$$

(Chevalier 1983). Note that γ_s goes smoothly from $\frac{5}{3}$ to $\frac{4}{3}$ as w goes from 0 to 1. The post-shock values of the density, etc., are given by (9.12)–(9.14) with γ replaced by γ_s. Since $\gamma_s < \frac{5}{3}$ for $w > 0$, the compression in a strong shock will exceed 4. In the presence of cosmic-ray acceleration, the post-shock temperature is reduced by a factor

$$\frac{T_s(w)}{T_s(0)} = \frac{1 - w^2}{(1 + 3w/4)^2}, \tag{9.34}$$

which is $(0.86, 0.61, 0.40)$ for $w = (0.1, 0.3, 0.5)$, respectively.

In the post-shock relaxation layer, the ratio of the cosmic-ray pressure to the gas pressure will change because the radiation drains energy only out of the gas and because the cosmic rays have a different ratio of specific heats. If the cosmic rays are tightly coupled to the gas, then they act like the magnetic field in limiting the post-shock compression in radiative shocks. The cosmic-ray pressure behind the shock is $p_r = p_{rs}(\rho/\rho_s)^{4/3}$, and the maximum allowable density n_{cr} corresponds to $p_r = \rho_0 v_s^2$, or

$$\frac{n_{cr}}{n_0} \approx \frac{(4 + 3w)^{7/4}}{[3w(1 + w)]^{3/4}} - \frac{3}{4}, \tag{9.35}$$

which is $(28.7, 13.6, 10.0)$ for $w = (0.1, 0.3, 0.5)$, respectively. Since we are considering fast shocks with $v_{s7} > 1$, this condition is often more stringent than that imposed by the magnetic field (9.20). However, in contrast to the magnetic field, the cosmic rays can escape from the compressed gas at a velocity v_A while the gas is ionized, and at a higher velocity when the gas recombines; hence the cosmic rays only delay the final, magnetically limited compression.

9.3   J-shocks
9.3.1   *Structure of the shock front*
The shock jump conditions determine the effects of the shock without reference to the complex dissipative processes occurring in the shock front. If the dissipation is due to particle collisions, then the thickness

of the shock front is of order $(n_s\sigma)^{-1}$, where n_s is the post-shock number density and σ is the collision cross section. In many astrophysical plasmas, however, this length is much greater than the lengths characterizing the collisionless behavior of the plasma, such as the ion Larmor radius $r_{Li} = 1.0 \times 10^9 (v_7/B_{-6})$ cm (where v is perpendicular to B) and the ion inertial length $c/\omega_{pi} = (v_A/v)r_{Li} = 2.3 \times 10^7 n_e^{-1/2}$ cm, where ω_{pi} is the ion plasma frequency. In this case it is possible for the dissipation to be due to turbulent E and B fields generated by the unstable plasma in the shock front, and the shock is said to be collisionless.

The interaction between the solar wind and the Earth's magnetosphere produces a bow shock with a thickness of order hundreds of kilometers, far smaller than the mean free path, which is of order 1 AU. The Earth's bow shock is an ideal laboratory for studying collisionless shocks at moderate Mach number $M_A \lesssim 10$ (Greenstadt & Fredricks 1979). The structure of the bow shock depends sensitively on θ_B, the angle between the magnetic field direction and the shock normal. For 'quasi-perpendicular' shocks ($\theta_B \gtrsim 45°$) the shock front has a thickness of a few times c/ω_{pi}, and the magnetic field overshoots the final downstream value by a factor of order 2. Theoretical calculations (e.g., Leroy *et al.* 1982) are in reasonably good agreement with the observations and demonstrate that the dissipation is due to the effects of ions which are reflected at the shock and are then magnetically deflected back downstream. 'Quasi-parallel' shocks (those with $\theta_B \lesssim 45°$) are much broader, and are less well understood. There is little observational knowledge or theoretical understanding of high Mach number shocks ($M_A \gg 10$): the Earth's bow shock is generally confined to $M_A \lesssim 10$, and the calculations of Leroy *et al.* break down for $M_A \gtrsim 12$. Bow shocks around the outer planets can be stronger than that around the Earth, reaching $M_A = 21$ in one observation of Saturn's bow shock (Scarf, Gurnett & Kurth 1981). Observations of young SNRs such as Cas A and Tycho indicate that collisionless shocks can exist at much higher Mach numbers, $M_A > 100$.

From the point of view of astrophysical spectroscopy, the most important unresolved questions about the structure of the shock front are: Does the acceleration of relativistic particles at the shock significantly alter the compression and heating of the thermal plasma? (This issue has been discussed in Section 9.2.4 above.) What is the electron/ion temperature ratio behind the shock front? The answer to this second question is crucial since the emission spectrum of a shock in an ionized gas is determined primarily by the electron temperature T_e. In a collisional shock front, collisions enforce $T_e = T_i$, but there is no such requirement for a collisionless shock. Two extremes may be envisioned: the turbulent electric fields in the collisionless shock front heat the electrons to approximate equipartition with the ions (McKee 1974); or, collisionless heating is negligible so that $T_e \ll T_i$ just behind the shock front. In the latter case equipartition is effected by collisions in a time $t_{eq} = 8.4\, T_e^{3/2}/n_e$ s (Spitzer 1978). Observations of electron heating at the Earth's bow shock, which is of moderate Mach number, are generally consistent with simple adiabatic compression of the electrons. Bow shocks around the outer planets have higher Mach number and show definite evidence for collisionless heating of the electrons. For very high Mach number shocks, McKee & Hollenbach (1980) used the data of Pravdo & Smith (1979) to argue that electrons in the young SNRs Cas A and Tycho have been collisionlessly heated to $T_e > 0.2\, T_s$. Thus the observational evidence suggests that collisionless heating of electrons is significant in high Mach number shocks, but the post-shock value of T_e/T_i remains unknown.

9.3.2 *Radiative J-shocks: maximum velocity*

Interstellar shocks are J-type if they occur in ionized gas, neutral atomic gas, or at velocities above about 50 km s^{-1} in molecular gas. When J-shocks become radiative, they emit primarily in the optical and uv, providing a beacon for astronomers. Most optical identifications of SNRs have been made on the basis of observations of radiative J-shocks, and analysis of these spectra allows a determination of the energetics of the SNR, the physical conditions in the ambient circumstellar and interstellar matter, and the elemental abundances in this matter. Radiative J-shocks have also been observed in HH objects (Schwartz 1983) and in H II regions (Lasker 1977). Radiative J-shocks may occur in molecular clouds, where they provide a signature for energetic outflows from young stellar objects and for embedded SNRs, but since there are no detailed observations of such shocks at present we shall not discuss them here (see McKee, Chernoff & Hollenbach 1984).

J-shocks can become radiative only if the age of the shock exceeds the cooling time. Rewriting (9.18) in terms of the shock velocity in an ionized gas (9.15) gives

$$t_c = 2.5 \times 10^{10}\, v_{s7}^3/n_0 \quad \text{s.} \tag{9.36}$$

For a steady shock, this corresponds to a column density

$$N_c = n_0 v_s t_c = 2.5 \times 10^{17}\, v_{s7}^4 \quad \text{cm}^{-2}. \tag{9.37}$$

Since the cooling rate drops dramatically for $T < 10^4$ K, this is an estimate for the column density from the shock front to the point at which $T \approx 10^4$ K. Comparison with detailed numerical calculations of shock structure (Raymond 1979, Shull & McKee 1979) shows that this estimate is accurate to within a factor 2 for $0.6 \lesssim v_{s7} \lesssim 2$; it should remain approximately valid up to $v_{s7} \approx 10$.

This result allows one to determine the maximum velocity for a steady radiative J-shock. First, consider the case of two clouds which collide. After impact, shocks are driven into each cloud; the shocks will become radiative if the cloud column densities exceed N_c, corresponding to a shock velocity v_{s7} less than $(N/(2.5 \times 10^{17})\,\mathrm{cm}^{-2})^{1/4}$. Next consider a cloud engulfed by an SNR blast wave. The age of an SNR expanding into an intercloud medium of density n_0 is related to its expansion velocity v_s as

$$t = 4.46 \times 10^{12}(E_{51}/n_0)^{1/3}v_{s7}^{-5/3}\ \mathrm{s}, \tag{9.38}$$

where E_{51} is the energy of the SNR in units of 10^{51} erg (e.g., McKee & Hollenbach 1980). The shock driven into the cloud is in approximate pressure equilibrium with the SNR blast wave, so that $n_0 v_s^2 = n_{co} v_{cs}^2$, where n_{co} and v_{cs} are the initial cloud density and the cloud shock velocity, respectively. In order for the cloud shock to become radiative, its cooling time must be less than the time for the pressure on the cloud to change significantly, which can be shown to be about $t/4$; with the aid of equations (9.36) and (9.38), this criterion becomes

$$v_{cs7} < 2.2(n_0^2 E_{51})^{1/4}(n_{co}/n_0)^{1/28}. \tag{9.39}$$

This condition on the cloud shock velocity depends only weakly on the SNR energy E and the ambient density n_0, and it is nearly independent of the cloud density n_{co}.

9.3.3   *Radiative J-shocks: structure and spectrum*

The earliest studies of the structure of interstellar J-shocks were made by Pikel'ner (1954) and by Field *et al.* (1968). Cox's (1972) pioneering study of the structure of fast ($v_{s7} \approx 1$) shocks in ionized gas revealed that the emission spectrum of such a shock is dominated by two distinct regions: a hot, collisionally ionized region just behind the shock front and a cooler ($T \approx 5000$ K), photoionized region farther downstream. The hot region produces a flux of ionizing photons $\Phi n^0 v_s$, where $\Phi \approx 1$ for $v_{s7} \approx 1$. The effect of the ionizing radiation on the cooler gas downstream can be characterized by the ionization parameter Γ (sometimes denoted U), the ratio of the ionizing photon density to the hydrogen density n. If the recombination region is magnetically supported (9.28) and if scattering of the ionizing radiation is neglected, then one finds

$$\Gamma = \frac{\Phi n_0 v_s}{nc} = 4 \times 10^{-6}\Phi b. \tag{9.40}$$

This is typically of order 10^{-5}, much lower than the values inferred for H II regions or active galactic nuclei. Hence, the signature of a radiative J-shock in an ionized gas is the combination of a *hot, collisionally ionized region*, characterized by high values of temperature-sensitive ratios such as [O III] $\lambda4363/([$O III$]\ \lambda4959+\lambda5007)$, and a *low excitation, photoionized recombination region*, characterized by strong [O I] and [S II] lines.

Extensive grids of shock models have been calculated by Raymond (1979), Dopita (1976, 1977), Shull & McKee (1979) and Dopita *et al.* (1984). Shull and McKee showed that the ionizing radiation from the hot, collisionally ionized region generates an ionizing precursor which self-consistently determines the ionization of the gas entering the shock front. Complete pre-ionization of the H I and He I is attained for $v_{s7} > 1.1$. (At sufficiently high shock velocities, where the shock is non-radiative, the ionizing photon flux from the shocked gas drops below the value needed for full pre-ionization; McKee & Hollenbach (1980) estimate this occurs for $v_{s7} > 4.1(n_0^2 E_{51})^{0.07}$.) A number of charge exchange reactions were included by Shull & McKee (1979) and, with a more complete set of reaction rates, Butler & Raymond (1980) showed that charge exchange reactions could significantly alter the strengths of a number of lines, especially low excitation ones. Dopita *et al.* (1984) have demonstrated that emission line intensities from all but the highest stages of ionization depend only weakly on the shock velocity for $v_{s7} > 1$, so that spectroscopy of such shocks provides a powerful probe of abundances in the interstellar medium.

The gas phase abundances of many elements in the interstellar medium is less than that in the Sun, and this depletion is attributed to the elements being locked up in interstellar grains (Black 1984). These grains are subject to destruction in interstellar shocks by sputtering and by grain–grain collisions (Cowie 1978; Shull 1978; Draine & Salpeter 1979), and observations show significantly less depletion in high-velocity gas than in low-velocity gas (Barlow & Silk 1977). The rate at which grains are destroyed in shocks is enhanced by the betatron acceleration they undergo in the compressing magnetic field behind the shock front (Spitzer 1976). The destruction is most effective for the largest grains, which are accelerated to the highest velocities because they are least affected by plasma drag. A 100 km s^{-1} shock destroys about half the grains of radius $> 10^{-5}$ cm, while leaving those of radius $< 10^{-5.5}$ cm virtually unaffected (Seab & Shull 1983). Much of the grain destruction occurs in the transition between the hot, collisionally ionized region of the shock and the recombination region, so that these regions may have different gas-phase abundances. In principle, detailed observations should allow one to infer both the gas-phase abundances in the pre-shock medium and the total abundances of the elements.

Table 9.1 gives preliminary results for shock emission spectra allowing for grain destruction (Shull, Seab & McKee: personal communication). Initially the depletion factors (ratios of gas phase to total abundance, taken as solar) were $\delta(C) = 0.2$, $\delta(O) = 0.75$, $\delta(Si) = 0.03$, and $\delta(Fe) = 0.02$. Seab & Shull (1983) describe the methods used to calculate grain destruction in the shock; the final depletions are listed in the table. The assumption of a steady shock is satisfied only if the shocked column density exceeds the values listed: for lines emitted at $T > 10^4$ K, a column $N(10^4$ K) is sufficient, whereas for the infrared [O I], [Si II], and [Fe II] lines, the much larger $N(500$ K) is required. For comparison with the theoretical calculations, Table 9.1 also gives the line strengths observed at Miller's (1974) position 1 in the northeast part of the Cygnus Loop SNR. Overall, the agreement between the observations and a shock at 100 km s^{-1} is reasonably good. The agreement could be improved by allowing for several effects: (1) The uv data of Benvenuti, Dopita & D'Odorico (1980) do not include any correction for interstellar extinction because they argued it was negligible, but optical data (Fesen, Blair & Kirshner 1982) suggest otherwise. (2) The observations probably cover a region of the SNR containing a range of shock velocities (Benvenuti *et al.* 1980). (3) The shock is optically thick in resonance lines such as C IV $\lambda 1549$, so that the intensity ratio of a resonance line to a nonresonance line decreases as the angle between the line of sight and the shock normal increases (Raymond *et al.* 1980). (4) The shock is probably not old enough to have a complete recombination layer, so the observed intensities of lines such as the [S II] doublets fall below the theoretical prediction for steady shocks (Raymond *et al.* 1980, Dopita & Binette 1983).

Extensive spectroscopic studies of the Cygnus Loop have shown that a range of shock velocities and shock ages are present in different regions of the SNR (Fesen *et al.* 1982, Hester, Parker & Dufour 1983). High spatial resolution observations with simultaneous optical and uv coverage, such as will be possible with the Space Telescope, should enable one to isolate individual shocks and overcome the difficulties cited above.

There are several active galactic nuclei which exhibit shock-like spectra, and, in view of the violent events known to be occurring there, it is reasonable to suppose that shocks are indeed present. However, it has recently been shown that a low-excitation, power-law photoionization spectrum mimics a shock spectrum, the only difference being the absence of the hot, collisionally ionized region in the photoionized gas (Halpern & Steiner 1983). Furthermore, the high densities encountered in active galactic nuclei can populate the lower level of the [O III] $\lambda 4363$ line, strengthening the line and eliminating it as a temperature diagnostic (Filippenko &

Table 9.1. *Emission lines from J-shocks in atomic gas*[a]

		Models				Cyg Loop[b]
		30	60	100	130	
v_s	(km/s)	30	60	100	130	
$N(T = 10^4$ K)	(cm^{-2})	7.9(17)	6.3(16)	4.0(17)	6.2(17)	
$N(T = 500$ K)	(cm^{-2})	2.1(19)	2.2(18)	6.6(18)	8.9(18)	
n_m	(cm^{-3})	6.6(2)	1.4(3)	2.3(3)	3.1(3)	
δ(C)		0.21	0.24	0.28	0.28	
δ(O)		0.76	0.85	0.88	0.87	
δ(Si) ≈ δ(Fe)		0.08	0.41	0.54	0.49	
I(Hβ) (erg/cm² s sr)		7.9 (−8)	1.4 (−6)	7.1 (−6)	1.4 (−5)	
Lα λ1216		2.3(4)	1.0(4)	4.6(3)	4.0(3)	
Hα λ6563		456	364	306	299	308
Hβ λ4861		100	100	100	100	100
C II λ2326		67	13	115	67	62
C III] λ1909		—	0.36	188	112	188
C IV λ1549		—	—	117	283	86
[O I] λ6300 + 6363		685	86	102	150	
[O I] λ63 μm		2800	98	126	126	
[O II] λ3726 + 3729		52	121	960	871	1300
O III] λ1663		—	—	142	217	62
[O III] λ4959 + 5007		—	—	268	324	440
[O III] λ4363		—	—	19	26	25
O IV] λ1403		—	—	7	60	
[Ne II] λ12.8 μm		—	5	134	135	
[Si II] λ34.8 μm		734	368	428	315	
Si IV] λ1397		—	—	26	111	60
[S II] λ6716 + 6731		1200	209	420	508	284
[S III] λ9069 + 9532		—	6.8	84	101	
[Fe II] λ1.28 μm		132	97	270	297	

[a] Shull, Seab & McKee (1986). Assumes $n_0 = 10$ cm^{-3}, $B_0 = 10^{-6}$ G. Notation: 1.0(3) means 1.0×10^3. Line intensities are relative to Hβ.

[b] Optical lines from Miller (1974), position 1; uv lines from Benvenuti et al. (1980) at the same position.

Halpern 1984). Hence, at present there is no definitive spectroscopic evidence for shocked regions in active galactic nuclei.

9.3.4   *Emission from non-radiative J-shocks*

Shocks with velocities in excess of $200 \, \text{km s}^{-1}$ are generally non-radiative in that radiative losses are not important in determining the temperature and density structure of the shock. Such shocks produce the X-ray emission from SNRs, much of which is in emission lines of highly ionized atoms (see Chapter 10 of this volume by McCray). Here we shall focus on the optical emission from non-radiative shocks.

Most of the emission from radiative shocks comes from an 'over-ionized' plasma: the recombination tends to lag the cooling of the shocked gas, so that the ionization state of the plasma corresponds to a temperature higher than the gas temperature. Just behind the shock front, however, the gas is underionized because it takes a finite amount of time for the ionization of the shocked gas to reach the level corresponding to the post-shock temperature. In a non-radiative shock, it is the emission from this underionized plasma which produces the observed spectrum.

An extreme case of an underionized plasma occurs when a fast non-radiative shock advances into a partially neutral gas (Chevalier & Raymond 1978, Chevalier, Kirshner & Raymond 1980). The spectrum of such a shock is dominated by emission in the Balmer lines, as seen in Tycho's SNR and in several other SNRs. The shock front propagates in the ionized gas, leaving the neutral gas behind. The 'slow' neutrals are subject to charge exchange with the 'fast' shocked ions, forming a population of fast neutrals. Both populations of neutrals may be excited several times before being ionized. As a result, the line profile exhibits a characteristic two-component structure: a broad component with a line width of order the post-shock thermal velocity, and a narrow component with a width characteristic of the unshocked gas.

Since the emission from a non-radiative shock comes from the vicinity of the shock front, observations of such shocks have the potential of answering the basic questions about shock front structure raised in Section 9.3.1 above. Raymond *et al.* (1983) have used optical and uv observations of a Balmer line filament in the Cygnus Loop to argue that the electron temperature T_{es} just behind the shock front is substantially less than the ion temperature there. The lower electron temperature in this case compared with that in the equipartition case leads to more copious production of N V and Ne V, in accord with observation. Uncertainties in the modeling prevent setting a strict upper limit on T_{es}/T_{is}, and in fact equipartition is not definitively ruled out, but it is clear that this method holds great promise for answering a question of fundamental importance. Raymond *et al.* (1983) also show how observations of non-radiative shocks can be used to measure the fraction of the shock energy going into relativistic particles: for a face-on shock, the broad Balmer lines would show a shift of at least $\frac{3}{4}v_s$, corresponding to a compression of at least 4, but a line width reduced (by a factor $(T_s(w)/T_s(0))^{1/2}$) from that expected in the absence of relativistic particles. However, the interpretation of the results would be complicated by the fact that the neutral H atoms producing the emission would tend to damp the hydromagnetic waves which couple the cosmic rays to the gas and allow the Fermi acceleration to occur. The study of the structure of non-radiative shocks has only just begun because the emission features are often faint, but they offer great promise. With the advent of Space Telescope it will be possible to study their structure using absorption line spectroscopy as well (Cowie *et al.* 1979).

9.4   C-shocks

Shocks in weakly ionized, magnetized, molecular gas with $v_s \lesssim 50$ km s^{-1} are C-shocks rather than J-shocks. The hydrodynamic variables change continuously over a length scale determined by the mean free path for a neutral to hit a charged particle, either an ion or a grain, which is much greater than the neutral–neutral mean free path because of the low ionization. (We shall not consider the possibility of cosmic-ray mediated C-shocks in this section – see 9.2.4.) Since a jump in these variables occurs when there is a transition from supersonic to subsonic flow, the essential condition for a pure C-shock is that the neutral gas remain supersonic ($v_n^2 > \gamma c_{sn}^2$, where v_n is the flow velocity of the neutrals and c_{sn} their isothermal sound velocity) throughout the shock. At first sight, this seemingly contradicts the fundamental requirements that the flow behind a shock front be subsonic relative to the front; however, it is the large Alfvén velocity associated with the compressed field, rather than the thermal velocity of the neutral gas, which renders the signal velocity large enough to satisfy this requirement. In order for the neutral gas to remain supersonic, it must remain cool (typically $T_n \lesssim 5000$ K). Hence strong C-shocks must have radiative shock fronts. Since an atomic gas of cosmic abundance is inefficient at cooling at these temperatures, strong C-shocks must occur in molecular gas. Hence C-shocks are non-dissociative; similarly, they must be non-ionizing. Because C-shocks are cool throughout, they emit primarily infrared molecular emission lines. Such lines are readily observed in highly obscured regions such as regions of active star formation and provide a wealth of information on the physical condition there.

Historically, shocks with C-type precursors were first analyzed by Mullan

(1971). He ignored radiative losses, which is approximately valid for the atomic shocks he considered, but this prevented him from finding strong C-shocks. Weak C-shocks ($M_A \leqslant 2.5$) can exist even in the absence of radiation and he did find these. Draine (1980) introduced the distinction between C- and J-shocks and laid the foundation for the study of the structure of C-shocks. Draine, Roberge & Dalgarno (1982) presented an extensive study of the properties of molecular C-shocks assuming constant fractional ionization. As discussed below, models of C-shocks are in good accord with observations of intense molecular emission lines from the BN-KL region of Orion (Draine & Roberge 1982; Chernoff, Hollenbach & McKee 1982, who included ionization chemistry). Infrared emission from C-shocks has been reviewed by McKee, Chernoff & Hollenbach (1984).

9.4.1   *Physical conditions in C-shocks*

When a C-shock approaches, the ions are the first to take notice: they compress to $M_A\sqrt{2}$ times their original density (if no chemistry occurs) and, in the shock frame, decrease in velocity from v_s to $v_s/M_A\sqrt{2}$. For M_A larger than a few and for low ionization levels, the neutrals have no time to react and instead slip with respect to the field and the charged particles, which are tightly coupled. The maximum slip, or drift velocity, v_d, occurs at the beginning of the shock; there $v_d = v_s(1 - 1/M_A\sqrt{2})$. Throughout the rest of the shock, the neutrals are gradually decelerated by collisions with charged particles.

The length scale l_B over which the magnetic field and charged particles slip through the neutrals is determined by balancing the magnetic force on the ions with the collisional drag due to the neutrals (Spitzer 1978)

$$B^2/8\pi l_B = n_i\rho(v_n - v_c)\langle\sigma v\rangle. \tag{9.41}$$

Since the magnetic pressure gradient drives the charged particles in the upstream direction, the neutral drag on the charged particles, which varies as $v_n - v_c$, must be in the downstream direction. Hence, the charged particles must be decelerated more than the neutrals: the charged particles and magnetic fields are compressed *before* the neutrals. This can be visualized as the compressed field behind the shock propagating ahead as a damped hydromagnetic wave. During the compression, n_i/B remains constant, so, in terms of the ion–neutral mean free path, $\lambda_{in} = 1/n_0\sigma_{in}$, the length scale l_B becomes

$$l_B = \lambda_{in}(B/B_0)/(M_A^2 x_{i0}), \tag{9.42}$$

where $x_{i0} = n_{i0}/n_0$ is the initial fractional ionization, and where we have set

$\langle \sigma_{in} v \rangle v \approx \sigma_{in} v_s^2/2$. Note that

$$M_A^2 x_{i0} = 4\pi \rho_{i0} v_s^2/B_0^2 \equiv M_{Ai}^2 \tag{9.43}$$

is the square of the Alfvén Mach number relative to the ionized gas.

When M_{Ai} exceeds unity, (9.42) shows that the field must initially increase on a scale less than the ion–neutral mean free path: a collisionless J-shock occurs in the ionized components of the gas. Physically, such a jump must occur because signals cannot propagate upstream faster than the ion Alfvén velocity. The field must be fairly weak in order to have $M_{Ai} > 1$; (9.28) gives $b < 0.017 \, v_{s7}(x_{i0})^{1/2}$. Chernoff (1985) has discussed shocks with $M_{Ai} > 1$.

The temperature of each species may be estimated by balancing the most important heating and cooling processes for each species. Ions are heated and cooled primarily by elastic collisions with neutrals and reach temperatures of

$$T_i = m_n v_d^2/3k, \tag{9.44}$$

where m_n is the mean mass of a neutral. The electron temperature is lower than the ion temperature since the small quantity (m_e/m_n) enters in the rate of elastic heating and cooling, but not in the rate of cooling by excitation of vibrational and rotational transitions in H_2. Collisionless plasma instabilities contribute to the heating also, so that the electrons typically reach temperatures of $T_e \approx (0.1–0.2)T_i$. At very high temperatures, they cool also by dissociation and ionization of H_2. Grains in the shock waves are heated primarily by collisions with neutrals and are cooled by emission of far-infrared photons; typically they equilibrate at $T_{gr} < 100$ K.

The situation for heating neutrals is more complicated. Far upstream of the shock, only ions significantly heat the neutrals, but as the flow approaches the strong downstream magnetic field, heavier charged particles (grains) are slowed by the field and contribute to the heating. Grains dominate the heating if the plasma is only slightly ionized, $x_i < 3 \times 10^{-7}$, and if the electrons are sufficiently hot to charge them ($v_d > 3$–5 km s^{-1}); otherwise, ions are more important. The neutral temperature may be estimated, if charged grains are the dominant heat source, by balancing heating and cooling:

$$n_{gr} n \sigma_{gr} \mu_H |v_n - v_{gr}|^3 = n^2 \Lambda, \tag{9.45}$$

where v_{gr} is the grain velocity, including gyration in the magnetic field. The maximum temperature T_{max} occurs near the beginning of the shock where the drift velocity v_d multiplied by the grain density is a maximum. As T_{max} and v_d increase, the ionization of the gas can increase. When it rises above 3×10^{-7}, the temperature is determined by ion–neutral scattering:

$$x_i \langle \sigma v \rangle_{in} m_n v_d^2 = \Lambda. \tag{9.46}$$

The cooling of the gas in a C-shock depends on the chemical composition of the predominantly neutral gas which, in turn, depends on the gas temperature. Hollenbach (1982), Hollenbach & McKee (1979), and McKee & Hollenbach (1980) have summarized the chemical reactions which are important in C-shocks. The neutral chemistry is determined by the ratio $x_2 = n(H_2)/n$. For $x_2 \approx \frac{1}{2}$ (molecular gas) and for an oxygen abundance exceeding that of carbon, all carbon becomes CO, and the left-over oxygen becomes H_2O, nitrogen becomes NH_3, sulfur becomes H_2S, OCS, and SO, and Si goes to SiO. For $x_2 \ll 1$, chemical reactions destroy the remaining molecules and the gas becomes atomic; in this case the shock would become J-type. For $x_2 \approx \frac{1}{4}$ a significant fraction of O is in the form of OH. We note that fast ($v_s \approx 50$ km s^{-1}) C-shocks incident upon molecular gas can produce quite intense OH infrared emission because of the dissociation of a fraction of the H_2 inside the shock wave. The ion chemistry, which is unimportant to the cooling of a C-shock but is important to the heating, is dominated by the collisional dissociation of molecular gas ions by H and H_2 for observable C-shocks ($v_s > 20$ km s^{-1}). The most abundant ion in these shocks is therefore H^+.

In C-shocks the primary coolants are O I, H_2, CO, OH, and H_2O. The column density of warm gas increases with density because collisional de-excitation at high densities reduces the cooling efficiency. The results are only weakly dependent on the pre-shock abundances of $x(O)$, $x(CO)$, $x(H_2O)$, $x(OH)$, etc., since shock chemistry rapidly equilibrates the initial distribution and since the total cooling at low temperatures and high densities depends primarily on the density of dipolar molecules.

As discussed above, a necessary condition for strong C-shocks is $v_n^2 > \gamma c_{sn}^2$. Since the characteristic length scale for coupling of the field to neutrals varies as the inverse of the ionization x_i, at low ionization levels small increases in x_i can drive the heating length below the cooling length, increase $c_{sn}^2 \propto T_n$, and possibly violate this condition, thereby producing a J-shock. There are a number of ionization mechanisms which become energetically feasible as the drift velocity increases. It is possible for hot, drifting ions and electrons to ionize the neutrals. Drifting grains reflect neutrals, thereby producing a high-velocity population of neutrals which then ionize other neutrals. Finally, hot neutrals can ionize other neutrals. The considerable uncertainties in the forms of the ionization cross sections near threshold are ameliorated, in part, by a sudden avalanching effect: as the gas heats, it ionizes itself and destroys molecular coolants, both of which cause it to heat even more quickly. The resulting ionization breakdown occurs between about 40 and 50 km s^{-1} for $n_0 \approx 10^6$ cm^{-3} and a range of values of B, and at 10–20 percent higher velocity for $n_0 \approx 10^4$ cm^{-3} (Chernoff *et al.* 1982; McKee *et al.* 1984).

9.4.2 *Application to Orion*

Computer modeling of a shock produces a 'fingerprint' which can be compared with observations to unravel the physical parameters. The input for a model consists of the parameters n_0, v_s, B, the grain opacity, $n_{gr}\sigma_{gr}$, the mean grain size, and the gas-phase abundances of atoms and molecules. Before comparing the resulting 'fingerprint' with observations, one must allow for extinction by dust and geometrical effects. In order to match observations with a model one would like to employ quantities which are sensitive to only a few of these parameters and insensitive to the rest. Thus, for example, far-infrared lines are insensitive to extinction, H_2 lines and line ratios are insensitive to elemental abundances, and line ratios are insensitive to geometrical effects. Lines of H_2, CO, and H_2O dominate the cooling and are thus best for inferring the total intensity of the emission and hence $n_0 v_s^3$ (see (9.29)). The 2-μm lines of H_2 are a thermometer for the shock, whereas the high J lines of CO allow one to infer both temperature and density. The 12-μm 0–0 S(2) line of H_2 measures the shocked column density.

Just as SNRs are the primary sites for the study of J-shocks, so star-forming regions in molecular clouds offer sites for studying C-shocks. Supersonic gas motions, hot molecular gas, and large-scale outflows have all been observed in the BN-KL region of Orion. The outflows center on IRc2, a luminous embedded protostar. Within a projected distance of 2×10^{17} cm from IRc2, a number of infrared H_2 lines indicating gas temperatures of ≈ 2000 K are seen (Nadeau & Geballe 1979; Nadeau, Geballe & Neugebauer 1982; Scoville *et al.* 1982). The emission and the line profiles which show supersonic wings extending to ± 100 km s^{-1} are highly suggestive of shocks. In addition, high J rotational transitions in CO (Goldsmith *et al.* 1981; Storey *et al.* 1981; Watson 1982) and the ground state OH rotational transition (Storey, Watson & Townes 1981) have been seen.

An instructive example of shock modeling comes from comparing two theoretical models of the BN–KL region of Orion which 'match' the observations yet arrive at somewhat different values for the fundamental physical parameters n_0, b, $x(CO)$, and A_v. The results of Draine & Roberge (1982) and Chernoff *et al.* (1982) are: $n_0 = (7 \times 10^5, 2 \times 10^5)$ cm^{-3}, $B = (1.5, 0.45)$ mG, $v_s = (38, 36)$ km s^{-1}, $x(CO) = (7 \times 10^{-5}, 3 \times 10^{-4})$, and $A_v = (4.0,$

2.5) mag at 2 μm, respectively. A comparison of Chernoff *et al.*'s model to the data appears in Table 9.2.

Draine and Roberge attempted to match the H_2 lines emitted at Peak 1, north of BN. Here, Knacke & Young (1981) observed large intensities of high pure rotational lines of H_2 (S(9)–S(15)), and Draine and Roberge infer high values of n_0 to insure LTE populations at 2000 K. Chernoff *et al.* averaged all intensities over the entire shock region since the CO beam does not resolve the region and they therefore ignored these high J H_2 transitions, which are localized and incompletely mapped. The high values of n_0 used by Draine and Roberge increase the CO 30–29/21–20 ratio to unacceptably high levels; however, they argue that the excitation cross sections for high J transitions in CO are incorrect. Recent calculations by Schinke *et al.* (1985) in fact support this contention. By ignoring the H_2 emission from the higher rotational levels at Peak 1, Chernoff *et al.* are able to match the CO observations with a lower density ($n_0 \approx 2 \times 10^5$ cm^{-3}) and the standard CO excitation rates (McKee *et al.* 1982).

Since the shock intensity varies as $n_0 v_s^3$ and both models have a similar value of v_s, the higher value of n_0 used by Draine and Roberge means that they must assume larger values of A_v in order that the 2-μm H_2 absolute intensities match those observed. Recent observations (Scoville *et al.* 1982; Beck & Beckwith 1983) support the lower extinctions found by Chernoff *et al.* In addition, Draine and Roberge must assume more extinction at 12 μm than do Chernoff *et al.*; the latter were driven to higher values of the CO abundance because of their lower densities. The CO lines are unattenuated by grains and therefore the absolute intensities of CO relative to extinction-corrected H_2 emission give a measure of the CO abundance. The abundance found by Chernoff *et al.* is greater than that observed elsewhere in the ISM.

A noteworthy feature of both shock models is the low value of $(n_{gr}\sigma_{gr}/n)_0 \approx 10^{-22}$ cm^2 required to produce viable shock models. Although grain mantles are destroyed in these C-shocks, this value of grain area per hydrogen atom is nearly an order of magnitude lower than that appropriate to diffuse clouds, where grains may also be free from mantles. These low values of $(n_{gr}\sigma_{gr}/n)_0$ were required in order to spread the heating over a larger column density, thereby keeping T_{max} at an acceptable level. Coagulation of grains in molecular clouds could lead to this reduction in grain surface area.

Overall, the differences between the models are relatively minor, and the models are remarkably successful in accounting for the strengths of the large number of molecular lines which have been observed in Orion. Both models require an ambient magnetic field of order a milligauss, large enough to be dynamically significant. Both predict large column densities of high-velocity

Table 9.2. *Comparison of predicted and observed line intensities for BN–KL*

Line	Intensity[a] or line ratio		Refs.	Notes
	Predicted	Observed		
H_2				
1–0 S(1)	1.0 (-1)	1.0 (-1)	1	
0–0 S(2)	6.4 (-3)	5.0 (-3)	2	
2–1 S(1)/1–0 S(1)	1/12	1/12	3	
3–2 S(3)/1–0 S(1)	1/200	1/64	4	d
CO				
6–5	1.0 (-3)	1.0 (-3)	5	
21–20	9.0 (-3)	1.0 (-2)	6	
27–26	4.0 (-3)	6.5 (-3)	6	
30–29	2.3 (-3)	2.4 (-3)	6	
34–33	9.8 (-4)	6.0 (-4)	7	
OH				
$^2\pi_{3/2}(5/2–3/2)$	3.7 (-3)	4.8 (-3)	8	
$^2\pi_{3/2}(7/2–5/2)$	2.2 (-3)	—		
H_2O				
$9_{55}–9_{46}$	$\approx 10^{-3}$	—		
$10_{56}–10_{47}$	$\approx 10^{-3}$	—		
$11_{57}–11_{48}$	$\approx 10^{-3}$	—		
OI				
63 μm	1 (-4)	3.3 (-2)	9	e
HI				
21 cm	4(20)	—		f

Notes:

[a] In $erg\,cm^{-2}\,s^{-1}\,sr^{-1}$.

[b] Fit to the observed average surface brightness and line ratios of OMC-1 with $n_0 = 2 \times 10^{-5}\,cm^{-3}$, $v_s = 36\,km\,s^{-1}$, and $B_0 = 4.5 \times 10^{-4}\,G$.

[c] Surface brightness averaged over a 1' beam, corrected for 2 μm and 12 μm extinction of 2.5 and 0.94 mag, respectively. The 1–0 and 2–1 S(1) intensities have been adjusted upward by an additional factor of 2 to allow for the more heavily attenuated blue component.

[d] Observed value at peak 1 only.

[e] Oxygen intensity depends sensitively on preshock O abundance, assumed to be 2.4×10^{-4}. The observed emission has been attributed to a J-shock by Werner *et al.* (1984).

[f] Atomic hydrogen column density through OMC-1, with 75 km s^{-1} width.

Table references:

(1) Beckwith *et al.* (1978).
(2) Beck, Lacy & Geballe (1979).
(3) Scoville *et al.* (1982).
(4) Beck & Beckwith (1983).
(5) Goldsmith *et al.* (1981).
(6) Storey *et al.* (1981).
(7) Watson (1982).
(8) Storey, Watson & Townes (1981).
(9) Werner *et al.* (1984).

H I, which may have been seen in NGC 2071 (Bally & Stark 1983). Chernoff *et al.* suggest that emission from J-shocks may also be present, either from the shock which decelerates the outflow from the central source or in regions where the velocity of the shock in the ambient gas is larger than average. Observation of [O I] $\lambda 63 \mu$m emission (Werner *et al.* 1984) well above that predicted for C-shocks but consistent with that from J-shocks (Hollenbach 1982) suggests that such shocks have indeed been seen in Orion.

Acknowledgments

The material in Section 9.4 is taken from McKee, C. F., Chernoff, D. F. & Hollenbach, D. J., in *Galactic and Extragalactic Infrared Spectroscopy*, ed. by M. Kessler and J. Phillips, p. 103, copyright © 1984 by D. Reidel Publishing Company, Dordrecht, Holland. I am indebted to my collaborators D. Chernoff, D. Hollenbach, G. Seab, and M. Shull for their many contributions to the work described here. This work is supported in part through NSF grant AST 83-14684.

References

Axford, W. I. (1981). In *Proceedings of the 17th International Cosmic Ray Conference, Paris*, **12**, 155.

Axford, W. I., Leer, E. & Skadron, G. (1977). Paper presented at 15th International Cosmic Ray Conference, Plovdiv, Bulgaria.

Bally, J. & Stark, A. A. (1983). *Astrophys. J. Lett.* **266**, L61.

Barlow, M. J. & Silk, J. (1977). *Astrophys. J. Lett.* **211**, L83.

Beck, S. C. & Beckwith, S. (1983). *Astrophys. J.* **271**, 175.

Beck, S. C., Lacy, J. H. & Geballe, T. R. (1979). *Astrophys. J. Lett.* **234**, L213.

Beckwith, S., Persson, S. E., Neugebauer, G. & Becklin, E. E. (1978). *Astrophys. J.* **223**, 464.

Bell, A. R. (1978a). *Mon. Not. Roy. Astron. Soc.* **182**, 47.

Bell, A. R. (1978b). *Mon. Not. Roy. Astron. Soc.* **182**, 443.

Benvenuti, P., Dopita, M. & D'Odorico, S. (1980). *Astrophys. J.* **238**, 601.

Black, J. (1984). This volume, ch. 12.

Blandford, R. D. & Ostriker, J. P. (1978). *Astrophys. J. Lett.* **221**, L29.

Blandford, R. D. & Ostriker, J. P. (1980). *Astrophys. J.* **237**, 793.

Butler, S. E. & Raymond, J. C. (1980). *Astrophys. J.* **240**, 680.

Chernoff, D. F. (1985). Ph.D. Thesis, University of California, Berkeley.

Chernoff, D. F., Hollenbach, D. & McKee, C. F. (1982). *Astrophys. J. Lett.* **259**, L97.

Chevalier, R. A. (1983). *Astrophys. J.* **272**, 765.

Chevalier, R. A., Kirshner, R. P. & Raymond, J. C. (1980). *Astrophys. J.* **235**, 186.

Chevalier, R. A. & Raymond, J. C. (1978). *Astrophys. J. Lett.* **225**, L27.

Cox, D. P. (1972). *Astrophys. J.* **178**, 143.

Chu, C. K. & Gross, R. A. (1969). In *Advances in Plasma Physics*, ed. A. Simon & W. B. Thompson, **2**, 139. New York: Wiley.

Cowie, L. L. (1978). *Astrophys. J.* **225**, 887.

Cowie, L., Laurent, C., Vidal-Madjar, A. & York, D. G. (1979). *Astrophys. J. Lett.* **229**, L81.

Dopita, M. A. (1976). *Astrophys. J.* **209**, 395.

Dopita, M. A. (1977). *Astrophys. J. Suppl.* **33**, 437.

Dopita, M. A. & Binette, L. (1983). In *Supernova Remnants and their X-ray Emission*, ed. J. Danziger & P. Gorenstein, p. 221. Dordrecht: Reidel.

Dopita, M. A., Binette, L., D'Odorico, S. & Benvenuti, P. (1984). *Astrophys. J.* **276**, 653.

Draine, B. T. (1980). *Astrophys. J.* **241**, 1021.

Draine, B. T. & Roberge, W. G. (1982). *Astrophys. J. Lett.* **259**, L91.

Draine, B. T., Roberge, W. G. & Dalgarno, A. (1982). *Astrophys. J.* **264**, 485.

Draine, B. T. & Salpeter, E. E. (1979). *Astrophys. J.* **231**, 438.

Drury, L. & Volk, H. (1981). *Astrophys. J.* **248**, 344.

Fesen, R. A., Blair, W. P. & Kirshner, R. P. (1982). *Astrophys. J.* **262**, 171.

Field, G. B., Rather, J. D. G., Aanestad, P. A. & Orszag, S. A. (1968). *Astrophys. J.* **151**, 953.

Filippenko, A. V. & Halpern, J. P. (1984). *Astrophys. J.* **285**, 458.

Goldsmith, P. F., Erickson, N. R., Fetterman, H. R., Clifton, B. J., Peck, D. D., Tannenwald, P. E., Koepf, G. A., Buhl, D. & McAvoy, N. (1981). *Astrophys. J. Lett.* **243**, L79.

Greenstadt, E. W. & Fredricks, R. W. (1979). In *Solar System Plasma Physics*, ed. L. J. Lanzerotti, C. F. Kennel & E. N. Parker, p. 3. Amsterdam: North Holland.

Halpern, J. P. & Steiner, J. E. (1983). *Astrophys. J. Lett.* **269**, L37.

Hester, J., Parker, R. A. R. & Dufour, R. I. (1983). *Astrophys. J.* **273**, 219.

Hollenbach, D. (1982). *Ann. N.Y. Acad. Sci.* **359**, 242.

Hollenbach, D. J. & McKee, C. F. (1979). *Astrophys. J. Suppl.* **41**, 555.

Kahn, F. D. (1976). *Astron. Astrophys.* **50**, 145.

Knacke, R. F. & Young, E. T. (1981). *Astrophys J. Lett.* **249**, L65.

Krimsky, G. F. (1977). *Dokl. Akad. Nauk. USSR*, **234**, 1306.

Lagage, P. O. & Cesarsky, C. J. (1983). *Astron. Astrophys.* **125**, 249.

Lasker, B. M. (1966). *Astrophys. J.* **146**, 471.

Lasker, B. M. (1977). *Astrophys. J.* **212**, 390.

Leroy, M. M., Winske, D., Goodrich, C. C., Wu, C. S. & Papadopoulos, D. (1982). *J. Geophys. Res.* **87**, 5081.

Mathews, W. G. & O'Dell, C. R. (1969). *Ann. Rev. Astron. Astrophys.* **7**, 67.

McKee, C. F. (1974). *Astrophys. J.* **188**, 335.

McKee, C. F. (1982). In *Supernovae: A Survey of Current Research*, ed. M. J. Rees & R. J. Stoneham, p. 433. Dordrecht: Reidel.

McKee, C. F., Chernoff, D. & Hollenbach, D. (1984). In *XVI ESLAB Symposium, Galactic and Extragalactic Infrared Spectroscopy*, ed. M. Kessler & J. Phillips, p. 103. Dordrecht: Reidel.

McKee, C. F. & Hollenbach, D. J. (1980). *Ann. Rev. Astron. Astrophys.* **18**, 219.

McKee, C. F., Storey, J. W. V., Watson, D. M. & Green, S. (1982). *Astrophys. J.* **259**, 647.

Miller, J. S. (1974). *Astrophys. J.* **189**, 239.

Mouschovias, T. C. (1976). *Astrophys. J.* **207**, 141.

Mullan, D. J. (1971). *Mon. Not. Roy. Astron. Soc.* **153**, 145.

Nadeau, D., Geballe, T. R. & Neugebauer, G. (1982). *Astrophys. J.* **253**, 154.

Nadeau, D. & Geballe, T. R. (1979). *Astrophys. J. Lett.* **230**, L169.

Pikel'ner, S. B. (1954). *Izv. Krym. Astrofiz. Obs.* **12**, 94.

Pravdo, S. H. & Smith, B. W. (1979). *Astrophys. J. Lett.* **234**, L195.

Raymond, J. C. (1979). *Astrophys. J. Suppl.* **39**, 1.

Raymond, J. C., Black, J. H., Dupree, A. K., Hartmann, L. & Wolff, R. S. (1980). *Astrophys. J.* **238**, 881.

Raymond, J. C., Blair, W. P., Fesen, R. A. & Gull, T. R. (1983). *Astrophys. J.* **275**, 636.

Raymond, J. C., Cox, D. P. & Smith, B. W. (1976). *Astrophys. J.* **204**, 290.

Russell, C. T. & Greenstadt, E. W. (1979). *Space Sci. Rev.* **23**, 3.

Scarf, F. L., Gurnett, D. A. & Kurth, W. S. (1981). *Nature*, **292**, 747.

Schinke, R., Engel, V., Buck, U., Meyer, H. & Diercksen, G. H. F. (1985). *Astrophys. J.* **299**, 939.

Schwartz, R. D. (1983). *Ann. Rev. Astron. Astrophys.* **21**, 209.

Scoville, N. Z., Hall, D. N. B., Kleinmann, S. G. & Ridgeway, S. T. (1982). *Astrophys. J.* **253**, 136.

Seab, C. G. & Shull, J. M. (1983). *Astrophys. J.* **275**, 652.

Shull, J. M. (1978). *Astrophys. J.* **226**, 858.

Shull, J. M. & McKee, C. F. (1979). *Astrophys. J.* **227**, 131.

Spitzer, L. (1976). *Comments Astrophys.* **6**, 177.

Spitzer, L. (1978). *Physical Processes in the Interstellar Medium.* New York: Wiley.

Stahler, S., Shu, F. & Taam, R. (1981). *Astrophys. J.* **248**, 727.

Storey, J. W. V., Watson, D. M. & Townes, C. H. (1981). *Astrophys. J. Lett.* **244**, L27.

Storey, J. W. V., Watson, D. M., Townes, C. H., Haller, E. E. & Hansen, W. L. (1981). *Astrophys. J. Lett.* **247**, L136.

Tidman, D. A. & Krall, N. A. (1971). *Shock Waves in Collisionless Plasmas.* New York: Wiley.

Watson, D. M. (1982). Ph.D. Thesis, University of California, Berkeley.

Werner, M. W., Crawford, M. K., Genzel, R., Hollenbach, D. J., Townes, C. H. & Watson, D. M. (1984). *Astrophys. J. Lett.* **282**, L281.

Zel'dovich, Ya. B. & Raizer, Yu. P. (1966). *Physics of Shock Waves and High Temperature Hydrodynamic Phenomena*, Vols. 1, 2. New York: Academic Press.

10

Coronal interstellar gas and supernova remnants

RICHARD McCRAY

10.1 Introduction

The study of coronal ($T \gtrsim 10^6$ K) interstellar gas is a relatively new branch of astronomy. Before the 1970s, there was little direct evidence for such gas, although theoretical models predicted that it should be found in the interiors of supernova shells. In 1956, Spitzer made the prescient suggestion that the galaxy would likely possess a hot corona much like the solar corona. By the early 1970s, a series of rocket experiments had shown that the Milky Way was glowing in soft X-rays, indicating that coronal gas was pervasive in the interstellar medium; this interpretation was supported by observations by the *Copernicus* satellite of the interstellar absorption line O VI $\lambda 1035$, showing that this tracer of high-temperature gas was extensively distributed throughout the galaxy.

We now have good maps of the brightness and temperature distribution of the soft X-ray emission from the Milky Way. With X-ray telescopes we have seen emission from coronal gas in elliptical galaxies and between the galaxies in clusters. As a result of these observations, the theory of coronal interstellar gas has advanced rapidly. The atomic processes that determine the local temperature, ionization, and spectral emissivity of the gas have been studied in detail. We have also learned much about the energy sources and macroscopic processes that control the global properties of the interstellar gas. It is now clear that the coronal gas in the Milky Way is produced mainly by the blast waves from supernova explosions, although stellar winds and compact X-ray sources may dominate in specific locales. It is also clear that the same physical processes control the properties of coronal gas wherever it may be – between galaxies and in the nuclei of galaxies as well as in the Milky Way.

Interstellar gas can be heated to coronal temperatures by three sources: energetic particles, X-rays, and shock waves. The first two heat sources may be steady; if so, the resulting state of the gas is a function of a single parameter, the ratio of the gas pressure to the flux of particles or radiation. On the other hand, it is certainly possible or likely that the flux of particles or radiation is transient, and heating by shock waves is intrinsically transient. In those cases, the problem of interpreting or modeling the state of the gas becomes considerably more complicated.

The plan of this chapter is as follows. In Section 10.2 I describe the atomic processes that determine the local state and spectral emissivity of the gas. Then, in Section 10.3, I describe the stationary equilibrium states that result from steady sources of heating or ionization and discuss qualitatively how these results will be modified when the source is transient. In Section 10.4 I describe the physics of the interfaces between hot and cool interstellar gas, where electron thermal conduction may play an important role. Finally, in Section 10.5, I show how these concepts may be used to describe the structure and evolution of expanding interstellar shells caused by the action of supernovae and stellar winds.

10.2   Thermally ionized gas: coronal approximation

In the absence of ionizing radiation, the local state of hot interstellar gas is controlled by collisions of ions with thermal electrons. The dominant processes are electron impact excitation, electron impact ionization, radiative recombination, dielectronic recombination, and bremsstrahlung.

Electron impact excitation, in which an ion X^z (charge state z of element X) is excited by a passing electron to a state X^{z*} and subsequently radiates a photon with energy ε, is the primary mechanism by which the heat of a coronal gas is converted to radiation in emission lines. The thermal excitation rate coefficient, $q_\varepsilon = \langle \sigma_\varepsilon v_e \rangle$, is conventionally written

$$q_\varepsilon(T) = 8.63 \times 10^{-9} \, T_6^{-1/2} \Omega_\varepsilon(T) \omega_g^{-1} \exp(-\varepsilon/kT) \quad \text{cm}^3 \, \text{s}^{-1}, \quad (10.1)$$

where $T_6 = T/(10^6 \text{ K})$, ω_g is the statistical weight of the ground state and Ω_ε is the 'collision strength', a dimensionless function of temperature that incorporates all the details of the atomic physics. For resonance transitions Ω_ε is a weak function of temperature and has a value of order unity.

Electron impact ionization, in which $e + X^z \to 2e + X^{z+1}$, has a rate coefficient that may be written (for $kT < I_z$)

$$q_z(T) = 1.3 \times 10^{-5} \, T_6^{1/2} I_z^{-2} r_z F_z \exp(-I_z/kT) \quad \text{cm}^3 \, \text{s}^{-1}, \quad (10.2)$$

where I_z is the ionization threshold in eV, r_z is the number of electrons in the outer shell, and F_z is another dimensionless number ≈ 1–3.

Radiative recombination, in which $e + X^{z+1} \to X^z + \varepsilon$, has a net (to all bound states) rate coefficient that may be written

$$\alpha_{r,z}(T) = 2.06 \times 10^{-14}(z+1)^2 T_6^{-1/2}\phi_z(T_6) \quad \text{cm}^3 \text{ s}^{-1}, \tag{10.3}$$

where the function $\phi_z(T_6)$ embodies the details of the atomic physics, and $\phi_z(T_6) \approx 0.63(z+1)^{1/2} T_6^{-1/4}$ for hydrogenic ions. This expression provides a fairly good approximation to the radiative recombination coefficient for any ion as z becomes large.

In dielectric recombination, a fast incoming electron simultaneously excites an ion and is trapped into a weakly bound autoionizing state, which then stabilizes by the emission of a resonance line photon, viz.: $e + x^{z+1} \rightarrow X^{z*} \rightarrow X^z + \varepsilon_{ij}$. The second step, needed to complete the recombination, is rare; usually the first step is followed by the reverse reaction. The rate coefficient for dielectronic recombination is given by the expression:

$$\alpha_{d,z}(T) = 3.0 \times 10^{-12} T_6^{-3/2} B(z)$$
$$\times \sum_j f_{ij} A(x) \exp(-E_{ij}/kT) \quad \text{cm}^3 \text{ s}^{-1}, \tag{10.4}$$

where f_{ij} is the oscillator strength of the resonance transition $i \rightarrow j$, $E_{ij} = \varepsilon_{ij}/[1 + 0.015 z^3(z+1)^{-2}]$, $B(z) = z^{1/2}(z+1)^{5/2}(z^2 + 13.4)^{-1/2}$, $x = \varepsilon_{ij}/(z+1)$, and $A(x) = x^{1/2}/(1 + 0.105 x + 0.015 x^2)$. One can see, by comparing (10.3) and (10.4), that dielectronic recombination may dominate radiative recombination in high-temperature plasmas by factors $\approx 10^2$, especially for those ions with partially filled L or M shells, which have low-lying excited states and therefore can form autoionizing states with $\varepsilon_{ij} \lesssim kT$.

In the 'coronal model' approximation, the elements are ionized only by thermal electrons. The master equation describing the rates of change of the fractions, f_z, of element X in each ion stage z is:

$$\frac{df_z}{dt} = n_e \sum_{z'} M_{z,z'} f_{z'}, \tag{10.5}$$

where the matrix $M_{z,z'}$ is tri-diagonal and has elements

$$M_{z,z-1} = q_{z-1}$$
$$M_{z,z} = -(q_z + \alpha_{r,z-1} + \alpha_{d,z-1}) \quad . \tag{10.6}$$
and
$$M_{z,z+1} = \alpha_{r,z} + \alpha_{d,z}$$

If the ionization is stationary, $df_z/dt = 0$, equation (10.5) can be recast into a set of equations connecting pairs of adjacent ionization states:

$$q_z f_z = (\alpha_{r,z} + \alpha_{d,z})f_{z+1}. \tag{10.7}$$

Note that the electron density, n_e, does not appear in equation (10.7) because it is a common factor of all ionization and recombination processes. Therefore, in stationary coronal ionization balance the ion fractions, f_z, are functions only of temperature. (This statement is not strictly true at high electron densities, for which $\alpha_{d,z}$ can be suppressed.)

A potential source of confusion is the common use of the term 'coronal equilibrium' to describe the stationary ionization balance represented by (10.7). Equation (10.7) does not describe a state of thermodynamic equilibrium because it does not represent the result of a detailed balance of time-reverse processes (e.g., photoionization vs. radiative recombination, electron impact ionization vs. 3-body recombination). Indeed, according to the Saha equation, the ion fractions in thermodynamic equilibrium depend on electron density as well as temperature (at fixed T, $f_{z+1}/f_z \propto n_e^{-1}$).

Fig. 10.1 is a typical result of a coronal ionization balance calculation. Note that the temperature, T_z, at which the ionization state shifts from state z to $z+1$ is typically about $kT_z \approx 0.2$–$0.3\,I_z$. The effects of dielectronic recombination are apparent in Fig. 10.1. For example, the He-like Fe XXV and Ne-like Fe XVII are very durable and span large temperature ranges because they have low dielectronic recombination rates ($\alpha_d/\alpha_r < 1$) and high ionization potentials, whereas ions with partially filled L or M shells span relatively narrow temperature ranges because their dielectronic recombination rates are high ($\alpha_d/\alpha_r > 10$).

Fig. 10.1. Coronal model fractional abundances, $f_z(T)$, of ionization stages of iron. (Courtesy of Dr. J. M. Shull).

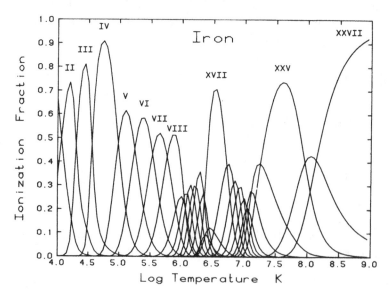

Given the ionization fractions, $f_z(T)$, one can calculate the emission spectrum by summing the spectral emissivities due to the various radiative processes. The line spectrum is dominated by radiative decays following electron impact excitation; the volume emissivity of a given line is given by:

$$\eta_\varepsilon(T) = n_e n_H A_X f_z(T) \varepsilon q_\varepsilon(T) \quad \text{erg cm}^{-3}\,\text{s}^{-1}, \tag{10.8}$$

where A_X is the abundance of element X relative to hydrogen (atomic number density n_H) and $q_\varepsilon(T)$ is given by (10.1). The emission line spectrum of the gas includes the sum of all these collisionally excited lines, plus a relatively small contribution of recombination lines. The continuum emissivity is dominated by bremsstrahlung and recombination, and has a spectral shape given approximately by $d\eta_\varepsilon/d\varepsilon \propto \exp(-\varepsilon/kT)$.

The total power radiated by a unit volume of coronal gas is calculated by integrating the line plus continuum emissivity over the spectrum; it may be written $P = n_e n_H \Lambda(T)$. Fig. 10.2 shows the radiative cooling function $\Lambda(T)$ and indicates the contributions of different elements to the emission line cooling in particular temperature ranges. For $T > 10^5\,\text{K}$ it may be

Fig. 10.2. Radiative cooling function $\Lambda(T)$, for coronal model. Contributions to cooling from individual elements are indicated. (From Gaetz & Salpeter (1983); courtesy of *Astrophysical Journal*.)

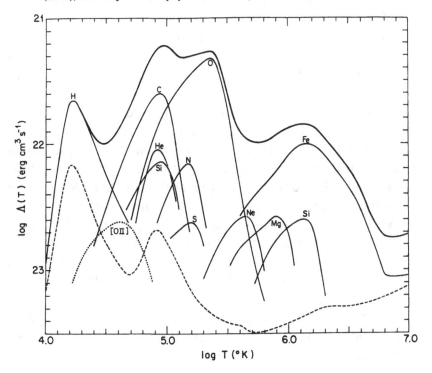

approximated by

$$\Lambda(T) = 1.0 \times 10^{-22} T_6^{-0.7} + 2.3 \times 10^{-24} T_6^{0.5} \quad \text{erg cm}^3 \text{ s}^{-1}, \qquad (10.9)$$

where the former term accounts for line emission and is accurate within a factor 2 and the latter term accounts for bremsstrahlung. For $T > 5 \times 10^6$ K the gas radiates most of its power in the X-ray ($\varepsilon > 0.5$ keV) band. The emission line spectrum is dominated by Lyman-α transitions of hydrogenic ions and the $1s^2 \to 1s2p$ transitions of He-like ions of abundant elements such as O, Si, S, and Fe. The spectrum also has a rich cluster of emission lines near 1 keV due to L-shell transitions of Fe ions. When the temperature drops below $\approx 5 \times 10^6$ K, the spectrum changes radically. The emission line spectrum is dominated by many strong EUV ($13.6 < \varepsilon < 100$ eV) lines from ions of O, Si, S, with L-shell electrons and from Fe ions with M-shell electrons. (The strongest emission lines with $\varepsilon < 13.6$ eV are Lyman-α, C IV $\lambda1550$ and O VI $\lambda1035$). In the temperature range $10^4 < T < 3 \times 10^6$ K the power in this EUV line emission exceeds the continuum power by factors $\approx 10^2$ and is much greater than the power radiated at higher or lower temperatures.

The peak in $\Lambda(T)$ forms a 'thermal barrier' that causes interstellar gas to avoid this range of temperatures. One way to understand this effect is to consider the thermal relaxation of a gas that has been heated suddenly to some initial temperature T_0. Then, if the gas cools at constant pressure, its temperature will relax according to the equation

$$\frac{d}{dt}\left(\frac{5}{2}kT\right) = -n_e\Lambda(T). \qquad (10.10)$$

It follows from (10.10) that the thermal relaxation time scale is roughly $t_r \approx 5kT/[2n\Lambda(T)]$. For typical interstellar pressures, $nT = 10^4$ cm^{-3} K, we may estimate from (10.10) that $t_r \approx 10^7 T_6^{2.7}$ yr. We see from this example that the gas temperature will linger near T for a time $\approx t_r$ and then will decrease at an accelerating rate to $T \lesssim 10^4$ K, at which point the radiative cooling drops rapidly and the temperature may be stabilized by the heating of ultraviolet starlight. We also see that radiative cooling is unimportant for interstellar gas with $T \gtrsim 10^7$ K. This result, which depends only on atomic physics and is independent of the details of the heating mechanism, is the main reason why interstellar gas is usually found with $T \lesssim 10^4$ K or with $T \gtrsim 10^6$ K, but rarely with intermediate temperatures.

The above example is not self-consistent, because we used the radiative cooling function $\Lambda(T)$ that was derived with the assumption of a stationary ionization balance in order to estimate timescales for relaxation of a nonstationary gas. In fact, fairly substantial errors arise from this approximation, because changes in ionization will lag changes in temperature. Consequently, gas which has been cooling down will be 'over-ionized' compared with a stationary gas at a given temperature, while gas which has recently been heated impulsively will be 'under-ionized'. Since lower ionization stages of a given element usually have more low-lying bound states that may be excited by thermal electrons, the radiative cooling at a given temperature is enhanced compared with a stationary gas if the gas is under-ionized, and suppressed if it is over-ionized. The enhancement or suppression factors may be substantial, say, ≈ 3. As will be discussed later, these nonstationary effects can be important for the interpretation of the spectra of supernova remnants. However, they do not obviate the qualitative conclusions of the previous paragraph.

10.3   Photoionized gas: nebular approximation

The ionization state and radiative cooling of interstellar gas is very different if the gas is illuminated by X-rays from a compact galactic X-ray source or an active galactic nucleus. In that case photoionization dominates ionization by thermal electrons for most ions. The X-rays may also dominate the heating of the gas. As a result, the ionization and temperature of the gas may reach a stationary state that is determined by the gas pressure and the flux and spectrum of radiation. Models for the ionization and spectral emissivity of gas around a compact X-ray source are called 'X-ray nebular models', by analogy with models for planetary nebulae.

Suppose that a parcel of gas with given pressure $P = 2.3 \, n_H kT$, is located a distance r from a compact isotropic X-ray source with luminosity L and spectrum $dL/d\varepsilon = Lg(\varepsilon)$. Assume further that the primary spectrum is not significantly attenuated by photoabsorption. Then the photoionization rate for electrons in shell, s, of an ion, X^z, is given by:

$$J_{z,s} = \frac{L}{4\pi r^2} \int_{I_{z,s}}^{\infty} d\varepsilon \, \frac{1}{\varepsilon} \, g(\varepsilon) \sigma_s(\varepsilon) \quad s^{-1}, \tag{10.11}$$

where $I_{z,s}$ is the photoionization threshold. The photoionization cross section, $\sigma_s(\varepsilon)$, is a rapidly decreasing function of ε; for highly stripped ions or inner shells it may be approximated fairly well by

$$\sigma_s(\varepsilon) = 7.9 \times 10^{-18} (13.6 \, \text{eV}/I_{z,s}) r_s F_s (I_{z,s}/\varepsilon)^3 \quad \text{cm}^2, \tag{10.12}$$

where r_s is the number of electrons in shell s and the numerical factor $F_s \approx 1$ for hydrogenic ions, $F_s \approx 0.8$ for filled K shells, and $F_s \approx 0.5$ for L-shell electrons.

Photoionization of inner shell electrons is often followed by the emission of one or more Auger electrons. For example, if an ion has two or more L-shell electrons, photoionization of a K-shell electron will leave the resulting ion in an autoionizing state that can relax by ejecting one or more L-shell electrons as another L-shell electron drops to the K-shell. Likewise, the L-shell vacancy may result in further Auger electrons if the ion has two or more M-shell electrons. Thus, there is a set of probabilities, $p_{z,z'}(s)$, that photoionization from shell s of ion X^z will result in ion $X^{z'}$. Given these probabilities, we may write an expression for the rate to produce ions $X^{z'}$ by photoionization of ion X^z:

$$J_{z,z'} = \sum_s J_{z,s} p_{z,z'}(s) = \frac{L}{4\pi r^2} j_{z,z'},\tag{10.13}$$

where we have removed the explicit dependence on $L/(4\pi r^2)$ in defining the quantities $j_{z,z'}$, which are numbers depending on the spectral shape function $g(\varepsilon)$.

As in the coronal case, we may write the master equation (10.5) describing the rates of change of the ion fractions, f_z, and look for stationary solutions such that $df_z/dt = 0$. However, the matrix $M_{z,z'}$ is no longer tri-diagonal because the Auger emission populates ionization stages z from stages with $z' < z - 1$. The matrix elements including photoionization and Auger emission may be written:

$$M_{z,z'} = j_{z,z'}, \quad z' < z - 1$$

$$M_{z,z-1} = q_{z-1} + \xi j_{z,z-1},$$

$$M_{z,z} = -(q_z + \alpha_{r,z-1} + \alpha_{d,z-1} + \xi j_{z,z'}),\tag{10.14}$$

and

$$M_{z,z+1} = (\alpha_{r,z} + \alpha_{d,z}),$$

where we have defined an 'ionization parameter', $\xi = L/(4\pi r^2 n_e)$. The stationary ionization fractions are found by solving the matrix equation $\sum_{z'} M_{z,z'} f_{z'} = 0$, with the normalization $\sum_{z'} f_{z'} = 1$. We see that the solutions depend on two parameters: $f_z = f_z(T, \xi)$. Of course, if $\xi = 0$ we recover the coronal model, and with increasing ξ the ionization at a given temperature shifts to higher stages.

As with the coronal model, the spectral emissivity of the photoionized gas is calculated by summing the line and continuum emissivities due to all radiative transitions as in (10.8). As before, the total power radiated per unit volume can be written $P = n_e n_H \Lambda(T, \xi)$. Generally, the radiative cooling function $\Lambda(T, \xi)$ is a decreasing function of ξ for fixed T because as elements become more highly ionized they tend to have fewer low-lying resonance transitions. Indeed, as ξ becomes very large the peak in $\Lambda(T, \xi)$ for $10^4 < T < 10^7$ K due to line emission vanishes, leaving only bremsstrahlung cooling.

In the nebular model it is assumed that the gas temperature obtains a stationary state in which this radiative cooling is balanced by heating due to photons from the central source. Thus, T is no longer an arbitrary parameter of the model but is determined as a function of ξ. The photoelectric heating per ion is given by an expression similar to (10.11):

$$\Gamma_{x,z} = \frac{L}{4\pi r^2} \sum_{s} \int_{I_{z,s}}^{\infty} d\varepsilon \frac{1}{\varepsilon} g(\varepsilon) \sigma_s(\varepsilon) E_{z,s}(\varepsilon), \qquad (10.15)$$

where $E_{z,s}(\varepsilon)$ is the net kinetic energy of photoelectrons and Auger electrons resulting from an ionization from shell s. The net heating per hydrogen atom is calculated by summing over all species:

$$\Gamma_p(T, \xi) = \sum_{x} A_x \sum_{z} f_z(T, \xi) \Gamma_{x,z}. \qquad (10.16)$$

In addition, the electrons are heated directly by Compton scattering at a net rate (per hydrogen atom):

$$\Gamma_c = \frac{L}{4\pi r^2} \frac{n_e}{n} \sigma_T \int d\varepsilon g(\varepsilon) \left[\frac{\varepsilon}{m_e c^2} - \frac{4kT}{m_e c^2} \right] \qquad (10.17)$$

where the second term in brackets represents cooling of hot electrons by scattering off soft photons. This Compton heating term vanishes when the $kT \approx \varepsilon_{max}$, where ε_{max} is a characteristic maximum X-ray photon energy. Since $L/(4\pi r^2)$ is a common factor of all heating terms, the net heating per unit volume can be written in the form

$$\Gamma_{TOT} = \frac{L}{4\pi r^2} \gamma(T, \xi). \qquad (10.18)$$

Now, by equating this radiative heating to the radiative cooling, we find:

$$\xi \gamma(T, \xi) = \Lambda(T, \xi). \qquad (10.19)$$

Equation (10.19) is a complicated nonlinear equation that depends implicitly as well as explicitly on ξ and T through the ionization fractions $f_z(T, \xi)$. However, it can be solved on a computer, giving the solution $\xi = L/(4\pi r^2) = \xi(T)$.

It is convenient to define a new function,

$$\Xi(T) = \frac{L}{4\pi r^2 cP} = \frac{\xi(T)}{2.3 \, ckT}, \qquad (10.20)$$

which is the dimensionless ratio of radiation pressure to gas pressure. This function determines the possible stationary thermal states of gas that is heated by a cosmic X-ray source. It depends on cosmic abundances and on the shape, $g(\varepsilon)$, of the X-ray source spectrum; a typical example, with standard cosmic abundances and $g(\varepsilon) \propto \varepsilon^{-0.7}$, is plotted in Fig. 10.3. For points to the right of the curve, heating exceeds cooling and the temperature must rise; to the left, the temperature must fall. Thus, specifying gas pressure, P, source luminosity, L, and radial distance, r, determines the value of Ξ, and given that value we can find the corresponding value of T from Fig. 10.3. Far from the source, Ξ has a low value, say, Ξ_1, and the stationary temperature will be $T_c \approx 10^4$ K, where the strong cooling due to emission of EUV resonance lines balances the X-ray photoelectric heating. Close to the source, Ξ has a high value, say, Ξ_3, and most elements are fully ionized so that there is very little line emission and the stationary temperature will obtain a high value, T_h, for which the net Compton heating (10.17) is almost zero. In this limit the equilibrium temperature is determined solely by the spectral shape function $g(\varepsilon)$.

A very interesting phenomenon occurs when $\Xi = \Xi_2$, where Ξ_2 falls in the

Fig. 10.3. Gas temperature, T, as a function of ionization parameter, Ξ, in the nebular model. Source spectrum is power law, $dL/d\varepsilon \propto \varepsilon^{-0.7}$, with high energy cut-off at $\varepsilon = 500$ keV. (Courtesy of Dr. T. Kallman.)

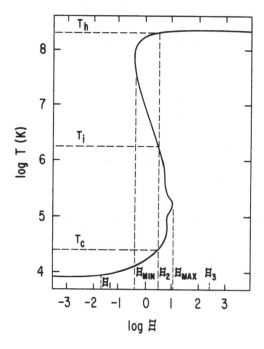

intermediate range $\Xi_{min} < \Xi_2 < \Xi_{max}$. In this case there are three temperatures, T_c, T_i, and T_h, for which heating equals cooling. However, the intermediate state, $T = T_i$, is thermally unstable: a perturbation to higher (or lower) temperature at fixed Ξ will grow because it puts the state of the gas to the right (or left) of the curve where heating is greater than (or less than) cooling. By the same argument, both the cool ($T = T_c$) and hot ($T = T_h$) states are stable. Thus, in the region where $\Xi_{min} < \Xi < \Xi_{max}$, relatively dense 'clouds' with $T = T_c \approx 10^4$ K can coexist in pressure equilibrium with low-density coronal gas with $T = T_h \approx 10^7 - 10^8$ K.

Although the stationary nebular model discussed here is very different from the time-dependent cooling model discussed in Section 10.2, the qualitative conclusion is the same: interstellar gas is rarely found in the intermediate temperature range $10^4 < T < 3 \times 10^6$ K where the atomic cooling function $\Lambda(T)$ peaks.

The detailed shape of the function $\Xi(T)$ is sensitive to the shape of the source spectrum shape $g(\varepsilon)$. Generally, $\Xi(T)$ doubles back on itself if the source spectrum is fairly flat up to the X-ray band, i.e., $n < 2$ for $g(\varepsilon) \approx \varepsilon^{-n}$. The range from Ξ_{min} to Ξ_{max} increases with decreasing n and if the source spectrum is deficient in low-energy photons. However, if the source spectrum is steep, $n > 2.5$, $\Xi(T)$ does not double back on itself and there is no range of Ξ for which two phases may coexist.

In this two-phase model the mass fractions of gas in the cool and hot phases are indeterminate. However, one can imagine a scenario that would have the gas pressure at any distance from the source rising to that value for which $\Xi = \Xi_{max}$. Suppose that there is initially a distribution of clouds with $T = T_c$, but at low pressure so $\Xi > \Xi_{max}$. Then the X-rays will heat the gas at the surface of the clouds to T_h, causing it to expand and fill the volume between the clouds. This process will continue until the pressure of the hot intercloud medium is great enough that $\Xi = \Xi_{max}$, at which point the clouds become stable.

This multiphase behavior plays an important role in the dynamics of gas near any cosmic X-ray source. For example, the model predicts that hot coronal gas will surround the dense gas associated with an accretion disk and stream near a compact galactic binary X-ray source. Another prediction is that the dense clouds responsible for the broad emission lines in active galactic nuclei and quasars ('AGN') should be confined by the pressure of a hot coronal gas. If this model for the broad-line clouds is correct, the parameter Ξ should be restricted to the fairly narrow range, $\Xi_{min} < \Xi < \Xi_{max}$. The value of Ξ for the broad-line clouds can be inferred by fitting theoretical models for the emission line spectra to the observed spectra, and it indeed seems to be in the right range. The model also provides a possible explanation for the absence of broad emission lines in active galactic nuclei, such as BL Lacertae objects, which have steep (spectral index $n > 2$) X-ray spectra; for such sources the curve $\Xi(T)$ does not double back on itself, so there is no region in which dense clouds can coexist with coronal gas.

In the description of the nebular model up to this point, we have neglected the effects of photon absorption and have assumed that the source spectrum is attenuated only by the $1/r^2$ geometrical dilution. This approximation is adequate for determining whether broad-line clouds may exist, provided that the Compton optical depth of the hot gas is small. However, the approximation is not usually adequate for calculating the emission line spectrum of broad-line clouds. Since the cooler emission line clouds have much greater density at constant pressure, $n_c T_c = n_h T_h$, they are likely to have photoelectric optical depths, $\tau(\varepsilon) > 1$. To take account of the photoelectric absorption of the source spectrum, one must replace $g(\varepsilon)$ by $g(\varepsilon) \exp[-\tau(\varepsilon)]$ in (10.11) and (10.15) and one must add the diffuse emission of ionizing radiation to $g(\varepsilon)$. Then it becomes necessary to solve a radiative transfer problem, in which the system of equations (10.11)–(10.19) must be solved at every depth point in the cloud.

Several authors have constructed such radiative transfer models for the emission line spectra of AGN. Since the line emitting regions must be relatively cool, $T = T_c \approx 10^4$ K, and dense, $n_e \approx 10^9$–10^{11} cm^{-3}, we require $\Xi < \Xi_{max}$ at the illuminated surface of the cloud. The line emitting region is thin, $d \approx 10^{12}$ cm, compared with its radial distance, $\approx 10^{18}$–10^{19} cm from the source, so Ξ is almost constant throughout the cloud. Generally, the surface of the cloud is an H II zone in which most of the ionizing uv luminosity is converted into uv emission lines, the strongest of which are C IV $\lambda 1549$ and O VI $\lambda 1035$. Beneath the H II region is a deeper, partially ionized H I zone, in which the luminosity of the more penetrating X-rays is converted primarily to hydrogen (L_α, H_α, H_β, etc.), Mg II $\lambda 2798$, and Fe II line emission. It is remarkable, and possibly fortuitous, that theoretical emission line spectra agree fairly well with the observed spectra even for models consisting of a single slab at a given distance from the source.

10.4   Thermal conduction

In Section 10.3 we described how relatively cool ($T \approx 10^4$ K) gas might coexist in pressure equilibrium with hot ($T \approx 10^7$–10^8 K) gas, and we pointed out that gas at intermediate temperatures would be thermally unstable and therefore improbable. However, this statement cannot be strictly true at the interfaces between the cool and hot gas, where the temperature must change continuously. Otherwise the heat flux due to conduction, $F = K(T) \nabla T$, would be infinite. As we shall see, thermal conductivity at these interfaces produces observable amounts of intermediate temperature gas and can play an important role in the thermodynamics of the interstellar medium.

According to the classical kinetic theory, the thermal conductivity of a fully ionized plasma is given by $K(T) = CT^{5/2}$, where the constant $C \approx 6 \times 10^{-7}$ erg cm^{-1} s^{-1} K$^{-7/2}$. We may write the second law of thermodynamics for a compressible gas, including thermal conduction, as follows:

$$\frac{D}{Dt}\left(\frac{3}{2}\frac{P}{\rho}\right) - \frac{P}{\rho^2}\frac{D\rho}{Dt} = \frac{1}{\rho}\nabla\cdot[K(T)\nabla T] - \frac{n_e n}{\rho}\Lambda(T) \tag{10.21}$$

where the gas pressure $P = \rho kT/\mu$, $\mu = 0.6\, m_H$, and D/Dt represents the time derivative in Lagrangian coordinates. The terms on the right-hand side represent heat loss due to conduction and radiation, respectively. The dynamics and thermodynamics of interstellar gas may be described by (10.21) and Euler's equation, a very complicated set of equations in general.

Here we shall describe only a highly simplified model problem, which nevertheless illustrates the main effects of thermal conduction. The model is as follows. Assume that there is a spherical cloud of cold ($T = T_c \approx 0$) gas with radius R embedded in hot gas with temperature T_h at very large distance. Assume further that there is a stationary flow of mass from the cloud, $\dot{M} = 4\pi r^2 \rho v$. (For truly stationary flow, there must be a hypothetical mass source within the cloud.) Assume that the gas flow is everywhere subsonic, in which case Euler's equation may be replaced by the condition $P = $ const. Finally, assume that the radiative cooling term may be neglected. (The validity of these assumptions will be discussed below.) Then, replacing D/Dt by $v\, d/dr$, we obtain:

$$\frac{5}{2}\frac{\dot{M}}{4\pi}\frac{k}{\mu}\frac{dT}{dr} = \frac{d}{dr}\left[r^2 CT^{5/2}\frac{dT}{dr}\right]. \tag{10.22}$$

The solution of (10.22), subject to the boundary condition $T(R) = 0$, is

$$T(r) = T_h(1 - R/r)^{2/5}. \tag{20.23}$$

The existence of such a stationary solution requires that

$$\dot{M} = \frac{16\pi}{25}\frac{\mu C}{k}RT_h^{5/2} \approx 4 \times 10^{-7}\, M_\odot\, \text{yr}^{-1}\, R_{pc}T_6^{5/2}, \tag{10.24}$$

where $T_6 = T_h/(10^6\, \text{K})$. Thus, we see that the heat conducted from hot gas into cool clouds causes them to 'evaporate', with mass loss rate proportional to the cloud radius. From (10.24) we may estimate the time in which a cloud of atomic density n_c and radius $R_{pc} = R/(1 \text{ pc})$ will be completely evaporated:

$$t \approx 3 \times 10^5 \text{ yr } n_c R_{pc}^2 T_6^{-5/2}. \tag{10.25}$$

Equation (10.23) tells us the distribution of intermediate temperature ($10^5 < T < 10^7$ K) gas in the interfaces between hot interstellar gas and cool clouds. Such gas can be detected through narrow ultraviolet absorption lines of ions such as C IV, N V, and O VI that predominate at these temperatures (cf. Section 10.2). However, such ions are not unambiguous tracers of intermediate temperature gas because they can also be produced in cooler gas by photoionization. The most reliable tracer of such gas is probably the O VI $\lambda 1035$ line, because (except near a cosmic X-ray source) there is probably very little interstellar radiation at photon energies $\varepsilon > 114$ eV, the ionization potential of O V. Assuming that the O VI fractional abundance, $f_{OVI}(T)$, is given by the stationary coronal approximation, its column density in the conduction front may be derived from (10.23) and the condition of pressure equilibrium, $n(r)T(r) = n_h T_h$. The result is

$$N_{OVI} = \frac{5}{2} X_O n_h T_h^{-3/2} R \int_0^{T_h} \frac{f_{OVI}(T)T^{1/2} \, dT}{[1 - (T/T_h)^{5/2}]^2}, \tag{10.26}$$

where we have written $n_{OVI} = n(r)X_O f_{OVI}(T)$. We can evaluate the integral in (10.26) [$f_{OVI}(T)$ peaks at $T \approx 3 \times 10^5$ K] to obtain $N_{OVI} \approx 10^{14} \, n_h T_6^{-3/2} R_{pc}$ cm^{-2}, where we have assumed a cosmic oxygen abundance $X_O = 6.8 \times 10^{-4}$. Observed column densities of interstellar O VI are typically in the range $N_{OVI} \approx 10^{13} - 10^{14}$ cm^{-2}. These values can be understood if the line of sight passes through several clouds with, say, $R \approx 1$ pc, $T_6 \approx 1$, and $n_h \approx 10^{-2}$ cm^{-3}.

This order-of-magnitude agreement is appealing and suggests that the observed O VI column density does come mainly from conduction fronts. However, this interpretation must be regarded as tentative for a variety of reasons. First, the predicted column densities are sensitive to uncertain parameters such as n_h, T_6, R_{pc}, and number of interfaces along the line of sight. Second, the observed O VI column density may contain a significant but very uncertain contribution due to photoionization by soft X-rays. Third, there is a great deal of uncertainty in the theory itself.

One may rightly question the assumptions that led to (10.22)–(10.26). First, consider the assumption that the flow is everywhere subsonic, $v/c_s = v/(kT/\mu)^{1/2} < 1$. One finds for the stationary flow above that the ratio v/c_s reaches a maximum for $r = 1.1 R$, and that the condition $v/c_s < 1$ at that point can be written approximately $R_{pc} \gtrsim 10^{-2} T_6^2 n_h^{-1}$, where n_h is the density of the hot intercloud gas at large distance. Thus, for typical parameters of the hot gas, $T_6 \approx 1$ and $n_h \approx 10^{-2}\,\mathrm{cm}^{-3}$, this condition is violated for clouds with radius $R < 1$ pc. This is the regime of 'saturated conductivity', in which the heat flux is limited to a few times $\rho_h c_h^3$, where $c_h = (kT_h/\mu)^{1/2}$. Mass evaporates from the cloud surface supersonically with a flux of a few times $\rho_h c_h$, and the rocket force due to this supersonic expansion raises the cloud pressure substantially above that of the hot intercloud gas.

The other assumption that one must check is that of neglecting the radiative cooling term in (10.21). This may be done by using the solution (10.23), in which radiative cooling is neglected, to evaluate the ratio of the radiative cooling term to the advected enthalpy term in (10.21). Using (10.9), we find that this ratio is small for $r \approx R$ only if $R_{pc} < 0.1\,T_6^2 n_h^{-1}$. If this condition is violated, radiative losses compete effectively for the heat that is conducted toward the clouds and suppress the evaporative mass flux. Indeed, the mass flux may even reverse, causing gas to condense onto large clouds while it is evaporating off the small clouds. Note that the domain for which (10.22)–(10.26) are valid is only a factor ≈ 10 in cloud radius.

When one attempts to include the effects of radiative losses and saturation, the theory of conduction fronts becomes much more complicated and is not yet fully understood. A further major uncertainty arises from the possible effects of magnetic fields. Since magnetic fields can greatly suppress thermal conductivity across field lines, the structure of conduction fronts may depend strongly on the topology of interstellar magnetic fields at the interfaces between the hot and cool gas.

The conclusions that we may draw from this section are rather unsatisfying. The simple classical theory indicates that thermal conductivity may play a major role in the thermodynamics of coronal interstellar gas, cooling the hot medium and evaporating the smaller clouds. Furthermore, the observed column densities of O VI are roughly consistent with the theory. However, we know that the classical theory is valid only over a very limited range of parameters, and that the theory itself becomes very complicated and uncertain outside this range or if magnetic fields are present.

10.5 Supernova remnants, interstellar bubbles, and superbubbles

10.5.1 *Supernova remnants*

In a supernova explosion, a star ejects a mass $M_{ej} \approx 1 - 10\,M_\odot$ into interstellar space with terminal velocity $V_{ej} \approx 10^4\,\mathrm{km\,s^{-1}}$ and kinetic energy $E_0 \approx 10^{51}$ ergs. The ejecta expand at nearly constant velocity until they encounter a comparable mass of ambient interstellar gas (density $\rho_0 = n_0 m$, where $m = 1.4\, m_H$). This occurs at a time $t_0 \approx 200\ \text{yr}\ (M_{ej}/M_\odot)^{1/3} n_0^{-1/3}$ and radius $R_0 \approx 2\ \text{pc}\ (M_{ej}/M_\odot)^{1/3} n_0^{-1/3}$. From the surface of impact a shock propagates outward into the interstellar gas and another shock propagates inward into the expanding ejecta. After a few times t_0 the ejecta have slowed down and have transferred most of their kinetic energy into the energy of the outgoing shock. This blast wave continues to expand for $> 10^4$ yr, sweeping the interstellar gas into a thin shell and leaving a hot rarefied cavity of coronal gas that may linger for millions of years before it is once again filled with 'normal' $(T < 10^4\ \text{K})$ interstellar gas.

Suppose, as a first approximation, that the ambient interstellar gas (temperature T_0) has uniform density ρ_0. In that case a detailed theory has been developed for the hydrodynamics of the supernova blast wave. Here we present a simplified approximate version of that theory, which illustrates the main physical effects and is surprisingly accurate. We assume that the mass of the system resides in a thin expanding shell of radius $R(t)$, and that it is dominated by the mass of swept-up interstellar gas, $M(t) = \frac{4}{3}\pi\rho_0 R(t)^3$. We assume that the shell is expanding hypersonically, $\dot{R} \gg (kT/\mu)^{1/2}$. Then, according to the jump conditions for a monatomic gas, the shocked gas in the shell will have density $\rho_1 = 4\rho_0$ and temperature $T_1 = \frac{3}{16}\mu\dot{R}^2/k$. The thickness of the shell, $\Delta R \approx R/12$, may be estimated from the condition $4\pi R^2\,\Delta R\rho_1 = \frac{4}{3}\pi R^3\rho_0$. The dynamics of the shell follows from Newton's Second Law applied to a conical section of solid angle $\Delta\Omega$: $(d/dt)(\Delta M\dot{R}) = \Delta F$, where $\Delta M = \frac{1}{3}R^3\rho_0\,\Delta\Omega$ and the driving force $F = PR^2\,\Delta\Omega$. Thus, we have the general equation for an expanding spherical system:

$$\frac{d}{dt}\left[\frac{1}{3}R^3\rho_0\dot{R}\right] = R^2 P. \tag{10.27}$$

To solve (10.27) we need an expression for the interior pressure, P. In the early part of the expansion, the total energy of the blast wave is conserved. Assume that this energy is mostly thermal energy and that the interior pressure is uniform: then $P = \frac{2}{3}E_0/(\frac{4}{3}\pi R^3)$. Inserting this result into (10.27), we find the solution,

$$R(t) = \left[\frac{25E_0 t^2}{4\pi\rho_0}\right]^{1/5} \approx 13\ \text{pc}\left(\frac{E_{51}}{n_0}\right)^{1/5} t_4^{2/5}, \tag{10.28}$$

where $E_{51} = E_0/(10^{51}\ \text{ergs})$ and $t_4 = t/(10^4\ \text{yr})$. Equation (10.28) agrees with the exact result within 1 percent. From it we may also derive the expansion velocity, $\dot{R}(t)$, and the temperature of the shocked gas,

$$T = 3.3 \times 10^6\ \text{K}\left(\frac{E_{51}}{n_0}\right)^{2/5} t_4^{-6/5}. \tag{10.29}$$

Note that the shell remains hot enough ($T > 10^6$ K) to emit observable soft X-rays for $\approx 3 \times 10^4$ yr. The X-ray spectrum is dominated by continuum for $\varepsilon > 1.5$ keV and by strong emission lines of C, N, O, Ne, Mg, and Fe for $\varepsilon < 1.5$ keV.

The instantaneous power radiated by the shell is given by

$$L(t) = 4\pi R^2 \, \Delta R n_1^2 \Lambda(T) = \frac{16\pi}{3} R^3 n_0^2 \Lambda(T). \qquad (10.30)$$

If we use (10.28), (10.29), and the approximate formula (10.9) for $\Lambda(T)$, we find that $L(t)$ rises slowly at first, $L(t) \approx t^{0.6}$ for $T > 10^7$ K, then rapidly, $L(t) \approx t^{2.04}$ for $T < 10^7$ K, when strong line emission sets in. Actually, $L(t)$ rises even more rapidly than this. Because the shocked gas is 'under-ionized' (cf. Section 10.2), the cooling due to line emission is enhanced by factors ≈ 3. The X-ray emission spectra of supernova shells clearly show enhanced line emission due to these non-equilibrium effects.

The energy conserving phase ends at a time t_1 when $\int L(t) \, dt \approx 0.3 \, E_0$. From (10.28)–(10.30) and (10.9) we obtain

$$t_1 \approx 3 \times 10^4 \text{ yr } E_{51}^{0.22} n_0^{-0.55}, \qquad (10.31)$$

and

$$R_1 = 20 \text{ pc } E_{51}^{0.29} n_0^{-0.42}, \qquad (10.32)$$

where we have increased $\Lambda(T)$ by a factor of 3 to account for the enhanced line emission. After this time the shell collapses, becoming cool, dense, and very thin, and ceases to emit observable X-rays. However, the shell continues to expand, pushed by very hot low-density interior gas, which does not radiate significantly. The pressure of the interior gas decreases according to the adiabatic law, $PV^{5/3} = \text{const.}$, or $P = P_1(R/R_1)^{-5}$, where $P_1 = \frac{2}{3}E_0/(\frac{4}{3}\pi R_1^3)$. Inserting this expression into (10.27), we obtain the expansion law for late times:

$$R(t) = \left[\frac{147}{4\pi} \frac{E_0 R_1^2 t^2}{\rho_0} \right]^{1/7} \approx 19 \text{ pc } E_{51}^{0.23} n_0^{-0.26} t_4^{2/7}, \qquad (10.33)$$

where we have used (10.32) to derive the second expression. The supernova shell will continue to expand for $t > 10^6$ yr, reaching a radius > 60 pc. Finally, after several million years, relatively cool and dense gas will flow back into the cavity.

The simplified theory presented above agrees very well with more-detailed theoretical models. However, the theory might have limited applicability to actual supernova explosions because a key assumption of the model, that of uniform density ambient gas, may be invalid. In fact, the interstellar medium is known to be intrinsically nonuniform, containing a chaotic distribution of relatively dense gas clouds and filaments embedded in a low-density substrate, much of which may be coronal ($T \approx 10^6$ K) gas. In such a complex region the development of a supernova blast wave will be very different from that described above. The blast wave will propagate rapidly through the low-density gas, passing and enclosing the dense clouds. The mass of the expanding system may be dominated by gas evaporated from the clouds rather than by the hot gas that is swept up. If so, one can show that the radius of the system increases as $R \propto t^{3/5}$ rather than $R \propto t^{2/5}$ as in (10.28).

Moreover, the ambient interstellar gas may be modified substantially by the pre-supernova star. Most supernovae in the galactic disk are thought to come from fairly massive stars, say, $> 6\,M_\odot$, corresponding to main-sequence spectral type B4. If the pre-supernova star (or any associated star) is somewhat more massive, say, $> 20\,M_\odot$, it would probably be luminous enough to ionize any interstellar clouds within a radius > 30 pc. If so, the clouds would expand and fill the region with gas of fairly uniform density. On the other hand, massive stars are also likely to have strong stellar winds, and such winds will create cavities of low-density gas around the pre-supernova star. Furthermore, since massive stars tend to be found in clusters and associations, there is a good chance that a given supernova will occur in a region of interstellar gas that has been disturbed by other supernovae during the past few million years. These possibilities are discussed below.

10.5.2 *Interstellar bubbles*

Early-type stars with bolometric luminosity $L_* > 10^5\,L_\odot$ are known to lose mass in strong stellar winds, with mass-loss rates $10^{-8} < \dot{M} < 10^{-5}\,M_\odot$ yr^{-1} and terminal velocities $1000 < V_w < 4000$ km s^{-1}. The mechanical luminosity of the wind, $L_w = \dot{M}V_w^2/2$, is a strongly increasing function of bolometric luminosity, $L_w \approx 10^{35}$ erg s^{-1} $[L_*/(10^5\,L_\odot)]^{1.7}$. Over its lifetime, such a star can eject a substantial fraction of its mass with such velocities and a net mechanical energy $\approx 10^{50}$ ergs. The result is a huge cavity in the ambient interstellar medium that resembles a supernova remnant. Here we sketch the theory of these expanding 'interstellar bubbles'.

As with supernova remnants, interstellar bubbles at first conserve most of their energy in a hot expanding shell, reach a radius at which radiative losses cause the shell to collapse, and thereafter expand more slowly. However, the energy-conserving phase of the bubble lasts longer, typically a few times 10^6 yr instead of a few times 10^4 yr. Thus, typical interstellar bubbles are likely to be somewhat larger than supernova remnants, even though they are formed with less total energy.

The expansion law for interstellar bubbles can be derived fairly accurately from the thin shell approximation. For the energy conserving phase one takes $P = \frac{2}{3} L_w t / (\frac{4}{3} \pi R^3)$ for the right-hand side of (10.27) and obtains

$$R(t) = \left[\frac{25}{14\pi} \frac{L_w t^3}{\rho_0} \right]^{1/5} \approx 31 \text{ pc} \left(\frac{L_{36}}{n_0} \right)^{1/5} t_6^{3/5}, \tag{10.34}$$

where $L_{36} = L_w / (10^{36} \text{ erg s}^{-1})$ and $t_6 = t / (10^6 \text{ yr})$.

The interior structure of an interstellar bubble differs significantly from that of a supernova blast wave. The thin shell of swept-up interstellar gas is driven by the pressure of the hot shocked stellar wind, which occupies most of the interior volume. (The wind itself is stopped in a shock at a radius $r_1 \ll R$.) There is a conduction front at the interface of the shell and the hot interior gas, where some of the swept-up interstellar gas in the shell is evaporated and mixed with the hot shocked wind.

Very early in the evolution of the bubble, $t \approx 10^4 \text{ yr } n_0^{0.14} L_{36}^{0.3}$, the outer shell collapses due to radiative losses, becoming dense and thin. However, the bubble continues to expand according to (10.34), because the hot interior retains a fixed fraction, $\frac{5}{11}$, of the total wind energy and continues to drive the shell. (The numerical coefficient in (10.34) drops by a factor $(\frac{5}{11})^{1/5}$ to 27 pc.) Eventually, the radiative losses of the hot interior become comparable with L_w, and the shocked wind region loses pressure and collapses. The radius, r_1, where the wind shocks moves out so that $r_1 \approx R(t)$. This occurs at a time

$$t_2 \approx 9 \times 10^6 \text{ yr } L_{36}^{0.3} n_0^{-0.7} \tag{10.35}$$

and radius

$$R_2 \approx 115 \text{ pc } L_{36}^{0.4} n_0^{-0.6}. \tag{10.36}$$

(In order to derive (10.35), one must solve for the evolution of the density and temperature of the interior using the evaporation theory of (10.24), then calculate the radiative loss rate.)

Subsequently, the bubble continues to expand, driven now by the ram pressure of the wind itself, $P = \rho_w V_w^2 = L_w / (2\pi R^2 V_w)$. Putting this into the right-hand side of (10.27), we obtain the law for the late expansion of the bubble:

$$R(t) \left[\frac{3}{\pi} \frac{L_w t^2}{\rho_0 V_w} \right]^{1/4} \approx 15 \text{ pc } \left[\frac{L_{36}}{n_0 V_{1000}} \right]^{1/4} t_6^{1/2}, \tag{10.37}$$

where $V_{1000} = V_w / (1000 \text{ km s}^{-1})$.

Interstellar bubbles are difficult to observe in emission because of their large size and low surface brightness. However, a few large faint ring-shaped emission nebulae have been seen around early-type stars. Another potentially observable characteristic of the system is the narrow blue-shifted O VI $\lambda 1035$ absorption line due to the conduction front inside the bubble. The predicted column density, $N_{\text{O VI}} \approx 5 \times 10^{13} \, \text{cm}^{-2}$, is in the range of observed column densities of O VI toward early-type stars. However, it may be difficult to distinguish the O VI in the interstellar bubbles from that in other conduction fronts that may lie along the line of sight to the star.

10.5.3 *Superbubbles*

There is now a substantial body of evidence for expanding shells of interstellar gas with radii exceeding 100 pc and sometimes even 1 kpc. These 'superbubbles' have been observed in various ways. Expanding H I shells have been seen in 21-cm emission line surveys of the Milky Way. Optical emission from giant shells of H II have been seen in interference filter photographs of the Milky Way and other nearby galaxies, particularly the Magellanic Clouds. Large-scale coherent structures of high-velocity gas have been observed through surveys of interstellar absorption lines in the ultraviolet spectra of hot stars. There is also one example, the 'Cygnus superbubble', of a region of soft X-ray emission in the Milky Way with radius ≈ 400 pc. For some of these systems the kinetic or thermal energy can be inferred from observations; often the net energy is tens or hundreds of times 10^{51} erg, the typical energy of a supernova remnant.

A natural explanation for these superbubbles is that they are formed by the combined action of tens or hundreds of supernovae from a cluster of massive stars. Theory and observations indicate that stars with initial masses $> 6 \, M_\odot$ will terminate their evolution as type II supernovae. Type II supernovae are associated with the massive stars and interstellar gas of the galactic disk (Population I); they account for roughly half the supernovae that occur in our Galaxy. (The other half, type I supernovae, are associated with the galactic bulge and halo (Population II) and do not significantly affect the dynamics of the disk gas.) Since massive stars are normally born in clusters and have relatively short lifetimes, type II supernovae should occur in clusters of tens or hundreds, spaced out over $\approx 5 \times 10^7$ yr, the lifetime of a $6 \, M_\odot$ star. Since the interval between supernova explosions is less than the time for interstellar gas to fill the cavity created by previous explosions, the repeated explosions will hammer out a huge cavity in interstellar space.

The theory for the evolution of such superbubbles is very similar to that for a stellar wind bubble. For a first approximation, one may merely replace the quantity L_w in (10.34) by the quantity $r_{\text{SN}} E_0$, where r_{SN} is the rate of supernova explosions in the cluster and E_0 is the energy of each supernova. A simple idealized model for an OB cluster suggests that r_{SN} should be almost constant for $t < 5 \times 10^7$ yr. If the star cluster is formed with a typical initial mass function, $dN_*/d(\log M_*) \approx M_*^{-\alpha}$, and the lifetime of the star can be written $t_* \approx M_*^{-\beta}$, we find that $r_{SN} \approx t^\gamma$, where $\gamma = \alpha/\beta - 1$. For stars with $6\,M_\odot < M_* < 20\,M_\odot$, we have $\alpha \approx 1.6$ and $\beta \approx 1.6$, so $\gamma \approx 0$. Therefore, we may write $r_{SN} \approx N_*/(5 \times 10^7 \text{ yr})$, or $r_{SN} E_0 \approx 0.7 \times 10^{36}$ erg s^{-1} $N_* E_{51}$, where N_* is the total number of stars with mass $> 6\,M_\odot$ initially in the cluster. Substituting this expression in (10.34), we obtain the expansion law for the superbubble in the energy conserving phase:

$$R(t) \approx 116 \text{ pc} \left[\frac{N_* E_{51}}{n_0} \right]^{1/5} t_7^{3/5}, \qquad (10.38)$$

where $t_7 = t/(10^7 \text{ yr})$.

If the assumptions of this model held true indefinitely, a cluster would continue to grow according to (10.38) until $t_7 \approx 5$, when the last supernova occurs. Thus, for a cluster such as Cyg OB2, with $N_* \approx 200$, $E_{51} = 1$, $n_0 = 1$ cm^{-3}, (10.38) predicts a final ($t_7 = 5$) radius $R > 800$ pc! (Cyg OB2 is a very young ($t < 3 \times 10^6$ yr) cluster containing several very luminous stars with strong stellar winds. At this early time in its evolution, the Cygnus superbubble may be driven primarily by the combined action of the stellar winds rather than by supernovae.) However, there are several reasons why the assumptions of (10.38) should break down before the bubble grows to such a large radius.

The first is that radiative cooling may set in at some earlier time. As a first estimate, we may use $L_{36} = 0.7 N_* E_{51}$, $N_* = 200$, $E_{51} = 1$, $n_0 = 1$ cm^{-3}, (10.35) and (10.36) to obtain $t_2 = 5 \times 10^7$ yr and $R_2 = 800$ pc for the time and radius at which cooling becomes important. However, these values may be overestimates if, as seems likely, the superbubble overtakes and entrains interstellar clouds which subsequently evaporate and enhance the interior cooling. But even if cooling becomes important very early, the superbubble will grow to a minimum radius determined by the momentum imparted by the supernova ejecta, given by (10.37). Taking $V_w \approx 10^4$ km s^{-1}, the velocity of the ejecta, we obtain for this limit $R \approx 90$ pc $t_7^{1/2}$. Thus, although there is considerable uncertainty in the time and radius at which cooling becomes important, it seems clear that a star cluster with a few hundred OB stars will create an expanding superbubble that will grow to a radius $R > 90$ pc in less than 10^7 yr.

The massive hot stars in a young cluster can probably ionize and heat interstellar gas to large (> 100 pc) distances. This ionization and heating will cause filamentary gas to expand, homogenizing the interstellar density. The ionization will also cause the expanding shell to fluoresce as a filamentary H II region. However, after $t \approx 10^7$ yr, all the stars with large ionizing flux (spectral type earlier than B0) will have evolved off the main sequence. Subsequent supernovae will continue to hammer the expanding shell, causing it to grow until $t \approx 5 \times 10^7$ yr. If there are no other ionizing stars in the region, the shell will be invisible optically, although it may show up as an H I shell in 21-cm emission line surveys.

The most important fact that we have neglected in our discussion up to now is that the interstellar gas density decreases with vertical distance, z, above the galactic plane with a scale height $\Delta z \approx 100$ pc. When the superbubble grows to this dimension, the shell will break up and the interior energy will squirt into the galactic corona. Thus, the maximum radius of the superbubble in the Milky Way is limited to the scale height of the gas in the galactic disk (which increases with galactocentric distance).

This is probably the reason why the most spectacular superbubbles have been found in the Magellanic Clouds, where several giant H II shells have been observed with radii ranging from 300–600 pc. In contrast to the Milky Way, the Magellanic Clouds have low gravity and large gas scale height, so that a superbubble can contain its energy and grow to large dimension before it squirts into intergalactic space.

Additional evidence for superbubbles in the Magellanic Clouds comes from observations of their supernova remnants. In contrast to the Milky Way supernova remnants, they appear to expand freely, with $R \propto t$ rather than $R \propto t^{2/5}$, until they reach large radii, $R \approx 50$ pc. This phenomenon could be explained if most of the Magellanic Cloud supernovae occur inside superbubbles which have been almost evacuated by the action of previous supernovae.

Finally, we might speculate on the possibility that star formation may be induced in the expanding shell of a superbubble. If so, we would expect star formation to be most efficient in a system where the gas scale height is large. Indeed, the Magellanic Clouds are unusually rich in young massive stars. Perhaps they provide a modest example of the 'starburst' phenomenon that is seen on a more spectacular scale in some galaxies.

References

Section 10.1

References to observations of coronal interstellar gas are given by
McCray, R. & Snow, T. P., Jr. (1979). *Ann. Rev. Astron. Astrophys.* 17, 213.

Section 10.2

The formulae for atomic cross sections and rate coefficients given in this section are approximate. Extensive references to more accurate results can be found in the references below.

The rate coefficient for dielectronic recombination (10.4) was derived by

Burgess, A. (1965). *Astrophys. J.* **141**, 1588.

Ionization fractions for stationary coronal models have been calculated by

Shull, J. M. & Van Steenberg, M. (1982). *Astrophys. J. Supp.* **48**, 95; *errata*, **49**, 351.

Emission spectra of a coronal plasma and the atomic cooling function have been calculated by

Stern, R., Wang, E. & Bowyer, S. (1978). *Astrophys. J. Suppl.* **37**, 195.

Gaetz, T. J. & Salpeter, E. E. (1983). *Astrophys. J. Suppl.* **52**, 155.

Section 10.3

X-ray photoionized nebular models have been calculated by

Kallman, T. R. & McCray, R. (1982). *Astrophys. J. Suppl.* **50**, 263.

Kallman, T. R. (1984). *Astrophys. J.* **280**, 269.

Two-phase models for emission line regions of active galactic nuclei have been discussed by

McCray, R. (1979). In *Active Galactic Nuclei*, ed. C. R. Hazard & S. Mitton, pp. 227–39. Cambridge University Press.

Krolik, J. H., McKee, C. F. & Tarter, C. B. (1981). *Astrophys. J.* **249**, 422.

Guilbert, P., Fabian, A. C. & McCray, R. (1983). *Astrophys. J.* **266**, 466.

Models for the emission line spectra of active galactic nuclei are discussed by

Davidson, K. & Netzer, H. (1979). *Rev. Mod. Phys.* **61**, 715.

Kwan, J. & Krolik, J. H. (1981). *Astrophys. J.* **250**, 478.

Section 10.4

The theory of conduction fronts is discussed in more detail by

Cowie, L. L. & McKee, C. F. (1977). *Astrophys. J.* **211**, 135.

McKee, C. F. & Cowie, L. L. (1977). *Astrophys. J.* **215**, 213.

Balbus, S. A. & McKee, C. F. (1982). *Astrophys. J.* **252**, 529.

Section 10.5

More details of the theory of supernova blast waves and their influence on the interstellar medium are given by

McKee, C. F. & Ostriker, J. P. (1977). *Astrophys. J.* **218**, 148.

McKee, C. F. (1982). In *Supernovae, A Survey of Current Research*, ed. M. J. Rees & R. J. Stoneham. Cambridge University Press.

Model X-ray emission spectra of supernova remnants have been calculated by

Shull, J. M. (1982). *Astrophys. J.* **262**, 308.

Hamilton, A. J. S., Sarazin, C. L. & Chevalier, R. A. (1983). *Astrophys. J. Suppl.* **51**, 115.

Observations of stellar winds from hot stars are summarized by

Garmany, C. D., Olson, G. L., Conti, P. S. & Van Steenberg, M. E. (1981).
    Astrophys. J. **250**, 660.

The theory of interstellar bubbles is described by

Castor, J. I., McCray, R. & Weaver, R. (1975). *Astrophys. J. Lett.* **200**, L107.

Weaver, R., McCray, R., Castor, J. I., Shapiro, P. & Moore, R. L. (1977).
    Astrophys. J. **218**, 377.

Theory and observations of superbubbles are discussed by

McCray, R. (1983). In *Highlights of Astronomy*, Vol. 6, ed. R. M. West,
    pp. 565–79. Reidel Publishing Co.

Heiles, C. (1979). *Astrophys. J.* **229**, 533.

Cash, W., *et al.* (1980). *Astrophys. J. Lett.* **238**, L71.

Bruhweiler, F. C., Gull, T. R., Kafatos, M. & Sofia, S. (1980). *Astrophys. J. Lett.*
    238, L27.

Kafatos, M., Sofia, S., Bruhweiler, F. & Gull, T. (1980). *Astrophys. J.* **242**, 294.

Meaburn, J. (1980). *Mon. Not. Roy. Astron. Soc.* **192**, 365.

Tomisaka, K., Habe, A. & Ikeuchi, S. (1981). *Astrophys. Space Sci.* **78**, 273.

Abbott, D. C., Bieging, J. H. & Churchwell, E. (1981). *Astrophys. J.* **250**, 645.

Georgelin, Y. M., Georgelin, Y. P., Laval, A., Monnet, G. & Rosado, M. (1983).
    Astron. Astrophys. Suppl. **54**, 459.

Observations of supernova remnants in the Magellanic Clouds are discussed
by

Mathewson, D. S., *et al.* (1983). *Astrophys. J. Suppl.* **51**, 315.

11

Diffuse interstellar clouds

JOHN H. BLACK

11.1 Introduction

Proof of the existence of interstellar gas was first provided by observations of narrow absorption lines in the visible spectra of distant stars. In 1904, J. Hartmann identified lines of Ca II that did not share the periodic variations in Doppler shift exhibited by the principal stellar absorption lines in a spectroscopic binary star, and attributed these 'stationary lines' to foreground material outside the stellar system. Somewhat more than three decades later, the first interstellar molecules, CH, CH^+, and CN, were discovered in similar ways. There is a similarly long history of related investigations in laboratory spectroscopy and in theoretical interpretation.

The interstellar absorption lines tend to be very narrow compared with the photospheric absorption features in the spectra of the hot stars used as background light sources. In terms of line broadening by thermal motions and frequent atomic collisions, this suggests low densities and low temperatures for the absorbing material. In most cases, the observed lines arise only in the lowest states of atoms and molecules, indicating also that the densities and temperatures are too low to maintain significant excited state populations. As we will see, it is possible to infer from such observations quite a lot of information about element abundances, temperatures, densities, cosmic-ray fluxes, and intensities of radiation inside particular clouds.

The term 'diffuse interstellar cloud' has no precise denotation and distinctions between different kinds of interstellar clouds – e.g., diffuse, dark, and giant molecular – are somewhat poorly defined. For our purposes, it is appropriate to adopt a simple empirical notion of a cloud as a sample of interstellar matter that manifests itself as a distinct component of radial velocity in an interstellar spectral line. A cloud that is sufficiently opaque to be identifiable on a photograph as a region of enhanced obscuration (i.e., a decreased density of star images) is properly called a dark cloud. The dividing line between diffuse and dark clouds can be arbitrarily set in terms of the amount of extinction of background starlight that they produce; a reasonable value is $A_V = 3$, the extinction in magnitudes in the visual ($\lambda 5500$ Å) photometric band.

Interstellar matter exhibits a variety of motions ranging from the thermal motions of atoms and molecules to the systematic differential rotation of the Galaxy. Because the difference in Doppler shift between clouds in the same direction is often found to be rather larger than the internal velocity dispersion within a cloud, it is reasonable to suppose that clouds are localized condensations with separations that are larger than their sizes. However, it would be difficult, if not impossible, to recognize as distinct clouds two similar but widely separated absorbing regions in the same direction that happened by coincidence to have the same radial velocity with respect to the observer. Our knowledge of the shapes and sizes of diffuse clouds is in general very limited, because most of the detailed information about the clouds comes from absorption observations that are possible only where suitable background stars can be found. Densities and line of sight dimensions of particular clouds can be inferred from detailed fitting of theoretical models to data. Average densities and separations of clouds can be determined in a statistical sense from measured column densities and the numbers of cloud components toward stars of known distances. Outside the diffuse clouds, there is even more dilute, broadly distributed gas, which tends to have rather higher temperatures than that in the clouds. At the other extreme are large, opaque molecular clouds. For our purposes, the intercloud gas can be considered merely a source of boundary pressure for the diffuse clouds, and the opaque regions, a more extreme version of them.

A diffuse cloud can be described exclusively in terms of spectroscopically determined properties: the strengths and velocity profiles of spectral lines. The study of diffuse clouds thus exemplifies the application of spectroscopic techniques in astrophysics. To go on to describe diffuse clouds according to derived properties such as density, temperature, chemical composition, and so on, some degree of theoretical interpretation is required. It is instructive to compare the analysis of a diffuse cloud with the spectroscopic analysis of a stellar atmosphere. A trivial distinction is that a sphere of gas in hydrostatic equilibrium is always a good first approximation to the structure of a normal star, while the shape and size of a cloud are from the outset unknown. The treatment of radiative transfer in a diffuse cloud is generally much simpler than in a stellar atmosphere: the absorbing region and the background continuum light source are physically distinct in the former case while the formation of both lines and continuum are connected with the structure of the atmosphere in a complicated way in the latter case. On the other hand, thermodynamic equilibrium, in which the physical state of a gas can be characterized by its elemental composition and temperature alone, is a good approximation to the conditions in many stellar atmospheres. At the low densities, $1–1000\,cm^{-3}$, of diffuse interstellar clouds, where the effective temperatures of ultraviolet starlight and penetrating cosmic rays greatly exceed the local kinetic temperature, the conditions must be far from equilibrium. The best one can hope for is that the state of the gas is approximately time-independent and is governed by a steady state among a finite – if not small – number of microscopic processes.

Diffuse interstellar clouds are interesting objects in their own right, but they are also useful tools for the investigation of various other phenomena. They provide important tests of theories of molecule formation that can be applied to the study of more opaque star-forming regions. They can be used to estimate cosmic-ray fluxes outside the solar system. Spectroscopy of diffuse clouds can yield constraints on conditions at the beginning of the cosmic expansion through limits on their content of primordial deuterium and can confirm the black-body character and effective temperature of the cosmic background radiation. In some instances, diffuse clouds can even be used as laboratory absorption cells for the measurement of spectroscopic properties of atoms and molecules that are difficult to study under stable conditions in laboratories on Earth.

11.2 Observations: extinction

Diffuse interstellar clouds contribute to the general extinction of starlight. Extinction is the effect of scattering and absorption of light by solid particles whose diameters are typically less than or of the same order of the wavelength of the light. The intrinsic flux of starlight at wavelength λ, $f_{\lambda 0}$, is diminished by a factor $\exp(-\tau_\lambda)$ upon passing through a cloud of optical thickness τ_λ. The corresponding difference between the apparent and intrinsic (unextinguished) magnitudes of a star, m_λ and $m_{\lambda 0}$, is called the extinction, A_λ, and is related to the ratio of apparent and intrinsic fluxes by

$$m_\lambda - m_{\lambda 0} = A_\lambda = -2.5 \log_{10}(f_\lambda/f_{\lambda 0}) = 1.0857\,\tau_\lambda. \tag{11.1}$$

Interstellar extinction is wavelength dependent in the sense of being relatively stronger at shorter wavelengths and thus has the effect of reddening starlight. The reddening can be described as a color excess $E(\lambda - \lambda')$, which is the difference between the apparent and intrinsic color indices of a star measured at two wavelengths:

$$E(\lambda - \lambda') = (m_\lambda - m_{\lambda'}) - (m_{\lambda 0} - m_{\lambda' 0}) = A_\lambda - A_{\lambda'}. \tag{11.2}$$

The extinction is conventionally expressed as a function of wavelength in terms of the extinction curve, $E(\lambda - V)/E(B - V)$ where B and V refer to the standard blue and visual photometric bands with effective wavelengths 4400 and 5500 Å, respectively. The ratio of total to selective extinction, which normalizes the extinction curve, is found to have a value $R = A_v/E(B - V) \approx$ 3.1 in most directions. A typical galactic extinction curve shows broad absorption resonances at $\lambda = 19.5$ and $9.7\,\mu$m attributable to amorphous silicate particles and at $\lambda = 2200$ Å, usually identified with carbon particles. The strength of the 2200 Å feature and the steepness of the rise in extinction with decreasing λ in the ultraviolet vary somewhat from place to place in the Galaxy, but the basic features of the extinction curve are remarkably uniform. Wherever measurements of extinction and of the amount of interstellar gas can be made toward the same star, it is found that the gas and dust are strongly correlated with each other. On average, the total column density of hydrogen nuclei is

$$N_H = N(H) + 2N(H_2) = 4.77 \times 10^{21}\ E(B - V)\quad \text{cm}^{-2} \tag{11.3}$$

where $N(H)$ and $N(H_2)$ are the column densities of hydrogen atoms and molecules, respectively. This mean relation has been derived from data on 85 lines of sight through diffuse cloud material with $E(B - V) \lesssim 0.53$, and does not necessarily apply to more opaque regions. It can also be shown that the mean projected cross-sectional area of dust grains is probably as large as

$$n(g)\sigma_g/n \approx 1.5 \times 10^{-21}\quad \text{cm}^2 \tag{11.4}$$

per hydrogen nucleus, where $n(g)$ and n are the local number densities of grains and hydrogen nuclei, respectively.

Comparison of the observed interstellar extinction with models of the composition, optical properties and size distribution of dust particles indicates that the solid particles account for approximately 0.7 percent of the mass in interstellar clouds. Using the elemental composition of the Sun as a reference standard, we expect 1.7 percent of the total interstellar mass to be in elements other than hydrogen and the inert gases. Slightly less than half the available supply of these heavy elements then is in solid form in diffuse clouds. Dust grains are known to form in the atmospheres of cool giant stars, in novae and planetary nebulae, and possibly in protostellar nebulae and in supernova remnants. Perhaps only half the mass in existing grains comes directly from these sources, however; the remainder results from a gradual process of accretion by adsorption in atom–grain collisions in interstellar clouds. The grains thus travel from sources associated with highly evolved stars into clouds that provide raw materials for future generations of stars, and thereby participate in the overall chemical evolution of the Galaxy.

Those grains that are non-spherical and that have some preferential alignment in space can polarize the light from background stars. The light from distant stars in the galactic plane is in fact linearly polarized, and the fractional polarization is correlated with the amount of extinction. The direction of alignment is associated with the weak interstellar magnetic field.

11.3 Observations: atomic and molecular absorption lines

Most of the information about the composition, motions and physical conditions of diffuse clouds has come from studies of interstellar absorption lines superimposed upon the spectra of background stars. Table 11.1 lists the atomic and molecular species that have been identified in the most thoroughly studied diffuse cloud in the ultraviolet, visible and near infrared regions of the spectrum. Diffuse clouds are predominantly neutral regions. The observed species are neutral atoms and first ions of the same elements that are abundant in the Sun, plus some diatomic molecules composed of them. No polyatomic molecule has yet been observed in interstellar clouds by optical techniques. In a cold, dilute gas, only the lowest energy levels of atoms are populated; therefore, only species with resonance transitions at accessible wavelengths are detectable. In the visible region in particular, only a limited variety of minor constituents has such transitions: metals like Ca, Na, K, Ti, Fe, and Li. With the advent of high-resolution ultraviolet spectrographs in spacecraft came a deluge of new information on a large number of elements. It also became possible at last to confirm theoretical expectations that the hydrogen molecule, H_2, should be widespread and abundant compared with atomic hydrogen.

Fig. 11.1 illustrates some of the practical considerations that affect absorption line observations of interstellar atoms and molecules. Many of the lines of interest are both weak and narrow; thus they put extreme demands upon both the resolution and sensitivity of a spectrograph. One would expect stronger interstellar lines in clouds of larger total column density. At the same time, clouds of large column density necessarily produce greater extinction of the background starlight, thus making the measurements more difficult. Even though narrow interstellar lines can often be distinguished from stellar absorption lines easily on the basis of widths alone, accurate measurement of their strengths can still be made difficult if they happen to coincide with a stellar feature as shown in the figure.

The observable property in absorption line studies is the absorption relative to the adjacent continuum, $\exp(-\tau_\lambda)$, where the line optical depth τ_λ is a function of wavelength about the line center λ_0. Often, as in Fig. 11.1, the intrinsic width of the line profile is smaller than the resolution so that only the integrated absorption, called the equivalent width W_λ, can be measured. In terms of the intensity in the line, I_λ, relative to the intensity of the continuum nearby, I_0,

$$W_\lambda = \int \frac{I_0 - I_\lambda}{I_0} \, d\lambda = \int 1 - \exp(-\tau_\lambda) \, d\lambda. \qquad (11.5)$$

The equivalent width can be measured accurately only if the continuum varies slowly with λ in the vicinity of the line and if $W_\lambda \gg \Delta\lambda/(S/N)$, where $\Delta\lambda$ is the resolution and S/N is the ratio of continuum signal to noise achieved in the observation. Note in Fig. 11.1 that the interstellar lines that lie in otherwise featureless parts of the stellar spectrum are more prominent and

Fig. 11.1. A high-resolution spectrum of the star HD 80077 near 8750 Å obtained by E. F. van Dishoeck and J. H. Black using the 4-m telescope and echelle spectrograph with a CCD detector at the Cerro Tololo Inter-American Observatory. Narrow interstellar absorption lines in the $A^1\Pi_u - X^1\Sigma_g^+$ (2, 0) band of C_2 are indicated by the branch designation (P, Q, or R) and the rotational quantum number of the lower state. Note that the R(4), R(6), and R(8) lines are coincident with a strong, broad absorption due to the Paschen 12 line of atomic hydrogen in the atmosphere of the star. For reference, the triangle has an equivalent width of 10 mÅ. (The Cerro Tololo Inter-American Observatory is operated by the Association of Universities for Research in Astronomy, Inc., under contract with the US National Science Foundation.)

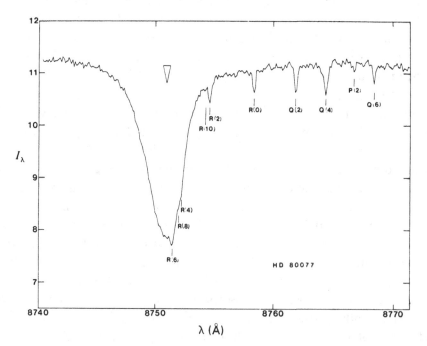

more reliably measured. The optical depth depends upon the column density N_1 of absorbers in the initial (lower) state 1 of the upward transition $1 \rightarrow u$, the oscillator strength of the transition, f_{lu}, and a line shape function $\theta(\lambda)$

$$\tau_\lambda = \frac{\pi e^2}{mc^2} \lambda_0^2 N_1 f_{lu} \theta(\lambda - \lambda_0). \tag{11.6}$$

The line shape function is normalized

$$\int \theta(\lambda - \lambda_0) \, d\lambda = 1 \tag{11.7}$$

in units consistent with those used for N_1, λ_0, and the fundamental constants e and m, the charge and mass of the electron, respectively, and c, the velocity of light. The column density N_1 in (11.6) contains an implicit integration of absorber concentration over path length along the line of sight. For the moment we assume a uniform distribution, but we will consider some consequences of stratified cloud structures later. It is worth noting that atomic transitions are often designated with the lower state first, while in molecular spectroscopy the opposite convention is used regardless of whether the transition is upward or downward.

When nature is being most cooperative, a single cloud (i.e., a single radial velocity component) without systematic velocity structure provides the observed absorption and the line shape function is a convolution of functions representing: (a) thermal Doppler motions of the absorbers, (b) macroscopic but random motions within the cloud (loosely called turbulence), and (c) natural broadening (radiation damping). The observed linewidth is usually larger than that expected for purely thermal motions. As long as the spectrum of 'turbulence' can be assumed to produce approximately Gaussian line shapes, terms (a) and (b) together can be characterized by a single Doppler parameter $b = (2 \ln 2)^{1/2} \Delta V$ where ΔV is the full-width at half-maximum of the Gaussian part of $\theta(\lambda - \lambda_0)$ in velocity units, $\Delta V = c \, \Delta\lambda / \lambda_0$. Term (c) contributes extended wings of Lorentzian shape, that become apparent only in the very strongest lines.

In the limit of very weak lines, $\tau_{\lambda_0} \ll 1$, the exponential function in (11.5) can be replaced by the first two terms in its series expansion

$$W_\lambda \approx \frac{\pi e^2}{mc^2} \lambda_0^2 N_1 f_{lu}, \tag{11.8}$$

and W_λ increases linearly with the column density of absorbers. As soon as $\tau_{\lambda_0} \gtrsim 1$, W_λ increases more slowly with increasing N_1, and the line is said to be saturated or to lie on the flat part of the curve of growth. When $\tau_{\lambda_0} \gg 1$, only the Lorentzian wings far from the saturated line core contribute additional absorption and $W \propto N_1^{1/2}$. Thus, for strong lines, W_λ must be compared with a direct integration of τ_λ over the line, and it will depend on both N_1 and b. The linear approximation (11.8) applies to an accuracy of 10 percent or better in derived N_1 only when

$$W_\lambda/\lambda \lesssim 1.66 \times 10^{-6} \, b \tag{11.9}$$

for b in km s^{-1}. Note that for lines on the flat part of the curve of growth, an uncertainty of 10 percent in the measured equivalent width can lead to an uncertainty in derived N_1 of a factor of 2 or more. The interstellar lines in Fig. 11.1 are just weak enough to satisfy the criterion of (11.9).

Interstellar absorption line spectra can be obtained with signal-to-noise ratio S/N $\gtrsim 100$ at a resolving power $\lambda/\Delta\lambda = 10^5$; therefore, lines as weak as $W_\lambda/\lambda \approx 10^{-7}$ are detectable. Combining equations (11.4) and (11.8), we find that trace constituents with relative abundances as small as

$$N_1/N_H \approx 1.6 \times 10^{-11}(3000 \text{ Å}/\lambda)(0.05/f_{lu})E(B-V)^{-1} \tag{11.10}$$

can be detected in such a spectrum.

In addition to the narrow absorption lines of atoms and molecules, there are also approximately 45 diffuse interstellar bands. The most prominent of these occur at wavelengths of 4428, 5780, and 6284 Å. They exhibit broad, shallow profiles whose equivalent widths or central depths correlate fairly well with extinction. The 6284 Å band, for example, has on average $W_\lambda = 0.2$ Å when $E(B-V) = 0.3$ magnitudes. The origin of the diffuse bands remains somewhat mysterious. Either intrinsically diffuse transitions in a large gas phase molecule or transitions in some species bound to grains could account for the breadth of the diffuse bands, but in any case their carrier must be a relatively abundant component of the interstellar medium.

The basic absorption line data on diffuse clouds can be summarized as follows. Hydrogen, the most abundant element, is almost entirely neutral and a significant fraction of it is in molecular form in diffuse clouds. The line of sight molecular fraction

$$f(H_2) = 2N(H_2)/(2N(H_2) + N(H)) \tag{11.11}$$

shows an abrupt threshold between $f(H_2) < 10^{-3}$ and $f(H_2) \gtrsim 0.1$ at $E(B-V) \approx 0.1$ mag. Measurements of ultraviolet lines arising in various rotational levels in the ground state of H_2 (J=0 through J=7) provide column densities for each level and these typically show two distinct population distributions. The J=0, J=1, and J=2 levels, which have excitation energies $\varepsilon/k = 0.0$, 170.5, and 509.9 K, are populated as though in equilibrium at temperatures $T \approx 50$–100 K. The relative populations of the higher levels are usually characterized by rather higher temperatures $T \approx 200$–400 K. Several molecules containing C, N, and O in combination with H or with each other are observed in diffuse clouds, but they comprise only small fractions of these elements.

Although some elements in diffuse clouds show abundances relative to hydrogen that are comparable with those in the Sun, most show evidence of depletion compared with solar abundances. In some cases the depletion is modest, factors of 0.25 to 0.5 for C, Mg, Cl, P, K, and Na; but for some metals, notably Fe, Ca, and Ti, the depletion factors can be less than 0.01. In terms of solar abundances, the missing elements sum to the same mass fraction as that inferred for the dust particles. For this and chemical reasons as well, selective depletion of the interstellar gas can be attributed to the condensation and accretion of dust particles.

11.4   Observations: radio lines

Some atoms such as H, D, and Na have radio-frequency transitions between ground-state hyperfine levels. The 21-cm wavelength line of H is seen from atoms distributed throughout the interstellar medium. Often the 21-cm line emission from a specific diffuse cloud will be hopelessly blended with all background and foreground emissions at the same radial velocity within the response pattern of the antenna. In a statistical sense, however, models of the H line emission from the galactic plane indicate an atomic cloud component with mean properties $T = 60$ K, a radius of 2.5 pc, and $N(H) = 3 \times 10^{20}$ cm^{-2} that are very similar to those derived from optical absorption line data on diffuse clouds.

Although some weak emission lines of CO and OH and perhaps some lines of polyatomic molecules like H_2CO can be attributed to diffuse clouds, radio molecular lines have not been extensively used to study the diffuse clouds. Typical diffuse clouds do not always have sufficiently high hydrogen densities for excitation of radio lines above the background and may not contain detectable quantities of complex molecules. Consider the widely observed CO molecule as an example. It can be shown that in a diffuse cloud of modest density, say, $n(H_2) \approx 300$ cm^{-3}, the excitation temperature that characterizes the populations of two adjacent rotational levels, J″ and J′ separated by an energy difference $\Delta\varepsilon$ is defined by

$$\frac{n_{J'}}{n_{J''}} = \frac{g_{J'}}{g_{J''}} \exp\left(-\Delta\varepsilon/kT_{ex}\right) \tag{11.12}$$

and is unlikely to exceed $T_{ex} \approx 10$ K. In equation (11.12) n_J is the concentration, cm^{-3}, of molecules in level J and $g_J = 2J + 1$ is the statistical weight of the level. With the assumption that $T_{ex} \lesssim 10$ K, a total column density of CO

$$N(CO) \gtrsim 3.0 \times 10^{13} \Delta V \quad cm^{-2} \qquad (11.13)$$

is required to produce a $J = 1-0$ line at 2.6 mm wavelength that has a detectable brightness temperature of 0.05 K in excess of the 2.7 K cosmic background radiation, where ΔV is the line width in $km s^{-1}$. For comparison, observations of the $A^1\Pi-X^1\Sigma^+$ $(1,0)$ Q(1) line of CO in the ultraviolet at the level implied in equation (11.10) are sensitive to an amount of CO 20 times smaller than this.

Radio and optical observations can offer complementary views of diffuse clouds. Optical absorption line studies are restricted to the locations where suitable background stars can be found. This is a handicap with some benefits: the star subtends a negligible solid angle on the sky so that interpretation of inhomogeneity within the cloud must deal with one dimension only. With radio-frequency emission lines, on the other hand, the entire extent of a cloud on the sky can in principle be mapped, although usually with only modest angular resolution. Limited angular resolution means that the antenna averages over considerable structure within the cloud in ways that can severely complicate the interpretation.

11.5   Theory and interpretation

Diffuse interstellar clouds afford numerous possibilities for the analysis of spectra in terms of atomic and molecular processes. Not only can we learn much about the physical state of a cloud and about interstellar chemistry generally, but we can also use the clouds as remote spectroscopic sensors of the galactic radiation field and cosmic-ray flux. In the following, we will discuss various aspects of the theoretical description of diffuse clouds required for a full interpretation of the observations.

Information about abundances can be inferred more or less directly from observations. Another direct limit is that mean cloud densities must be at least as large as $\langle n \rangle \approx N_H/D$, where D is the distance to the background star: typically $\langle n \rangle > 1 cm^{-3}$. Observed linewidths also place limits on temperature. For example, a CH absorption line having a width $\Delta V = 1 km s^{-1}$ can be formed in a gas no hotter than $T \lesssim 600$ K when the broadening is assumed to be entirely thermal.

To proceed in more detail, it is necessary to apply theoretical methods, for which some boundary conditions must be established. A cloud is exposed to the combined light of hot stars in the galactic plane and to some flux of cosmic rays. Ultraviolet starlight and 1–100 MeV cosmic-ray protons most severely affect the state of a cloud. To the extent that the solar system occupies a typical place in the galactic plane, the locally measured diffuse ultraviolet starlight and cosmic-ray flux (after correction for modulation by the solar wind) can be adopted as reference values. Ultraviolet photons with $\lambda < 912$ Å, the Lyman limit of atomic hydrogen, are largely confined to the H II regions around the hot stars that produce them. A typical reference value for the flux of 1000 Å photons is $\phi_\nu = 3 \times 10^{-8}$ photons $cm^{-2} s^{-1} Hz^{-1}$. Including the effects of secondary electrons, but excluding the unseen and putative low-energy cosmic rays, a cosmic-ray ionization rate of the order of $10^{-17} s^{-1}$ per hydrogen atom can be adopted. The small internal velocity dispersions of clouds compared with line of sight velocity gradients due to galactic rotation imply that the clouds must occupy a small fraction of the total volume of interstellar space. Most of this volume is filled by an even more dilute, hotter gas, with which the clouds are probably close to pressure balance.

If the physical conditions through a cloud remain fairly constant over a sound-crossing time, then a typical cloud will live long enough to approach closely a chemical steady state. In what follows, steady state among microscopic processes will be assumed. The description of conditions in a diffuse cloud then consists in evaluating the rates of all microscopic processes that form and destroy all constituent species and populate all their internal energy states. The ionization and excitation of atomic carbon can be taken as a specific example. The photon energy corresponding to the first ionization potential of carbon greatly exceeds the kinetic temperature, $h\nu_0 \gg kT$, but the interstellar radiation field includes photons with frequencies $\nu > \nu_0$. Therefore collisional ionization processes will be negligible, and production of C^+ will be dominated by photoionization of C at a rate

$$\Gamma = \int_{\nu_0}^{\nu_L} \phi_\nu \sigma_C \, d\nu \quad s^{-1} \tag{11.14}$$

where ν_L is the frequency of the Lyman limit of H (the upper cut-off of the interstellar radiation field) and σ_C is the ground-state photoionization cross section. The photon flux ϕ_ν is a function of both position within the cloud and of its total thickness: extinction by dust and competition with H_2 for the same photons attenuate the radiation. A typical unattenuated value is $\Gamma \approx 2 \times 10^{-10} \chi s^{-1}$, where χ is a scaling factor for the radiation field of order unity. Because C^+ reacts only very slowly at low T with the principal constituents H and H_2, it is removed primarily by radiative recombination

$$C^+ + e \rightarrow C + h\nu \tag{11.15}$$

which has a thermal rate coefficient $\alpha_C \approx 1.4 \times 10^{-10} \, T^{-0.607}$ cm^3 s^{-1}. In steady state, the concentrations of C, C$^+$, and e are therefore related by

$$n(C^+)n(e)/n(C) = 1.4 \, T^{0.607}\chi \tag{11.16}$$

where the radiation field is unattenuated. Hydrogen is not highly ionized, but most elements like carbon with first ionization potentials less than $h\nu_L$ exist predominantly as first ions in diffuse clouds. For the conditions of interest and with typical gas-phase abundances, (11.16) implies that

$$n(e) \approx 1.3 \, n(C^+) \approx 2 \times 10^{-4} n \tag{11.17}$$

where

$$n = 2n(H_2) + n(H) \tag{11.18}$$

is the total density of hydrogen nuclei. Because the photoionization rate, and hence the $n(C^+)/n(C)$ abundance ratio, vary with depth through a cloud, it is necessary to use self-consistent models of cloud structure to interpret the observable column densities, $N(X) = \int n(X)\, dr$, for all but the thinnest clouds. If carbon atoms and ions were in thermodynamic equilibrium, according to the Saha formula

$$n(C^+)n(e)/n(C) = 3.22 \times 10^{15} \, T_i^{3/2} \exp\left(-129625/T_i\right), \tag{11.19}$$

an ionization temperature $T_i = 3050$ K would be required to produce the same level of ionization as is given by (11.16) with $\chi = 1$. This temperature, not surprisingly, is approximately equal to the brightness temperature of the radiation field at ν_0, and is much higher than the kinetic temperature of the gas.

The ground term of C$^+$, ^2P_J, has two fine-structure levels $J = \tfrac{1}{2}$ and $\tfrac{3}{2}$, separated in energy by an amount $\Delta\varepsilon/k = 91.25$ K, which is comparable with the mean kinetic energy per particle in the gas. Higher excited states of C$^+$ are energetically inaccessible and too short-lived to be important in diffuse clouds. The relative populations of the two fine-structure levels are governed by spontaneous magnetic dipole transitions with a probability $A = 2.29 \times 10^{-6}$ s^{-1}, by inelastic collisions with e, H, and H$_2$, and by absorption of starlight and subsequent fluorescence in the resonance transitions at 1335 Å and 1037 Å. With the use of analytic fits to the rates of these processes, we find for the ratio of level populations

$$\frac{n(\tfrac{3}{2})}{n(\tfrac{1}{2})} = \frac{2nF \exp\left(-91.25/T\right) + 1.12\,\chi}{nF + 2.29 \times 10^4 + 0.56\,\chi} \tag{11.20}$$

where

$$F(T, x_e, f) = 3.91(10^4 \, x_e)T^{-1/2} + (1.5 - 0.6 \, f(H_2))\log T$$

$$+ 5 - 4.41 \, f(H_2) \tag{11.21}$$

is a slowly varying function of the electron fraction $x_e = n(e)/n$, of the local molecular fraction $f(H_2) = 2n(H_2)/n$, and of T. When x_e is given by (11.17), and with $f(H_2) = 0.5$, $\chi = 1$, and $T = 100$ K, the two levels will be equally populated at a density $n = 6300$ cm^{-3}. At lower densities, spontaneous transitions dominate the removal from $J = \frac{3}{2}$, creating infrared photons that carry an amount of energy $h\nu \gtrsim kT$ and that escape from the cloud with high probability. This is an important cooling mechanism that helps determine the steady state temperature of the gas.

We consider next the forms in which the most abundant element, hydrogen, can exist. Hydrogen molecules cannot form by direct binary collisions of hydrogen atoms under dilute interstellar conditions. The only efficient mechanism for producing H_2 is evidently the reaction of one H atom with another held captive on the surface of a dust grain. If an incident H atom has a probability κ of sticking and migrating around the grain surface long enough to meet another atom and form a molecule that escapes, then the rate of H_2 formation is related to κ and to the rate at which H atoms strike grains:

$$\text{rate of formation} = n(H)n(g)\sigma_g v_H \kappa \quad \text{cm}^{-3}\,\text{s}^{-1} \tag{11.22}$$

where

$$v_H = 1.45 \times 10^4\, T^{1/2} \quad \text{cm s}^{-1} \tag{11.23}$$

is the mean thermal velocity of a hydrogen atom. Combining (11.4), (11.22), and (11.23) we find a formation rate for H_2 of $2 \times 10^{-17} \kappa T^{1/2} nn(H)$ cm^{-3} s^{-1}. Experimental and theoretical studies suggest that $\kappa \approx \frac{1}{4}$ is not unreasonable. At $n \approx 100$ cm^{-3} and $T = 100$ K, for example, an average atom will survive almost 7 million years before becoming part of a molecule. A substantial molecular abundance can arise only if a typical H_2 molecule has a comparable or greater lifetime.

In diffuse clouds the most effective destruction process for H_2 is a two-step fluorescent photodissociation first proposed by P. M. Solomon in 1966. H_2 can absorb ultraviolet photons in discrete lines in 21 bands of the Lyman system and 7 bands of the Werner system in the wavelength interval 912–1108 Å. The resulting excited molecules decay by spontaneous transitions back down to various rotational vibrational levels of the ground state and to the vibrational continuum that lies above the ground-state dissociation limit. In the latter case, the molecule is destroyed. Depending upon the initial rotational population distribution and the character of the ultraviolet radiation field, a fraction 0.1 to 0.15 of the initial line absorptions results in dissociation of H_2 into two H atoms. In the unattenuated interstellar radiation, the effective rate of dissociation is approximately $3 \times 10^{-11} \chi$ s^{-1}.

In balance with the H_2 formation rate, this implies a molecular fraction

$$f(H_2) \approx 3.3 \times 10^{-5}(\kappa/0.25)(T/100)^{1/2}(n(H)/10)\chi^{-1} \qquad (11.24)$$

comparable with that observed in the thinnest clouds where $E(B-V) <$ 0.1 mag, but significantly less than that seen in thicker clouds. Referring to the earlier discussion of curve-of-growth effects on the integrated absorption in narrow lines, we can find a life-saving complication (for H_2) of the dissociation process. From equation (11.3), $f(H_2) = 3 \times 10^{-5}$ at $E(B-V) =$ 0.05 corresponds to a column density $N(H_2) = 4 \times 10^{15}$ cm^{-2}, which may be distributed among several rotational levels. Lower state column densities of the order of 10^{14}–10^{15} cm^{-2} are sufficient to produce central optical depths $\tau_0 \gtrsim 1$ in the Lyman and Werner lines of H_2. At greater column densities and optical depths, the absorption rate (and hence the dissociation rate) is drastically reduced as the Doppler cores of the lines become highly saturated and the only increment to the absorption with increasing depth comes from the weakly absorbing line wings. Because H_2 dissociation is initiated by line absorptions that can saturate readily rather than by continuous processes, the molecules in the outer part of a cloud effectively shield the interior whenever the H_2 column density exceeds the threshold value 10^{15} cm^{-2}. In the interior of a fairly thick cloud, the self-shielding can reduce the dissociation rate by a factor as large as 10^4 compared with its boundary value and allow $f(H_2) \approx 1$ in the interior.

The nondissociating ultraviolet absorptions produce vibrationally-excited molecules that subsequently relax to rotational levels of the $v=0$ ground state through a cascade of weak electric quadrupole transitions. This has the effect of redistributing the ground-state rotational populations and is most effective for the higher levels, $J \gtrsim 3$, for which the effective radiative excitation rates can be rather larger than the rates of collisional processes. The abundance and excitation of H_2 are coupled strongly in another sense. The details of the depth dependence of the ultraviolet absorption are affected by the rotational populations within the ground state: lines arising in the most populous levels become saturated earliest, while less populous excited levels contribute to dissociation and excitation deeper into the cloud.

Collisional processes affect the populations of the very lowest rotational levels of H_2. Symmetry properties of a homonuclear diatomic molecule like H_2 require that states of even and odd rotational quantum number have different alignments of nuclear spin. A transition within a given electronic state that involves the change of an odd number of rotational quanta must also involve a change of nuclear spin states, but such nuclear spin interactions are exceedingly weak both in radiative processes and in unreactive collisions. A molecule of H_2 in $J=1$, if left to itself, will live for 4×10^{12} yr before undergoing a spontaneous transition to $J=0$. It is therefore somewhat surprising that the $J=0$ and $J=1$ levels of H_2 in diffuse clouds appear to be populated thermally at a mean temperature $T_{01} = 77$ K, rather than at a higher value that would follow from the formation of independent even-J (para) and odd-J (ortho) molecules on grain surfaces. The explanation is that reactive collisions of the sort

$$H^+ + H_2(J) \rightarrow H_2(J \pm 1) + H^+, \tag{11.25}$$

can occur rapidly enough to compete with excitation, formation, and destruction processes. In this process, a free proton replaces and changes the nuclear spin orientation of one of the nuclei in the molecule. A small residual ionization of hydrogen can be maintained by penetrating cosmic rays through

$$H + CR \rightarrow H^+ + e + CR \tag{11.26}$$

$$H_2 + CR \rightarrow H_2^+ + e + CR \tag{11.27}$$

and related processes. Free protons are removed by

$$H^+ + e \rightarrow H + h\nu, \tag{11.28}$$

by charge transfer with oxygen

$$H^+ + O \rightleftarrows O^+ + H, \tag{11.29}$$

an accidentally near-resonant process, and by reactions with less abundant molecular species. The hydrogen molecular ion, H_2^+, is rapidly converted to H_3^+ by

$$H_2^+ + H_2 \rightarrow H_3^+ + H \tag{11.30}$$

wherever the molecular fraction is appreciable. This ion drives a great amount of chemical activity. For example,

$$H_3^+ + O \rightarrow OH^+ + H_2, \tag{11.31}$$

together with the reaction

$$O^+ + H_2 \rightarrow OH^+ + H \tag{11.32}$$

that follows (11.29), lead to a sequence of reactions

$$OH^+ + H_2 \rightarrow H_2O^+ + H \tag{11.33}$$

$$H_2O^+ + H_2 \rightarrow H_3O^+ + H \tag{11.34}$$

that terminates in the formation of H_2O and OH when H_3O^+ recombines with electrons. Another effect of the low-level ionization of hydrogen is the series of processes involving deuterium

$$H^+ + D \rightleftarrows D^+ + H \tag{11.35}$$

$$D^+ + H_2 \rightarrow HD + H^+ \tag{11.36}$$

that provides a more rapid formation mechanism for HD than direct association on grain surfaces. Both HD and OH then have production rates that are tied ultimately to the rate of cosmic-ray ionizations of hydrogen, and both are removed primarily by photodissociation in diffuse clouds. It is possible to use observed abundances of these molecules to infer the cosmic-ray ionizing rate provided that the depth-dependent abundance effects can be modeled accurately.

In diffuse clouds with a small molecular fraction, the measurement of column densities of H and D yields directly a deuterium abundance. In terms of standard theories of evolutionary cosmology and nucleosynthesis, the current deuterium abundance is a lower limit to the primordial abundance, which in turn is very sensitive to conditions in the early universe at the beginning of the cosmic expansion.

As we have just seen, deuterium is incorporated into molecules at a different rate than hydrogen itself. As a result, the abundance ratio HD/H_2 can be somewhat different from the overall deuterium abundance. This is an example of isotope fractionation, which can occur readily for isotopic varieties of many other interstellar molecules. In the case of deuterium, note that reaction (11.35) proceeds in both directions but with temperature-dependent probabilities and that (11.36) goes preferentially in one direction at interstellar temperatures. Owing to their vastly different abundances, the self-shielding against photodissociation also differs significantly for H_2 and HD. Their relative abundances are thus sensitive to local conditions and are related to the overall deuterium abundance in a complicated way. Similar mechanisms cause the abundances of deuterated species like H_2D^+ and DCO^+ to be enhanced by factors of 100 to 1000 relative to the overall deuterium fraction.

Although C^+ cannot react directly with H_2 at low temperature, it can initiate an active carbon chemistry through a radiative association reaction

$$C^+ + H_2 \rightarrow CH_2^+ + h\nu \tag{11.37}$$

in which the inefficient production of a photon during the reaction conserves energy and momentum. Subsidiary reactions form CH_3^+ and a variety of neutral species including CH, CH_2, C_2, and C_2H. The very important molecule, CO, can be formed following reactions of C and C^+ with OH and H_2O, and by an alternate sequence of reactions of O with several simple hydrocarbons. At the time of this writing, however, the photodissociation processes that control the CO abundance in diffuse clouds must be considered poorly understood. As with H_2 and HD, photodissociation of CO is dominated by ultraviolet line absorptions that lead by some means to repulsive states of the molecule. Not only are oscillator strengths of some crucial transitions not determined within factors of 2 or more, but even the issue of whether these transitions predissociate strongly at all is unsettled.

As outlined above, it is possible to build up elaborate networks of gas-phase reactions to try to account for the abundances of interstellar molecules. Because this chemistry is so far out of equilibrium, it is necessary to have rates for a large number of individual processes in order to solve for the steady state of the chemical system. Although many of the relevant reactions have been studied in laboratory experiments or by means of *ab initio* quantum chemical calculations, the rates of other crucial processes have been woven out of thin threads of guesswork.

A compelling case can be made that H_2 is formed on grain surfaces, but the formation of heavier molecules on grain surfaces is problematical. In diffuse clouds where chemical time scales are relatively short, grains might provide a minor general source of some molecules whose relative abundances are likely to be re-arranged rapidly by gas-phase processes. Deep inside darker clouds, the dominant effect of grains may be to remove almost all species from the gas phase by accretion.

The excitation of minor species in diffuse clouds can provide valuable diagnostic tools. We have already seen that the abundance and rotational population distribution in H_2 are related to the total density, temperature, and ultraviolet intensity, but in a severely depth-dependent manner. Another homonuclear molecule, C_2, shows measurable populations in highly excited rotational levels (see Fi. 11.1) and these populations are also determined by the competition between radiative excitation in electronic transitions and inelastic collision processes. In this case, however, C_2 is a minor constituent with generally unsaturated absorption lines and is probably confined only to the central regions of the denser, thicker diffuse clouds so that depth-dependent effects are not as severe as with H_2. With C_2 it is possible to obtain independent information on temperature and on the ratio of hydrogen density to photon density.

Another molecule of great diagnostic value is CN. It is observable by means of interstellar absorption lines both in the violet region and in the near infrared. Owing to a large permanent dipole moment, the radiative transitions between its lowest rotational levels at 1.32 and 2.64 mm have very large probabilities, $A = 1.7 \times 10^{-4}$ and $1.2 \times 10^{-5}\,\mathrm{s}^{-1}$, respectively. These rates are much larger than those of the competing collisional processes in diffuse clouds with $n \lesssim 1000\,\mathrm{cm}^{-3}$. As a result, the populations of the rotational levels of CN in diffuse clouds are controlled entirely by radiative processes, specifically by spontaneous emission and by stimulated emission and absorption in the cosmic background radiation. In this low density limit, the excitation temperature (see (11.12)) must approach the brightness temperature, T_b, of the background radiation. The excitation temperature can be determined from absorption measurements of lines arising in the different rotational levels. The most recent study of this phenomenon by D. M. Meyer and M. Jura has confirmed the value of $T_b = 2.73 \pm 0.04$ K and the black-body character of the radiation field at 1.3 and 2.6 mm wavelength near its peak.

Diffuse clouds have proven to be useful laboratories. Interstellar absorption line observations have been used to determine oscillator strengths from relative equivalent width measurements for lines of Si II, C I, and OH. Comparison of observed abundances with chemical models can sometimes provide estimates of rate coefficients of reactions that cannot be determined by other means, a rate coefficient of the order of $5 \times 10^{-16}\,\mathrm{cm}^3\,\mathrm{s}^{-1}$ for reaction (11.37) being one example. Although this exercise should not be pursued too far into the wilderness of ignorance, it can sometimes be useful if only because it stimulates someone to find a better answer by rigorous experimental or theoretical means.

As a laboratory, diffuse clouds may have their greatest utility as testing grounds for theories of molecule formation and excitation which can then be applied with more confidence to the more complicated dense, dark clouds. For example, the search for a completely successful description of the abundance and excitation of CO in diffuse clouds will be important for understanding the depth dependences of its abundance and isotope fractionation in dark clouds.

We have seen that the abundance of H_2 depends on density and on the ultraviolet radiation field in ways that are very sensitive to depth within a cloud. The distribution of H_2 among its ground state rotational levels is also sensitive to density, temperature, radiation field, and to the proton abundance, which in turn depends upon the cosmic-ray flux. The excitation of the C_2 molecule is likewise related to temperature, density and another part of the radiation field. The relative abundances of neutral atoms and singly-ionized species, and of various molecules depend upon density,

Table 11.1. *Interstellar atoms and molecules observed in the well-studied diffuse cloud toward ζ Ophiuchi*

Species		log (column density) (cm^{-2})	Reference
Atomic species			
H		20.72	1
Li		9.57	2
Be^+		< 10.5	3
B^+		< 10.5	3
C	$J=0$	15.30 ± 0.05	4
	$J=1$	15.00 ± 0.10	4
	$J=2$	14.30 ± 0.30	4
	total	15.50 ± 0.07	4
C^+	$J=1/2$	16.97 ± 0.17	1
	$J=3/2$	14.99 ± 0.13	1
	total	16.97 ± 0.17	1
N		$16.72^{+0.21}_{-0.42}$	5
O	$J=2$	17.85 ± 0.03	6
	$J=1$	< 12.88	1
Na		13.78	7
Mg		14.15	1
Mg^+		15.91	1, 8
Al		< 10.21	1
Al^+		13.01 ± 0.10	9
Si		$\leqslant 12.58$	1
Si^+		15.04 ± 0.10	9
P^+		13.46 ± 0.09	1
S		13.93	1
S^+		16.05 ± 0.19	1
Cl		14.03	1
Cl^+		13.35 ± 0.27	1
Ar		15.06 ± 0.16	1
K		11.84	7
Ca		9.72	1
Ca^+		11.75	7
Ti^+		11.48 ± 0.04	10
Mn^+		13.29 ± 0.04	1
Fe		11.59	1
Fe^+		14.56 ± 0.08	1
Ni^+		13.12 ± 0.16	1
Cu^+		11.93 ± 0.15	1
Zn^+		13.27 ± 0.05	1
Rb		< 9.7	11

continued

Table 11.1 (*cont.*)

Species			log (column density) (cm^{-2})	Reference
Molecular species				
H_2	$v=0$	$J=0$	20.46	1
		$J=1$	20.10	1
		$J=2$	18.56 ± 0.15	12
		$J=3$	17.07 ± 0.35	12
		$J=4$	15.68 ± 0.15	12
		$J=5$	14.63 ± 0.05	13
		$J=6$	13.69 ± 0.05	13
		$J=7$	13.55 ± 0.05	13
	$v=1$	$J=0$	< 12.84	1
		$J=1$	< 13.08	1
		$J=2$	< 13.00	1
	$v=2$	$J=0$	< 12.64	1
	total		20.62	
HD	$v=0$	$J=0$	14.26	14
		$J=1$	13.44	14
		$J=2$	< 13.41	14
	total		14.32	
CO	total		15.38	15
CH			13.35 ± 0.07	16
CH^+			13.52	17
$^{13}CH^+$			11.83	17
C_2	total		13.18 ± 0.05	18
CN			12.42	19
OH			13.71 ± 0.11	20

References:
 (1) Morton, D. C. (1975). *Astrophys. J.* **197**, 85.
 (2) Ferlet, R. & Dennefeld, M. (1984). *Astron. Astrophys.* **138**, 303.
 (3) York, D. G., Meneguzzi, M. & Snow, T. P. (1982). *Astrophys. J.* **255**, 524.
 (4) de Boer, K. S. & Morton, D. C. (1979). *Astron. Astrophys.* **71**, 141.
 (5) Lugger, P. M., York, D. G., Blanchard, R. & Morton, D. C. (1978). *Astrophys. J.* **224**, 1059; Hibbert, A., Dufton, P. L. & Keenan, F. P. (1985). *Mon. Not. Roy. Astron. Soc.* **213**, 721.
 (6) de Boer, K. S. (1981). *Astrophys. J.* **244**, 848.
 (7) Hobbs, L. M. (1978). *Astrophys. J. Suppl.* **38**, 129.
 (8) Hibbert, A., Dufton, P. L., Murray, M. J. & York, D. G. (1983). *Mon. Not. Roy. Astron. Soc.* **205**, 535.
 (9) Barker, E. S., Lugger, P. M., Weiler, E. J. & York, D. G. (1984). *Astrophys. J.* **280**, 600.
(10) Stokes, G. M. (1978). *Astrophys. J. Suppl.* **36**, 115.
(11) Jura, M. & Smith, W. H. (1981). *Astrophys. J. Letters*, **251**, L43; Federman, S. R., Sneden, C., Schempp, W. V. & Smith, W. H. (1985). *Astrophys. J. Letters*, **290**, L55.

(12) Spitzer, L., Cochran, W. D. & Hirshfeld, A. (1984). *Astrophys. J. Suppl.* **28**, 373.

(13) Spitzer, L. & Morton, W. A. (1976). *Astrophys. J.* **204**, 731.

(14) Wright, E. L. & Morton, D. C. (1979). *Astrophys. J.* **227**, 483.

(15) Wannier, P. G., Penzias, A. A. & Jenkins, E. B. (1982). *Astrophys. J.* **254**, 100.

(16) Lien, D. J. (1984). *Astrophys. J.* **284**, 578.

(17) Equivalent widths from Hawkins, I., Jura, M. & Meyer, D. M. (1984). *Bull. Amer. Astron. Soc.* **16**, 878. Column densities follow from the best-fitting $(1,0)/(0,0)$ band 'doublet ratio', based on oscillator strengths of Larsson, M. & Siegbahn, P. E. M. (1983). *J. Chem. Phys.* **76**, 175.

(18) van Dishoeck, E. F. & Black, J. H. (1985). Preliminary result.

(19) Meyer, D. M. & Jura, M. (1984). *Astrophys. J. Letters*, **276**, L1; adjusted for improved oscillator strength.

(20) Chaffee, F. H. & Lutz, B. L. (1977). *Astrophys. J.* **213**, 394.

temperature, and ultraviolet intensity. It is thus necessary to construct comprehensive theoretical models of diffuse clouds in order to interpret observed column densities of atoms and molecules. The densities, temperatures, radiation fields and cosmic-ray fluxes of the models that best fit the complete array of observed data for a particular cloud can then be taken to represent the conditions in that cloud.

Although many of the details remain to be refined, most of the available data on the ζ Oph cloud (see Table 11.1), for example, can be represented well by such models with temperatures $T = 20-100$ K, and total densities $n = 100-800$ cm^{-3}, which may vary systematically with depth, the cloud being denser and colder in the core than near the periphery. With allowances for some uncertain molecular processes, it is possible to reproduce both the abundances and population distributions of observed diatomic molecules as outlined above. There is one notable exception to this, namely the CH$^+$ molecule. The large abundance of CH$^+$ in diffuse clouds has been a major problem for theory for some 40 years. The only viable explanation for CH$^+$ at present seems to be formation by a temperature-sensitive reaction

$$C^+ + H_2 \rightarrow CH^+ + H \tag{11.38}$$

at elevated temperatures behind shock waves passing through diffuse clouds. Although shocked regions will not affect the abundances of most species over the full line of sight through a cloud, they can add another level of complexity to the description of clouds.

11.6 Summary

We have seen that diffuse interstellar clouds are rich sources of spectroscopic information at a variety of wavelengths. Rather elaborate theoretical models must be devised to interpret fully some of this information. Although this is an interesting endeavour in its own right, it also leads to results of wider significance such as the cosmic-ray flux elsewhere in the Galaxy and the character of the cosmic background radiation near its peak in intensity.

In the future, we can expect important observational developments to result from more sensitive observations in the ultraviolet and from additional attempts to take advantage of important infrared absorption lines. One can look forward to theoretical investigations that will clarify many microscopic details on the one hand and will answer questions about the evolutionary status of diffuse clouds in a global sense on the other.

References

Section 11.1

The early work on interstellar absorption has been reviewed by

Seeley, D. & Berendzen, R. (1972). *J. Hist. Astron.* **3**, 52–64, 75–86.

The first comprehensive theoretical discussion of interstellar matter is contained in a remarkable paper

Eddington, A. S. (1926). *Proc. Roy. Soc. (London),* **A111**, 424–56.

A very useful text is

Spitzer, L. (1978). *Physical Processes in the Interstellar Medium.* New York: Wiley.

Section 11.2

Useful reviews of interstellar dust, extinction and polarization are

Savage, B. D. & Mathis, J. S. (1979). *Ann. Rev. Astron. Astrophys.* **17**, 73–111.

Aannestad, P. A. & Purcell, E. M. (1973). *Ann. Rev. Astron. Astrophys.* **11**, 309–62.

The correlation between gas and dust expressed in (11.3) is based on data in

Savage, B. D., Bohlin, R. C., Drake, J. F. & Budich, W. (1977). *Astrophys. J.* **216**, 291–307.

Section 11.3

Two classic studies of interstellar absorption lines are the papers of

Herbig, G. H. (1968). *Zeits. f. Ap.* **68**, 243–77.

Morton, D. C. (1975). *Astrophys. J.* **197**, 85–115.

Recent reviews include

Spitzer, L. & Jenkins, E. B. (1975). *Ann. Rev. Astron. Astrophys.* **13**, 133–64.

Snow, T. P. (1980). In *Interstellar Molecules.* IAU Symposium No. 87, ed. B. J. Andrew, pp. 247–254. Dordrecht: D. Reidel.

Black, J. H. (1984). In *Molecular Astrophysics – State of the Art and Future Directions*, ed. G. H. F. Diercksen, W. F. Huebner & P. W. Langhoff, pp. 215–36. Dordrecht: D. Reidel.

Diffuse interstellar bands are discussed by

Herbig, G. H. (1975). *Astrophys. J.* **196**, 129–60.

Smith, W. H., Snow, T. P. & York, D. G. (1977). *Astrophys. J.* **218**, 124–32.
The column densities in Table 11.1 are based upon a critical review of the literature.

Section 11.4

For a discussion of the diffuse cloud contribution to the galactic 21-cm line emission, see
Baker, P. L. & Burton, W. B. (1975). *Astrophys. J.* **198**, 281–97.
Examples of radio line studies of specific diffuse clouds are
Liszt, H. S. (1979). *Astrophys. J. Lett.* **233**, L147–50.
Crutcher, R. M. & Watson, W. D. (1981). *Astrophys. J.* **244**, 855–62.
Crutcher, R. M. (1976). *Astrophys. J. Lett.* **206**, L171–74.
Crutcher, R. M. (1977). *Astrophys. J. Lett.* **217**, L109–112.
Crutcher, R. M. (1979). *Astrophys. J. Lett.* **231**, L151–3.

Section 11.5

Rates of ionization and dissociation by ultraviolet starlight are discussed by
Roberge, W. G., Dalgarno, A. & Flannery, B. P. (1982). *Astrophys. J.* **243**, 817–26.
For detailed treatments of dissociation processes in specific molecules and their relation to interstellar chemistry, see for example
van Dishoeck, E. F. & Dalgarno, A. (1984). *Astrophys. J.* **277**, 576–80.
van Dishoeck, E. F. (1984). Thesis, Rijksuniversiteit Leiden.
Equation (11.20) incorporates rates of radiative and collisional processes taken from
Nussbaumer, H. & Storey, P. J. (1981). *Astron. Astrophys.* **96**, 91–5.
Hayes, M. A. & Nussbaumer, H. (1984). *Astron. Astrophys.* **134**, 193–97.
Hayes, M. A. & Nussbaumer, H. (1984). *Astron. Astrophys.* **139**, 233–6.
Launay, J.-M. & Roueff, E. (1977). *J. Phys. B*, **10**, 879–88.
Flower, D. R. & Launay, J.-M. (1977). *J. Phys. B*, **10**, 3673–81.
The excitation of molecules like H_2 and C_2 is discussed by
Black, J. H. & Dalgarno, A. (1976). *Astrophys. J.* **203**, 132–42.
van Dishoeck, E. F. & Black, J. H. (1982). *Astrophys. J.* **258**, 533–47.
A few of the many discussions of interstellar molecule formation are
Leung, C. M., Herbst, E. & Huebner, W. F. (1984). *Astrophys. J. Suppl.* **56**, 231–56.
Prasad, S. S. & Huntress, W. T. (1980). *Astrophys. J. Suppl.* **43**, 1–35.
Dalgarno, A. & Black, J. H. (1976). *Rep. Prog. Phys.* **39**, 573–612.
A few examples of comprehensive models of interstellar clouds are
Viala, Y. P. & Walmsley, C. M. (1976). *Astron. Astrophys.* **50**, 1–10.
Black, J. H. & Dalgarno, A. (1977). *Astrophys. J. Suppl.* **34**, 405–23.
Federman, S. R., Glassgold, A. E. & Kwan, J. (1979). *Astrophys. J.* **227**, 466–73.
de Jong, T., Dalgarno, A. & Boland, W. (1980). *Astron. Astrophys.* **91**, 68–84.
van Dishoeck, E. F. & Black, J. H. (1986). *Astrophys. J. Suppl.* **62**, 109.
The most recent use of interstellar CN lines to measure the cosmic background radiation is
Meyer, D. M. & Jura, M. (1984). *Astrophys. J. Lett.* **276**, L1–3.
Molecule formation in shock-heated diffuse clouds is discussed by
Elitzur, M. & Watson, W. D. (1980). *Astrophys. J.* **236**, 172–81.

12

Laboratory astrophysics: atomic spectroscopy

W. H. PARKINSON

12.1   Introduction

A close relationship has always existed between the progress of atomic theory and laboratory research on the one hand, and the growth of astrophysics on the other. That such a relationship should exist is not in the least surprising. The vast agglomeration of atoms, molecules, ions, and electrons that comprise every star and every nebula may in fact be regarded as an enormous physical laboratory, where matter is subjected to the most unusual and the most varied of physical conditions. Atomic studies in the laboratory should therefore logically supplement those in stellar atmospheres, and vice versa.

Leo Goldberg, Thesis, Harvard University, 1938.

These words describe the enduring relationship between astronomy and atomic physics and, in particular, the wonderful unity of laboratory and astrophysical plasmas. In this chapter we will explore the laboratory part of this unity and will discuss the measurements of some parameters and processes which are important to astronomy and atomic physics.

The determination of chemical abundances and the calculation of model atmospheres for the Sun and stars have become increasingly sophisticated since the landmark work of Goldberg and colleagues. However, despite the impressive progress, calculations based on the best existing atomic data, chemical abundances, and model atmospheres do not reproduce the measured, high-resolution, ultraviolet, solar spectrum nor do they match the center-to-limb variations. When high-resolution, ultraviolet, solar spectra (see Fig. 12.1) are examined it is apparent that the complex of overlapping and nearby lines ('line blanketing') could be responsible for much of the discrepancy between observations and calculations.

Astronomers are handicapped by incomplete and low-quality oscillator strength data (*f*-values) in their efforts to calculate accurately and completely the contributions from line blanketing. This situation has fostered semi-empirical procedures for calculating line opacity which are dependent in turn upon semi-empirically derived *f*-values.

Spectral synthesis methods have been used to determine the details of solar and stellar spectra and the elemental abundances from weak features. The successful application of these methods relies on knowledge of the atomic data of the elements of interest in order to establish a continuum level and the contributions of each spectral feature.

The atomic process which results in the emission of radiation from the source can become the basis of diagnostic methods for providing the best empirical determination of physical conditions of the source. Diagnostic methods can be devised to determine temperature, particle densities, ionization state distributions, and chemical abundance.

Many plasma diagnostic techniques which are used to obtain values for the electron density and temperature of an astrophysical source depend on the measured emergent intensity of optically thin lines and all require the associated atomic parameters. The accuracies with which the physical conditions of the sources can be inferred depend directly on the precision and the completeness of the underlying atomic parameters.

Following the development in the next section of the basic formulae and

Fig. 12.1. A plot of a 5-nm section at 280 nm of the disk center and near limb solar spectra showing the Mg II *h* and *k* features. The axes are absolute specific intensity (erg cm^{-2} s^{-1} sr^{-1} nm^{-1}) and wavelength (nm). This plot is taken from the atlas of center and limb solar spectra between 225.2 nm and 319.6 nm (Kohl, Parkinson & Kurucz (1978)).

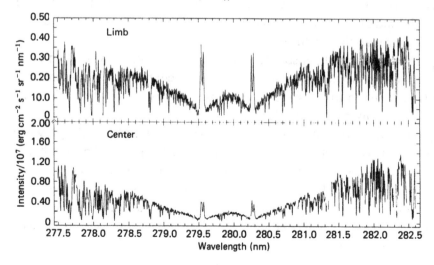

definitions necessary to describe emission and absorption of radiation, we will discuss the measurement of bound–bound radiative transition parameters. These include the measurement of radiative lifetimes, f-values and electron impact excitation cross sections. Next we will discuss the measurement of bound–free transition parameters. We will examine measurements of f-values of a series of lines approaching the series limit. Above the ionization limit, we will consider measurements of the adjoining photoionization and autoionization features. Autoionization will lead us to a discussion of dielectronic recombination and to the measurement of the cross section of this important process.

At the end of this chapter, a list of compilations of atomic data will be cited along with a selection of references which provide more detailed information on the subject of this chapter.

12.2   Basic formulae and definitions

In order to give some mathematical structure to the measurement procedures which will be discussed, we will develop very briefly the basic formulae and introduce some definitions that will allow us to discuss the processes of emission and absorption of radiation from atoms and ions. These formulae and definitions will also generally apply to neutral and ionic molecules, but would need modification and development in order to include the vibrational and rotational modes of excitation.

12.2.1   *Einstein probability coefficients*

Let us begin with the Einstein probability coefficients, which can be defined either in terms of the radiation density $\rho(v)$ with units, erg cm^{-3}, or in terms of the specific intensity of the radiation $I(v)$ where $\rho(v) = 4\pi/c\, I(v)$. Using the radiation density (Einstein's theory of radiation) we define the following probability coefficients:

A_{ji} is the probability per second (the transition probability) that an atom in state j will spontaneously emit a quantum of energy hv_{ij}. The number of such transitions to lower state i per sec per cm^3 is $A_{ji}N_j$ where N_j is the number of atoms per cm^3 on level j.

$B_{ij}\rho(v)$ is the probability per second that an atom in state i in the presence of radiation density $\rho(v)$ will absorb a quantum of energy hv_{ij}. The number of such transitions to upper state j per sec per cm^3 is $B_{ij}\rho(v)N_i$.

$B_{ji}\rho(v)$ is the probability per second that the atoms in state j in the presence of the radiation density $\rho(v)$ will undergo an induced transition to level i with the emission of a quantum of energy hv_{ij}. The number of such transitions to state j per sec per cm^3 is $B_{ji}\rho(v)N_j$.

The following relationships between the three Einstein coefficients can be derived by considering the thermodynamic equilibrium between radiation and atoms:

$$g_i B_{ij} = g_j B_{ji} \tag{12.1}$$

and

$$A_{ji} = (8\pi h \nu^3 / c^3) B_{ji} \tag{12.2}$$

where g_i and g_j are the statistical weights of the levels. If we had followed Milne's development of Einstein's theory of radiation (using specific intensity of isotropic radiation) then

$$B_{ji}(\text{Intensity}) = (4\pi/c) B_{ji}(\text{Density})$$

and

$$A_{ji} = (2h\nu^3 / c^2) B_{ji}(\text{Intensity}).$$

The radiative lifetime τ_j of the state j is related to the transition probability A_{ji}. The initial population of atoms decay according to

$$(\mathrm{d}N_j / \mathrm{d}t) = -N_j \sum_i A_{ji}$$

assuming that the population of level j is not being replenished or de-excited by collisions between atoms. The solution is

$$N_j(t) = N_j(0) \exp\left[\left(-\sum_i A_{ji}\right) t\right], \tag{12.3}$$

$\tau_j = (1/\sum_i A_{ji})$ is the mean life or lifetime of the state j. The sum is over all possible final states i.

Note that in order to use the knowledge of the lifetime of a given level to determine the transition probability of the line, we must know the relative strengths or intensities of all of the transitions from level j to all levels i. It is the simplest situation, when there is only one transition from level j, so that $A_{ji} = 1/\tau$. Otherwise, the intensity branching ratio must be measured. If the intensity branching ratio

$$B_r = \left(I_{ji} \bigg/ \sum_i I_{ji}\right) \tag{12.4}$$

is determined by measuring the intensity (photon s^{-1}) for all transitions, then we know

$$\left(I_{ji} \bigg/ \sum_i I_{ji}\right) = \left(A_{ji} \bigg/ \sum_j A_{ji}\right)$$

and thus

$$A_{ji} = (B_r/\tau_j).$$ (12.5)

The Einstein probability coefficients are also related to the absorption coefficient k_v. Consider a parallel beam of radiation of unit cross-sectional area passing through a layer of gas of length Δz in which the energy per unit volume is $\rho(v)\,dv$. The amount of energy passing through per second is $c\rho(v)\,dv$ and the amount absorbed per second is $c\,\Delta z \int_0^\infty k_v\rho(v)\,dv$. From the definition of B_{ij}, the number of quanta of energy hv_{ij} absorbed per second is $B_{ij}\rho(v)N_i\,\Delta z$ and the energy absorbed per second is $B_{ij}\rho(v)N_i\,\Delta z hv_{ij}$. We can equate the two expressions for the energy absorbed per second and, assuming that $\rho(v)$ is constant over the line profile,

$$B_{ij} = (c/N_ihv) \int_0^\infty k_v\,dv.$$ (12.6)

The absorption coefficient k_v is defined by the relationship $-\delta I_v(z) = k_vI_v(z)\,\delta z$ where I_v is the incident radiation and δI_v and δz are incremental energy and distance. For a thickness l, the relationship between the incident and emerging intensity is found by integration, so

$$I(l) = I_v(0) \exp\left(-\int_0^l k_v\,dz\right).$$ (12.7)

The $\int_0^l k_v\,dz$ is the optical depth or optical thickness of the medium and, if the medium is homogeneous so that k_v is not a function of z, then the optical depth is k_vl and

$$I_v(l) = I_v(0) \exp[-k_vl].$$ (12.8)

The shape of the curve of k_v versus v depends on the broadening mechanisms in the medium, but the total area of the curve $\int_0^\infty k_v\,dv$ depends on the atomic parameter B_{ij}. The atomic absorption coefficient or the atomic cross section, α_v (units of cm^2), is sometimes preferred and is related to k_v by

$$\alpha_v = (k_v/N_i).$$ (12.9)

If the curve k_v versus v is the profile of a line in which no broadening other than natural broadening (natural damping) is present then, from the uncertainty principal, the line width depends on the radiative lifetime of the upper level involved in the transition. As we will see later, δv_N is the natural width and

$$\tau = (1/2\pi\,\delta v_N).$$ (12.10)

12.2.2   *The oscillator strength or f-value*

An essential atomic parameter that we will be concerned with in the following sections is the oscillator strength or *f*-value. When we plan measurements of this parameter we naturally think of methods based on dispersion and absorption. As we will see, this plan is realized in measurements of anomalous dispersion (by the so-called 'hook' method) and by absorption measurements.

The real part of the refractive index *n* and the absorption coefficient *k* can be used to relate measurable quantities to the oscillator strength. The classical model of the absorption of electromagnetic energy at resonant, angular frequency ω_0 by $\mathcal{N}$ electron oscillators per cm^3 in a medium can be transferred to the quantum picture by identifying a photon at frequency ν_0 being absorbed with an oscillator strength *f* by *N* atoms per cm^3.

We begin with a relationship for the driving electric field with the electric vector in the *x*-axis

$$E_{tz} = E_{0,0} \exp\left[i\omega(t - z/v) \right] \tag{12.11}$$

where the phase velocity $v = (c/\tilde{n}) = (c/\tilde{\varepsilon}^{1/2})$; $\tilde{\varepsilon}$ is the complex dielectric constant and $\tilde{n}$ the complex refractive index, which can be written $\tilde{n} = n + \kappa\mathrm{i}$. When we substitute to obtain the fields in terms of refractive index and the phase of the wave, we obtain

$$E_{tz} = E_{0,0} \exp\left[-\omega(z/c)\kappa \right] \exp\left\{ i\omega[(z/c)n - t] \right\}. \tag{12.12}$$

The two exponential factors should be examined separately.

The intensity *I* of a wave front is proportional to the magnitude squared of the electric vector, and thus we have

$$I_z = I_0 \exp\left[-2\omega(z/c)\kappa \right]. \tag{12.13}$$

By comparing (12.13) with (12.8) we see that $2(\omega\kappa/c)$ is related to the absorption coefficient if we identify $2(\omega\kappa/c) = k$ and $z = l$.

Similarly the phase ϕ of this wave is

$$\phi = \omega[(z/c)n - t].$$

We can determine the properties of *n* and *k* by considering the classical model of the electric field E_{tz} driving an elastically bound electron of charge *e* and mass *m*. At a fixed value of *z*, the equation of motion of a one-dimensional oscillator with damping γ and resonant angular frequency ω_0, driven by the field is

$$m\ddot{x} + m\gamma\dot{x} + m\omega_0^2 x = eE_{0z} \exp\left[i\omega t \right]. \tag{12.14}$$

The solution of this equation is

$$x_{tz} = \frac{(e/m)}{\omega_0^2 - \omega^2 + i\gamma\omega} E_{tz}. \tag{12.15}$$

The electromagnetic wave, in traversing the medium, induces an electric dipole moment (i.e., it polarizes the atoms and molecules in the gas). The induced dipole moment is

$$P = ex_{tz} = \frac{(e^2/m)E_{tz}}{\omega_0^2 - \omega^2 + i\gamma\omega} \tag{12.16}$$

and if there are $\mathcal{N}$ oscillators or electrons per cm^3, the total dipole moment per cm^3 or the polarization for small field strengths is

$$P = \frac{\mathcal{N}(e^2/m)E_{tz}}{(\omega_0^2 - \omega^2 + i\gamma\omega)}. \tag{12.17}$$

We can define the complex polarizability by

$$\bar{\alpha} = \frac{(e^2/m)}{\omega_0^2 - \omega^2 + i\gamma\omega}$$

so (12.17) can simply be expressed

$$P = \bar{\alpha}\mathcal{N}E_{tz}.$$

We should note that the macroscopic, small field strength equivalent of the polarizability $\bar{\alpha}$ is the electric susceptibility χ_e of the medium or is the linear susceptibility of this gas $\chi^1(\omega)$. For higher fields, now available with lasers, the nonlinear properties of the medium can be studied. In this case the polarization is expanded as power series in the electric field. For large fields the polarization also depends on higher powers of E and the nonlinear susceptibility $\chi^n(\omega)$.

From Maxwell's constitutive relation, the displacement current D and the electric field can be expressed by

$$D = \bar{\varepsilon}(E)E = E + 4\pi P \tag{12.18}$$

$$= (1 + 4\pi\bar{\alpha}\mathcal{N})E$$

or

$$\bar{\varepsilon} = 1 + 4\pi\bar{\alpha}\mathcal{N} = \tilde{n}^2. \tag{12.19}$$

For small $\bar{\alpha}\mathcal{N}$ we can write

$$\tilde{n} = 1 + 2\pi\bar{\alpha}\mathcal{N}. \tag{12.20}$$

But if we write the complex refractive index in terms of n and κ, we have

$$n - 1 - i\kappa = 2\pi\bar{a}\mathcal{N}$$

and we can then separate the real and the imaginary parts of the dipole moment, (12.17).

The real part yields

$$n - 1 = \frac{2\pi e^2 \mathcal{N}}{m} \frac{(\omega_0^2 - \omega^2)}{(\omega_0^2 - \omega^2)^2 + \gamma^2 \omega^2}. \tag{12.21}$$

We can now introduce the oscillator strength f_{ij} or f-value, by defining that the number of oscillators $\mathcal{N}$ equals $N_i f_{ij}$, where N_i is the number density of atoms in the lower state of the transition at frequency v_{ij}. Equation (12.21) becomes

$$n - 1 = \frac{e^2 N_i f_{ij}}{2\pi m} \frac{v_{ij}^2 - v^2}{(v_{ij}^2 - v^2)^2 + (\gamma v / 2\pi)^2}. \tag{12.22}$$

In order to account for the more general case, we should write (12.22) as a sum over all discrete transitions and an integral over all continuum transitions from the ground state, where $(\partial f / \partial v_c)$ is the spectral density of oscillator strength in the continuum. Thus

$$n - 1 = \frac{e^2 N_i}{2\pi m} \left[\sum_j \frac{f_{ij}(v_{ij}^2 - v^2)}{(v_{ij}^2 - v^2)^2 + (\gamma v / 2\pi)^2} + \int \frac{\partial f}{\partial v_c} \frac{(v_c^2 - v^2)}{(v_c^2 - v^2)^2 + (\gamma v_c / 2\pi)^2} \right] dv_c. \tag{12.22a}$$

Similarly the imaginary part of the dipole moment, (12.17), leads to the absorption coefficient when we equate $2(\omega\kappa/c)$ with k and thus

$$k_v = \frac{e^2}{\pi m c} \frac{N_i f_{ij} \gamma v_{ij}}{(v_{ij}^2 - v^2)^2 + (\gamma v / 2\pi)^2}. \tag{12.23}$$

For a region near v_{ij} both (12.22) and (12.23) simplify somewhat:

$$n - 1 = \frac{e^2}{4\pi m v_j} N_i f_{ij} \frac{(v_{ij} - v)}{(v_{ij} - v)^2 + (\gamma/4\pi)^2} \tag{12.24}$$

and

$$k_v = \frac{e^2}{mc} N_i f_{ij} \frac{(\gamma/4\pi)}{(v_{ij} - v)^2 + (\gamma/4\pi)^2}. \tag{12.25}$$

When we can make the approximation in (12.24) that broadening can be neglected, i.e., $|v_{ij} - v| \gg c/4\pi$, it becomes the Sellmeier formula

$$n - 1 = \frac{e^2}{4\pi m v_{ij}} \frac{N_i f_{ij}}{(v_{ij} - v)}, \tag{12.26}$$

or in terms of wavelength

$$n - 1 = \frac{e^2}{4\pi m c^2} \frac{N_i f_{ij} \lambda_{ij}^3}{(\lambda - \lambda_{ij})}. \tag{12.27}$$

To express the oscillator strength in (12.25) in terms of measurable quantities, we integrate with respect to frequency from zero to infinity. Because the frequency v_{ij} is much larger than the full width $\gamma/2\pi$, the definite integral equals π. Therefore

$$\int_{\text{line}} k_v \, dv = (\pi e^2/mc) N_i f_{ij}. \tag{12.28}$$

It is important to note that the area of this profile, i.e., the integrated absorption coefficient, is independent of the shape of the line.

The oscillator strength can be expressed in terms of the Einstein probability coefficient and methods found to determine it.

By combining (12.6) and (12.28) we see that

$$B_{ij} = (\pi e^2/mh v_{ij}) f_{ij}. \tag{12.29}$$

In addition from (12.1) and (12.2) we have

$$A_{ji}/B_{ij} = (8\pi h v^3/c^3)(g_i/g_j), \tag{12.30}$$

and, after substituting for B_{ij} from (12.29), we have the resulting relation

$$g_i f_{ij} = [mc^3/8(\pi e)^2 v_{ij}^2] g_i A_{ji} = 1.499 \, \lambda_{ij}^2 g_j A_{ji} \tag{12.31}$$

where λ is in cm.

12.2.3 *Electron impact excitation cross sections*

Both transition probabilities and electron collision excitation rates are important in determining the intensity of transitions. The electron excitation rate C between initial and final states, depends on the electron impact excitation cross section, σ, the energy of the electrons E and the energy distribution function $F(E)$ of the colliding particles. The total electron collisional excitation rate is

$$C = \int_{E_0}^{\infty} E\sigma(E)F(E) \, dE \tag{12.32}$$

where E_0 is the threshold for excitation. Often $F(E)$ is a Maxwellian distribution characterized by a temperature T. The theoretical and experimental methods to determine accurate values of σ are complex and, as a result, there are few precise values available.

The theoretical methods used to calculate electron impact excitation cross sections range from a simple predictor formula, called the effective Gaunt factor formula, to highly-developed, close-coupling calculations.

The effective Gaunt factor G_f is related to the electron excitation cross section by

$$\sigma = \frac{8\pi}{\sqrt{3}} \frac{f}{EE_0} G_f \pi a_0^2 \tag{12.33}$$

where f is the oscillator strength of the transition, the energy is in Rydberg units (13.60 eV) and a_0 is the Bohr radius $[h^2/(4\pi^2 e^2 m)]$. It was first suggested, on the basis of comparisons with more sophisticated calculations, that with $G_f = 0.2$ the formula could be relied upon to within a factor of two or three for singly charged ions. The excitation cross section is often expressed in terms of a dimensionless and symmetrical parameter called the collision strength Ω. They are related by

$$\sigma = \pi a_0^2 \frac{\Omega}{g_i E} \tag{12.34}$$

where g_i is the statistical weight of the initial state of the atom or ion. If we equate (12.33) and (12.34), we see that we have removed the E dependence near threshold and

$$\Omega = \frac{8\pi}{\sqrt{3}} \frac{f}{E_0} g_i G_f \tag{12.35}$$

which is a constant if G_f is constant. It is also sometimes useful to express the collision rate in terms of the collision strength when the energy distribution function is known.

12.2.4 *Bound–free transition parameters*

An excited bound state has energy in eV

$$E_j = -R/(n-\mu)^2 = -R/(n^*)^2 \tag{12.36}$$

where R is the Rydberg (13.6 eV), n is the quantum number, μ the quantum defect and n^* the effective quantum number. The separation between the excited energy levels in the series is

$$\delta E_j = 2R/(n^*)^3. \tag{12.37}$$

Beyond the limit of the series is a region of continuous absorption, which corresponds to the photoionization of the atom. Quantum defect theory considers the series and its adjoining continuum as a channel. Channels are described by the states of the ion and the transition electrons. The analysis of spectra involving more than one channel can be done by means of multichannel quantum defect theory (MQDT). The MQDT, implemented with the aid of Lu–Fano plots of the quantum defects, is very successful in describing complicated spectra and series with a limited number of parameters.

To express the absorption per unit energy, or the differential oscillator strength, for each line of the series we can write

$$f_{ij}/\delta E_j = f_{ij}(n^*)^3/2R = df/dE. \tag{12.38}$$

We can express the atomic absorption coefficient or photoionization cross section in terms of the differential oscillator strength with the aid of (12.28)

$$\alpha(E) = (\pi e^2 h/mc)\, df/dE, \tag{12.39}$$

or

$$\alpha(E) = 1.098 \times 10^{-16}\, df/dE,$$

or

$$\alpha(v) = 2.654 \times 10^{-2}\, df/dv.$$

The bound–free, atomic absorption coefficient, for H-like ions, is

$$\alpha_n(v) = \frac{64\pi^4}{3\sqrt{3}} \frac{me^{10}}{ch^6} \frac{z^4}{v^3 n^5} G_n \tag{12.40}$$

where n is the quantum number, z the effective nuclear charge, and G_n the Gaunt factor.

For values of the Gaunt factor near unity and for large n, this relationship becomes Kramer's formula for a nucleus of charge Z

$$\alpha_n(v) = \frac{64\pi^4 e^{10} m Z^4}{3\sqrt{3}\, ch^6 v^3 n^3}. \tag{12.41}$$

If the photoionization cross section is known, then the reverse process – radiative recombination – can be determined. Just as (12.30) expresses the relationship between the Einstein coefficients for discrete transitions, the

Milne equation for a specific transition

$$\frac{\sigma(v)}{\alpha(v)} = \frac{h^2 v^2 g_k}{m^2 c^2 v^2 g_{\text{ion}}} \tag{12.42}$$

expresses the relationship between the photoionization and the radiative recombination cross sections, where v is the velocity of the free electron and g_k and g_{ion} are the statistical weights of the levels involved in the atom and ion, respectively.

12.3 Measurement of bound–bound radiative transition parameters

12.3.1 *Measurement of radiative lifetimes*

The lifetime of an excited state can be determined by measuring directly the radiative decay after excitation, or by measuring the energy width of the level involved. The method chosen to excite the levels determines the basic experimental approach used to measure the decay directly. The Hanle effect method employs the change in the polarization of the radiation by the magnetic field to infer the natural width of the energy level. There are two general means of excitation – modulated or pulsed optical or electron-impact excitation, and beam foil excitation. In all of these approaches there is no need to know the absolute population of the initial level, but accurate timing is required for the measurement of the direct radiative decay.

12.3.1.1 *Modulated excitation method*

Often the species to be examined will not exist under normal conditions as a collection of approximately independent atoms, molecules, or ions, and it will be necessary to create an appropriate sample with a high-temperature oven, electric discharge, or by collisions with charged particles or a laser beam. A vacuum chamber arrangement then confines the atoms, molecules, or ions at low densities in the form of a gas or a beam. The species are excited by optical radiation from a suitable light source or by electron impact from a beam of electrons. The source of excitation can either be pulsed or modulated at some frequency v_f.

When pulsed excitation is used, it is the radiative decay of the population $N_j(t)$ of a level j that is measured. Let us recall (12.3), $N_j(t) = N_j(0) \exp[-t/\tau_j]$, where $N_j(0)$ is the initial population at time $t=0$ and where we have neglected all other sources of population such as cascading from higher excited levels. The rate of change of intensity from any transition out of level j is given by

$$I_{jk}(t) = N_j(t) A_{jk}.$$

This exponential decay can be measured directly when the cut-off of the exciting pulse is short compared with the lifetime and when the time between pulses is much longer than the lifetime.

If the excitation source is not a short pulse but a modulated excitation with a period greater than the lifetime to be measured, then the phase shift between the periodic exciting radiation and the fluorescence radiation is measured. The shift θ between the two is due to the lifetime of the excited atom or molecule and can be expressed as $\tan \theta = 2\pi v_f \tau$. With optical excitation, the phase shift can be measured by observing both the excitation and fluorescence with the identical optical system and detector. However, for electron impact excitation the system must be calibrated with a known transition. Until the advent of lasers, optical excitation was restricted to resonance transitions which could be excited by traditional light sources. Electrons, however, are ubiquitous excitors and cover wide energy ranges. Their main advantage is that they excite forbidden as well as optically allowed transitions and can be used to excite highly ionized species. The modulation of electron beams is easier than photon beams.

Laser excitation is increasingly the method of choice for selective excitation in precise lifetime measurements. The selective feature of laser excitation eliminates the possibility of cascades from higher excited levels. Efficient optical filters and monochromators can be used for measurements of the fluorescence cascades to lower levels. With electron impact excitation, cascading is known to be a serious source of systematic error. The use of low-energy electrons, to limit excitation of higher levels, reduces the cascade problem, but threshold energies are low and the excitation cross sections are usually small so that there are technical difficulties in building a suitable electron source.

The pressure of test gas and background gas must be kept sufficiently low in the vacuum chamber to reduce the possibility of collisional depopulation. Collisions reduce $N_j(t)$ by non-radiative decay (quenching) of the excited states or, in the case of ions, by charge transfer. The apparent radiative lifetime is shortened. Conversely, the lifetime appears to be lengthened when the density of the test species in the cell is sufficiently great to allow photons emitted by an excited atom to be reabsorbed and re-emitted a number of times before being detected. The photon appears in this manner to be effectively delayed, and the measured lifetime is longer than that for the isolated atom.

A systematic error which is present with all measurements of long radiative lifetime is the movement of the excited atom out of the field of view of the detector during the course of the measurement. For example, during the measurement of the radiative lifetime of a level involved in a spin changing transition in O^{2+}, with $\tau = 1.25 \times 10^{-3}$ s, the ion could drift about three-quarters of a meter if at room temperature. In Section 12.3.2.1 we will consider means by which ions can be held or trapped in the field of view for times long compared with their radiative lifetimes.

12.3.1.2 *Beam foil excitation method*

This important method originated from and capitalized on the observational fact that optical radiation is emitted from an emerging beam of multiply-charged ions that has passed through a thin carbon foil. The ions of the emerging beam are in various stages of ionization and different states of excitation. As the ions move downstream from the foil, at speeds near 10^9 cm s^{-1}, the excited states decay. A measurement of the decrease of intensity of a particular line, as a function of linear distance from the foil, represents the radiative decay curve of the relevant excited state.

Typically, a Van de Graaff or a linear accelerator is used to produce the high-energy beam of ions which are mass analyzed before they pass through a thin carbon foil (5–10 μg cm^{-2}). The foil is located in an observation chamber where the radiation can be observed to decay. A monochromator is tuned to the emission line of interest, and the intensity of the decaying radiation is measured as a function of distance downstream from the foil by translating the foil or the monochromator. To take account of variations in beam current and foil characteristics with time, it is possible to normalize the intensities at several positions downstream from the point of excitation. The normalization can be done by monitoring the ion beam current using a Faraday cup or by monitoring the intensity of a line that can represent the line whose lifetime is being measured during the course of the experiment.

The method is best suited for lifetimes of about 10^{-8} s, with a useful range of plus or minus more than a factor of 10 about this value.

The characteristics that have made beam foil excitation methods successful are: (1) many elements can be studied, (2) high chemical and isotopic purity can be achieved in the beam before it impinges on the foil, (3) high stages of ionization can be reached, (4) large fractions of each species are in excited states and some are in double excited ones, (5) the high vacuum requirement of the beam chamber results in no collisional or optical depth problems in the beam, and (6) excitation is abrupt and coherent.

Some difficulties with the method are spectral line broadening and cascading. Because the particles in the beams are moving at a high speed (approximately 5×10^8 cm s^{-1}) and, because the spectrometers that are used have large light-gathering capability and thus large acceptance angles, the Doppler broadening usually limits the resolution ($\lambda/\Delta\lambda$) to 1000. Despite the use of various optical arrangements to reduce the effect of Doppler broadening, the limit to the spectral resolution is set by the inclined trajectories of the excited ions caused by small angle scattering within the foil. In favorable cases, the resolution can approach 10 000. Nevertheless, line blending from the large number of excited ionic species produced in the beam, together with limited spectral resolution, is a serious limitation in the beam foil technique.

Cascading from higher excited states is the major problem in interpreting the results from the beam foil method. Most decay curves are sums of many exponentials and there are few objective ways to separate them. However, beam foil experiments have produced a large amount of data on lifetimes for ionic species. In some cases it is the only experimental method that has been applied to highly ionized spectra. It has been particularly useful in the examination of systematic trends in oscillator strengths along isoelectronic sequences.

12.3.1.3   *The Hanle effect method*

It follows from (12.10), $\tau = \frac{1}{2}\pi \, \delta v_N$, that the lifetime of a state is related to the natural energy width of the level, and thus if the width can be determined, the lifetime can be found. One of the most successful methods used to determine the lifetime of atomic states originates from the early 1920s' experiments of R. W. Wood and A. Ellett and of W. Hanle who observed in the resonance fluorescence of mercury and sodium vapor how linear polarization varied with magnetic field strength, relative direction of propagation, and the polarization of the incident light. The effect known now variously as magnetic depolarization, zero field level-crossing, or the Hanle effect can be explained classically.

With reference to Fig. 12.2 consider a beam of linearly polarized radiation incident along the x-axis with the electric field vector in the y-direction, a magnetic field H available along the z-axis and atomic vapor confined at the origin of the coordinate system. With zero magnetic field the excited dipoles oscillate along the y-axis and radiate a sin^2 intensity pattern with zero intensity along and maximum intensity perpendicular to the y-axis. When the magnetic field is applied, the dipoles precess about the z-axis at the Larmor frequency until they radiate.

The Larmor frequency $\omega_L = g_J(\beta/\hbar)H$ where β is the Bohr magneton and g_J

is the Landé-*g* factor. In LS coupling,

$$g_J = 1 + \frac{J(J+1) - L(L+1) + S(S+1)}{2J(J+1)}.$$

Fig. 12.2. (*a*) Arrangement of mutually perpendicular axes containing the magnetic field and the incident and detected light directions for Hanle effect measurements; (*b*) sketch of the inverted Lorentzian-shaped curve of signal counts vs. applied magnetic field for the Hanle effect measurement of the lifetime of the upper level in the $4s^2\ ^1S_0 - 4s4p\ ^1P_1$ transition in Ca I (Kelly & Mathur (1980)).

(*a*)

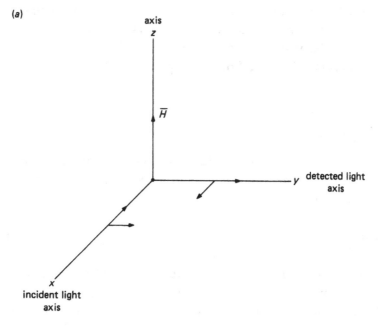

(*b*)

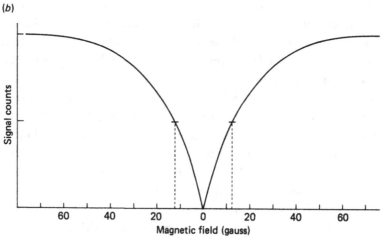

By combining expressions for the exponential radiative decay of the atoms with the dipole radiation pattern and integrating over time we obtain an equation for the observed intensity of scattered resonant radiation along the y-axis as a function of the strength of the magnetic field,

$$I = C \int_0^\infty e^{-t/\tau} \sin^2 \omega_L t \, dt,$$

$$= \frac{C}{2} \left[1 - \frac{(1/\tau)^2}{(1/\tau)^2 + (2\omega_L)^2} \right] \tag{12.43}$$

where C is a constant for the particular measurement and depends on the incident light intensity, the number of scatterers and the geometry.

We see that the variation of the fluorescence signal as a function of magnetic field H is an inverted Lorentzian profile (see Section 12.3.3.2) of the form

$$I \propto \left[\frac{(2\tau\omega_L)^2}{1 + (2\tau\omega_L)^2} \right]$$

with a minimum of zero when the field is zero, and with the full width at one-half the maximum of the profile equal to the mean life of the excited state. Therefore we obtain

$$\tau = \frac{h}{4\pi g_J H_{1/2}}. \tag{12.44}$$

Breit has given a more complete, quantum mechanical analysis of scattering of resonance radiation from atoms exhibiting degenerate excited states. However, the classical analysis will suffice for many simple situations, for example, in the alkaline earths where $J = 0 \rightarrow J = 1 \rightarrow J = 0$ and hyperfine structure is absent. In general the incident radiation may be unpolarized, in which case an analysis would be used to calculate each orthogonal component of polarization. However, the sensitivity will be highest when the incident radiation is polarized.

The Hanle effect is well suited for measurements of lifetimes of the order 10^{-8} s or shorter because the natural broadening is large compared with instrumental effects such as inhomogeneities in the magnetic field. The measured radiative decay can depend on the pressure through collisional redistribution which shortens the apparent lifetime or through radiation trapping (coherence narrowing of the natural linewidth) which lengthens the apparent lifetime. The width of the Lorentzian is usually measured as a function of pressure and a plot is made of the profile width versus a density parameter. When the plot is extrapolated to zero density the resulting width yields the radiative lifetime.

The Hanle effect method has been particularly successful for measurement of radiative lifetimes of levels of neutral atoms and molecules, which are easily produced and contained at low density.

12.3.2 *Measurement of radiative lifetimes of metastable levels*

When transitions occur in which the selection rules do not hold quite rigorously, we can observe electric dipole spin-changing 'intersystem' lines and lines forbidden as electric dipole radiation but allowed as electric quadrupole or magnetic dipole radiation. These transitions have longer radiative lifetimes than the corresponding electric dipole, spin-preserving ones. The upper levels for intersystem transitions in light ions of low charge, for example, have lifetimes of the order of 0.1–10 ms and are said to be metastable. The A-values for such transitions are of the order of $10^2 \, \text{s}^{-1} \leqslant A \leqslant 10^4 \, \text{s}^{-1}$, i.e., about four orders of magnitude smaller than those for allowed transitions. The A-values for forbidden lines are of the order of $1 \, \text{s}^{-1}$ or less.

12.3.2.1 *Ion trap method*

In order to determine the long lifetimes by direct observation of the spontaneous emission of a line from the excited level, the species must be held in the line of sight of the observer by some method which does not interfere with the radiative lifetime. Charged atoms and molecules can be contained with various configurations of electric and magnetic fields. In general there are three types of traps: the electrostatic trap which confines the ions in orbits around a central wire, the Penning trap which uses d.c. electric fields and a magnetic field, and the radio-frequency trap which uses the r.f. field to create a pseudo-potential well.

Let us consider the more useful Penning and r.f. traps in more detail. The electrodes of the traps are formed as hyperboloids of revolution, consisting of two endcaps and a ring. The ring is biased with respect to the endcaps by a d.c. voltage.

In the case of a Penning trap, a magnetic field is applied parallel to the z-axis, so that any ion that drifts radially will see crossed electric and magnetic fields and will be forced to precess about the z-axis. The ions will move in harmonic motion in the parabolic well along the z-axis, in cyclotron motion along the equatorial plane, and precess about the z-axis.

For a radio-frequency trap, instead of a magnetic field, an r.f. field is applied, so that the ring is biased with respect to the endcaps by d.c. plus an r.f. voltage, e.g., $U_0 + V_0 \cos \omega t$. With the r.f. fields, the potential is made to change sign before the ions can escape along any axis. Proper choice of the voltages U_0 and V_0 and the frequency ω allows preferential trapping of an ion of a single charge-to-mass ratio. Although hyperbolic traps have better charge-to-mass selectivity than the simpler shaped, cylindrical traps, the latter produce fields which are in close approximation to those with hyperbolic electrodes.

Typical voltages are d.c. bias of $U_0 = 10$ V, a.c. drive voltage of $V_0 = 150$ V, and r.f. frequency of 1 MHz. These produce potential well depths of the order of 10–20 eV.

Ions can be created within the trap by laser ablation of a solid target in or near the trap, which produces a tenuous plasma containing the required ion, or by electron impact ionization of a specific gas which contains the desired atom or molecule.

In an experiment in which ions are created in the trap by electron bombardment of a gas at pressures in the range 10^{-9}–10^{-6} Torr, a pulse of electrons of about 100 μA and 200–350 eV is used. After the creation and the storage of the ions, a delay period of at least 100 μs follows during which time those allowed transitions which have been excited radiate, and those species with unwanted charge-to-mass ratios and neutrals drift out of the trap. Following the delay period, photons from the intersystem transitions are counted for a period of about 10 lifetimes with a multichannel analyzer. The ions are ejected from the trap into a magnetic electron multiplier which can be used to monitor their number, and this completes one-half of a measurement cycle.

On the subsequent half cycle, the d.c. voltage is changed to a value that makes the trap repulsive for all ions and the create–delay–detect–eject cycle is repeated. On these alternate half cycles, the neutral background signal from, for example, decaying metastable levels in neutral species and photomultiplier dark counts is subtracted from that previously accumulated. The measurement proceeds for many full cycles. The magnetic electron multiplier signal is an indicator of trap performance and can be used to normalize data against fluctuations in the number of stored ions.

By using electron bombardment and excitation to produce the ion of choice, other charge states are produced and other levels excited. Therefore there is background radiation whose decays may be significant. These unwanted photons are reduced by using interference filters or, if the signal is sufficiently high, a spectrometer to isolate the transition of interest. The method chosen depends upon the signal-to-background ratio, the lifetime, and the time required to accumulate sufficient counts.

Although ion traps are operated at low pressures (10^{-9} Torr range), a source of systematic error is the continued presence of the source gas, which can lead to collisional redistribution of population among nearby energy levels in the ion and can be interpreted as a shortening in the radiative lifetime. In addition, charge transfer can remove metastable ions from the trap before they radiate. Measurements of the ion storage time and measurements of the radiative lifetime as a function of trap pressure (Stern–Volmer plots) must be carried out in order to determine whether these systematic errors are present. Conversely, ion traps can be used to study collisional processes such as charge transfer and dissociative recombination.

Fig. 12.3 is a semilogarithmic plot of the decay of the $^3P_1^0$ level of Si III as a function of time. The 3s3p $^3P^0$ term of Si III comprises three levels, but only one, $^3P_1^0$, decays by an electric dipole transition. The $^3P_2^0$ level has a radiative lifetime of 83 s, and the $^3P_0^0$ level decay is even slower. Consequently the only

Fig. 12.3. Semilogarithmic plot of the measured decay of the $^3P^0$ level of Si III as a function of time. The wavelength is 189.2 nm, the lifetime is 59.9 ± 3.6 μs, leading to a transition probability of $1.67 \pm 0.1 \times 10^4$ s^{-1} (Kwong, Johnson, Smith & Parkinson (1983)).

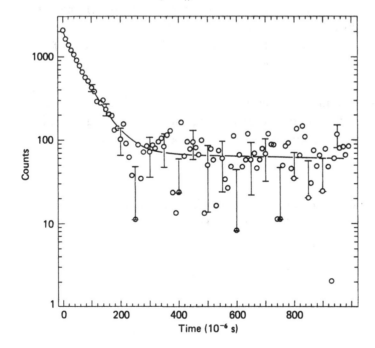

intersystem line of significance is $3s^2\ {}^1S_0-3s3p\ {}^3P_1^0$ at 189.2 nm. The ions for this measurement are produced by electron bombardment of SiH_4 in the ion trap at pressures in the 10^{-7} Torr range. The data for all measurements are summed to produce the decay curve shown. Representative uncertainties are given for each fifth datum. The lifetime is $59.9 \pm 3.6\ \mu s$, leading to a transition probability of $A = 1.67 \pm 0.1 \times 10^4\ s^{-1}$.

12.3.3 *Measurement of f-values*

12.3.3.1 *Anomalous dispersion and the 'hook' method*

The anomalous dispersion or 'hook' method for determining f-values was devised by Rozhdestvenskii in 1912. It uses the change in the refractivity in the region of a spectral line to determine the f-value of the transition.

A sample of length l and refractive index n of the absorbing gas of interest (Fig. 12.4) is contained in the test arm of an interferometer, usually of the Mach–Zehnder type, which is illuminated with collimated light from a background light source. The reference arm of the interferometer contains a compensation chamber and/or compensation plates of length l' and refractive index n'. The interferometer is adjusted at optical zero path difference, i.e., without the compensation plate, so that fringes are focused on the entrance slit of a stigmatic spectrograph and the spectrally dispersed fringe pattern recorded in the focal plane of the spectrograph. Higher-order fringes are created by changing the geometrical path length of one of the two arms or by inserting a compensation plate. This plate compensates for the phase shift caused by changing dispersion (anomalous dispersion) near each side of an absorption line. The resulting hook-like interference fringes around an absorption line (see Fig. 12.5) can be analyzed to determine Nlf, the product of the f-value of the transition, the column length, and the population density of the lower state in the vapor being studied.

Fig. 12.4. Mach–Zehnder interferometer arranged for hook method measurements with a stigmatic spectrograph.

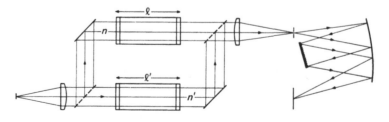

The equation which describes the behavior of a given fringe is

$$\phi y + (n-1)l - (n'-1)l' = p\lambda \tag{12.45}$$

ϕ is the angle between the test and reference beams at the spectrographic slit, y is the vertical position of the fringe at the slit, and p is the order number of the fringe.

The slope of a given fringe in the focal plane is, for an isolated spectral line,

$$(dy/d\lambda)_{dp=0} = (K/\phi)[1 + (1/K)(dn/d\lambda)]. \tag{12.46}$$

The hook constant K can be determined at conditions where the dispersion of the test beam is zero, so

$$K = -\lambda(\partial p/\partial\lambda) \approx (\lambda/\delta\lambda).$$

$\delta\lambda$ is the unperturbed horizontal fringe space (see Fig. 12.5) due to the compensation plate and in the absence of the test gas; or it is measured in regions without the influence of absorption lines. The locations of the hooks are the maximum and minimum along a given fringe near the line λ_{ij}. The basic relation of the hook method comes from combining the equation of the

Fig. 12.5. A sketch of the hook-like interference fringes about an absorption line at λ_{ij}. The spacings to be measured in using the hook and the hook vernier relationships (Huber (1971), Sandeman (1979)) are: Δ_{ij}, hook separation; $\delta\lambda$ and δy, spectral and vertical fringe spacing; and $\Delta\lambda$ and Δy, the spectral and vertical fringe shifts at the hook location.

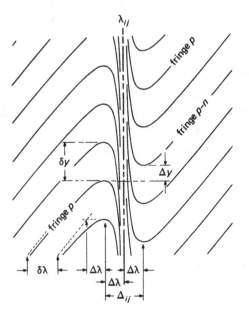

location of these extremes with (12.24). We note that the hooks are formed away from λ_{ij}, so we can make the approximation in (12.24) that broadening can be neglected, i.e., $|\nu_{ij} - \nu| \gg \gamma/4\pi$, and use the Sellmeier formula (12.27)

$$n - 1 = \frac{e^2}{4\pi mc^2} \frac{N_i f_{ij} \lambda_{ij}^3}{(\lambda - \lambda_{ij})}.$$

The basic hook method equation becomes

$$\Delta_{ij}^2 = \left(\frac{e^2}{\pi mc^2} \right) \left(\frac{\lambda_{ij}^3}{K} \right) N_i f_{ij} \tag{12.47}$$

where $\Delta_{ij} = 2(\lambda_{ij} - \lambda_h)$ is the hook separation and λ_h is the location of one of the hooks.

Recently the sensitivity of this method has been increased by almost a factor of 10 by recognizing the advantage of including the vertical fringe shift with the spectral fringe shift at the hook locations in the measurement and analysis of the hook spectrogram. The so-called hook 'vernier' relationship is

$$\frac{\Delta \lambda}{\delta \lambda} = \frac{1}{4} \left(a + \frac{\Delta y}{\delta y} \right) \tag{12.48}$$

where $\delta \lambda$ and δy are the spectral and vertical fringe spacing and $\Delta \lambda$ and Δy are the spectral and vertical fringe shifts at the hook location. With reference to Fig. 12.5, $\Delta \lambda$ is the distance of one hook from line center λ_{ij} and is one-half Δ_{ij} for an isolated line. The integer 'a' remains constant while $\Delta \lambda/\delta \lambda$ covers a range of 0.25 and then changes by unity (e.g., $a = 0$ for $0 \leqslant \Delta \lambda/\delta \lambda < 0.25$, and $a = 1$ for $0.25 \leqslant \Delta \lambda/\delta \lambda < 0.5 \ldots$). In practice, a measurement of Δ sets the value of a for the more accurate measurement in the y-direction.

In many cases where there are complex spectra involved and where neighboring lines interfere with each hook pattern, one must return to (12.22a), or the Sellmeier version, and sum over all relevant transitions in deriving the hook equation. Because there is a hook formed on each side of a single or a complex of lines whose wavelengths are all usually known, a computer assisted iterative calculation can solve the multi-line hook equation,

$$\sum_{i,j} \frac{N_i f_{ij} \lambda_{ij}^3}{\Delta_{ij}} = \frac{\pi mc^2}{e^2} K \tag{12.49}$$

for all of the lines involved.

The influence of nearby lines on the overall hook system must also be taken into account for hook vernier measurements. However, the spectral and vertical shifts of the hook locations are each affected in the opposite direction by neighboring lines, and there is a canceling effect. An average of the hook width and the hook vernier width will yield a good estimate of the isolated hook width.

The advantages of the hook method in determining Nlf are its large dynamic range and the fact that its accuracy does not depend on the line shape, on the resolution of the instrument, or on the strength of the line. With the advent of the hook vernier approach, hook patterns can now be measured for sufficiently small values of Nlf that they overlap the upper limit of the optically thin requirement of the absorption method. In addition, the linear distances of the hook pattern are easily measured parameters and the method is also insensitive to scattered light. The useful wavelength range over which the hook method can be used is determined by the transmission limit of the Mach–Zehnder interferometer and the background light source.

To make absolute f-value determinations from hook method measurements of Nlf, the column density Nl of the absorbing species must be known. Many of the pioneering determinations of absolute oscillator strengths by the hook method and absorption method depended upon the applicability of vapor pressure data for the elements of interest. However, this approach has been found unreliable, because the absolute column density is influenced by chemical reactions, losses at the walls, and nonuniformity in densities in the absorption chamber. The systematic losses in number density are particularly prevalent with the reactive iron group elements. The modern method is to make precise relative f-value measurements, which are then normalized using reference f-values that have been independently measured by other techniques. The few accurate absolute f-values, provided by atomic beam absorption measurements, by radiative lifetimes combined with branching ratio measurements or by Hanle-effect measurements, for example, can be used to normalize a much larger set of relative f-values obtained with the hook method. Thus, a standard transition at wavelength λ_{is} with a known f-value f_{is} and an observed hook separation Δ_{is} yields a column density Nl given by

$$N_i l = (mc^2 \pi / e^2)(K \, \Delta_{is}^2 / \lambda_{is}^3 f_{is}) \tag{12.50}$$

If we know the gas to be in equilibrium at temperature T, then from the Boltzmann equation

$$N_j / N_i = (g_j / g_i) \exp\left[-(E_j - E_i)/kT\right] \tag{12.51}$$

and from the partition function or state sum

$$U(T) = \sum_{i=0}^{\infty} g_i \exp\left[-E_i/kT\right] \tag{12.52}$$

we have

$$N_i/g_i = (N/U(T)) \exp[-E_i/kT] \tag{12.53}$$

where N is the total number density of atoms of the species.

With (12.53) we can write the general form of (12.50) as

$$\frac{N_i l}{g_i} = \frac{mc^2\pi}{e^2} \frac{K \Delta_{is}^2}{\lambda_{is}^3 g_i f_{is}} = \frac{Nl \exp[-E_i/kT]}{U(T)}. \tag{12.54}$$

Unknown f-values relative to f_{is}, for transitions from the same lower level with wavelengths λ_{ij}, follow from (12.47) and (12.54)

$$f_{ij} = \frac{mc^2\pi K \Delta_{ij}^2}{e^2 \lambda_{ij}^3 N_i l} = \left(\frac{\Delta_{ij}}{\Delta_{is}}\right)^2 \left(\frac{\lambda_{is}}{\lambda_{ij}}\right)^3 f_{is}$$

where the hook constant K is assumed to be unchanged over the wavelength range used.

12.3.3.2 *Absorption coefficient and equivalent width method*

A useful way to measure the rate at which energy is removed from the incident radiation $I_v(0)$ by an absorption line is to determine the equivalent width $W(v)$ of a rectangular strip which has the same area

$$\int_{\text{line}} 1 - [I_v(l)/I_v(0)] \, dv$$

as the absorption line. Furthermore, by recalling (12.8) for a homogeneous absorbing medium of length l, $I(l) = I_v(0) \exp[-k_v l]$, and writing it in terms of $W(v)$, we obtain

$$W(v) = \int_{\text{line}} \left(1 - \frac{I_v(l)}{I_v(0)}\right) dv$$

$$= \int_{\text{line}} (1 - \exp[-k_v l]) \, dv \tag{12.55}$$

a functional relationship between the equivalent width and the absorption coefficient. When the optical depth $kl \ll 1$ ('optical thin case'),

$$W(v) = \int_{\text{line}} k_v l \, dv = (\pi e^2/mc) N_i l f_{ij}. \tag{12.56}$$

If we plot the logarithm of (12.56) where $\log W(v)$ is the ordinate and $\log N_i l f_{ij}$ the abscissa, we develop the linear portion of a curve known as the curve of growth (Fig. 12.6). However, if $N_i l f_{ij}$ is increased so that $k_v l > 0.1$, then we should return to (12.25) and consider the influence of strong absorption on the absorption coefficient profile and the integral over the profile.

In the case so far considered, the curve of k_ν versus ν is the profile of a line whose broadening is due to the radiation damping γ. This so-called 'natural' damping or broadening gives the profile a full width at one-half peak of $\delta\nu_N = (\gamma/(2\pi))$. The width is related to the radiative lifetimes of the upper level involved in the transition at ν_{ij} so that $\tau = (1/(2\pi\,\delta\nu_N))$. The shape of the profile is called a damping profile or, because of the resonance denominator, a resonance profile. It is also referred to as a Lorentzian profile (after H. A. Lorentz).

When we consider thermal broadening of the line profile, due to the velocity of motion v_z of the emitting atoms or molecules in equilibrium and with a Maxwellian velocity probability distribution $P(v_z)$ described by the function

$$P(v_z)\,dv_z = (m/2kT)^{1/2}[\exp(-mv_z^2/(2kT))]\,dv_z \tag{12.57}$$

the resulting line has a full width of

$$\delta\nu_D = (2\nu_{ij}/c)(2kT\,\ell n\,2/m)^{1/2}. \tag{12.58}$$

The shape of the profile is called a Doppler or Gaussian profile and has the

Fig. 12.6. A sketch of a family of curves of growth.

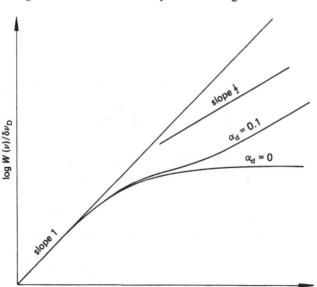

slope $\frac{1}{2}$

$\alpha_d = 0.1$

$\alpha_d = 0$

slope 1

$\log W(\nu)/\delta\nu_D$

$\log N\ell f/\delta\nu_D$

functional form $\exp(-y^2)$ where

$$y = [2(v_{ij} - v)/\delta v_D](\ell n\, 2)^{1/2}. \tag{12.59}$$

When the pressure in the medium is sufficiently high that collisions among the radiating atoms or molecules are the dominating source of damping, the pressure-broadened profile is Lorentzian of full-width δv_L. If, however, the Gaussian and Lorentzian components to the broadening are comparable in magnitude, the line profile is a combination or a convolution of both shapes and is known as a Voigt profile. The functional form of the Voigt profile is

$$\frac{\exp(-y^2)}{(x-y)^2 + \alpha_d^2} \tag{12.60}$$

where

$$\alpha_d = (\ell n\, 2)^{1/2} (\delta v_L / \delta v_D)$$

is called the damping ratio. Integration over a line profile of this shape cannot be done analytically but values of the function can be generated and are available in tables (e.g., Hummer, 1965).

If $|v_{ij} - v| \gg \delta v_L$, (12.25) becomes

$$k_v = \frac{e^2 N_i f_{ij}\, \delta v_L}{2mc(v_{ij} - v)^2}. \tag{12.61}$$

We can integrate (12.61) over the line profile, and with (12.28) and (12.55) we have

$$W(v) = \left(\frac{2\pi e^2}{mc}\,\delta v_L\right)^{1/2} (N_i l f_{ij})^{1/2}. \tag{12.62}$$

Here we see the square-root part of the curve of growth.

Sometimes it is advantageous in plotting the curve of growth to use a dimensionless quantity $W(v)/\delta v_D$. In this case the damping ratio α_d is required, and a family of curves of growth (Fig. 12.6) is generated for various values of α_d $(0 \leqslant \alpha_d \leqslant \infty)$. However, because of the difficulty in determining α_d and thus defining which one of the curves of growth is applicable to the absorption column, the absorption method is best suited for weak lines which are on the linear portion of the curve.

The main advantages of the absorption method are its sensitivity and accuracy for measuring weak lines and, in addition, the ease with which the area under the line profiles is obtained by the simple measurement of equivalent width. The line profile is not needed. Since relative intensity ratios $I_v(0)/I_v(l)$ are the measured quantities, the photometry is not a function of wavelength. Moreover, like the hook method, it avoids the difficulties and resulting errors of self-absorption.

12.3.3.3 *Example of f-value measurements by absorption and hook methods*

In order to get a clearer understanding of the uses and advantages of the absorption and hook methods to measure *f*-values, let us consider in some detail an experiment that applies both methods to the same column of absorbing atoms.

The purpose of the experiment is to measure the *f*-values of lines belonging to multiplet 1 and to multiplet 5 of Fe I (Fig. 12.7). The range in *f*-values covered by lines of these two multiplets spans nearly six decades, and the measurements offer a good test of combining the hook and the absorption techniques.

The experimental arrangement consists of a tungsten-ribbon-lamp light source which produces a background continuum for both the hook and the absorption measurements. Collimated light from this source enters a Mach–Zehnder interferometer containing in the test arm a graphite furnace and in the reference arm a compensation chamber and a mechanism which allows either a compensation plate or a completely opaque shutter to be inserted (Fig. 12.4). The light from the Mach–Zehnder interferometer is focused onto the entrance slit of a high-dispersion, stigmatic spectrograph. The focal plane

Fig. 12.7. Some energy levels and lines of multiplets 1 and 5 of Fe I. The bowtie-shaped array (Cardon, Smith & Whaling (1979)) has been emphasized.

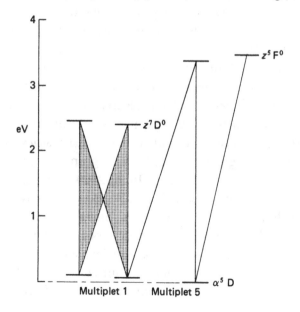

contains either a photographic plate for recording the hook spectrograms or a photomultiplier and exit slit over which the spectra can be scanned.

The procedure is to use the hook method to determine the column density with a line of the multiplet whose oscillator strength is accurately known. After the hook spectrum is obtained, the wavelength setting of the spectrograph is changed, the shutter is inserted into the reference beam, and absorption measurements are carried out on the same iron vapor column whose density is now known. The hook spectrum of the reference line is again photographed to obtain the number density, immediately following the absorption measurement. Any variation with time of the number density in the furnace is monitored in this way throughout the duration of the experiment. The combination of these alternating two techniques enables a nearly simultaneous measurement of very strong (hook method) and very weak (absorption) lines.

In the specific case addressed here the lower state of the line, at 371.99 nm, is also the ground state of atomic iron. The oscillator strength of this transition is known within ± 5 percent by measurements of the lifetime of the $z\,^5F^0_5$ level using four different methods with excellent agreement among them all. The mean log gf for the 371.99 nm line is -0.43. The hook measurement on this line enables us to determine the column density Nl for the Fe I atoms in the ground state (12.54).

Unknown oscillator strengths $g_i f_{ij}$ are determined from the equivalent width absorption measurement (12.56) and the column density Nl. Thus $g_i f_{ij} = (mc/\pi e^2)(U/Nl) \exp E_i/kT(W(\lambda)/\lambda_{ij}^2)$. T is the effective temperature of the Fe I vapor in the furnace and $W(\lambda)$ is the equivalent width.

Since the relative scale is made absolute by adopting the absolute gf-value for the 371.99 nm resonance line, the error in f-values, due to temperature error only can be calculated by considering the Boltzmann equation. The partition function is not a strongly varying function in this case. Therefore the relative error dT/T in the measurement of T leads to a relative error in the calculated number density of atoms dN/N, which is proportional to $E\,dT/T^2$. The temperature in a furnace can usually be measured *in situ* with an optical pyrometer, which is sighted on a series of graphite targets positioned along the optical axis inside the furnace. In the 2000 K range, the error in temperature usually does not exceed 7 or 8 K. However, serious error in comparison among lines of different excitation potential are caused by nonuniform temperature distributions along the vapor column. To account for this, it is usually necessary to construct a model which represents the measured temperature distribution through the furnace and at the boundary layers at the ends of the furnace. The equivalent width defined by the simple relation of (12.55) holds for conditions where emission from the line is negligible and all of the line profile is integrated. To account for line emission from the absorbing column, we must consider the equation of transfer for the column in the furnace. If the furnace is in local thermodynamic equilibrium and at uniform temperature T, we can write an equation which represents the radiation incident on the spectrometer from the combined background light source and the emitting vapor column:

$$I_\nu(0) = I_\nu(\text{lamp}) e^{-k_\nu l} + B_\nu(T)(1 - e^{-k_\nu l}) \tag{12.63}$$

where

$$B_\nu(T) = \frac{2h\nu^3}{c^2} \frac{1}{(e^{h\nu/kT} - 1)} \tag{12.64}$$

is Planck's function. Usually the emission can be measured separately and an emission equivalent width $W_E(\lambda)$ determined, so that the true equivalent width $W(\lambda)$ can be calculated from the measured apparent equivalent width $W^*(\lambda)$. The correction can be measured at one wavelength and temperature and then scaled to others, because the ratio of emission to absorption is proportional to $\exp[-hc/\lambda kT]$.

The second requirement of (12.55) is that all of the line profile be integrated. If the effective line shape is Gaussian, this important condition is readily fulfilled with a good diffraction grating spectrometer. However, to minimize any pressure-broadened Lorentzian component in the profile which would cause extensive wings, the total pressure of the absorption column must be kept sufficiently low (< 50 Torr). The shape of the observed line profile can be tested by comparing fitted Gaussian, Lorentzian, and Voigt profiles to assure that the measured equivalent width is the true equivalent width of the profile of the atomic line.

Fig. 12.8 is a graph which compared the Fe I relative *gf*-values of lines of multiplets 1 and 5 made by the combination of the hook and absorption method at Harvard College Observatory with the high-precision absorption method (described below) developed by the group at Oxford University. The figure shows that the deviation between the data from these two different techniques is on the average $< \pm 9$ percent over nearly six decades. Both sets of measurements were normalized to $\log gf = -0.43$ of the 371.99 nm iron resonance line of multiplet 5.

The Oxford absorption method for measuring *gf*-values is a modern, technically-improved version of the technique first used by King and King in the 1930s. The most important improvement, besides the aid of computers to control and to analyze the data, is the use of a pair of spectrometers equipped with modern, large echelle gratings of high quality. These gratings reduce the scattered light, which is a serious problem in the measurement of weak absorption lines. The spectrometers are equipped with low-noise, cooled, photoelectric detectors.

The ratio of the oscillator strengths of two lines is determined from the ratio of the measured equivalent widths of the Gaussian shaped line profiles. The two spectrometers are used simultaneously so that sets of ratios of relative *gf*-values are formed by measuring the equivalent widths of successive pairs of lines originating from a common energy level or, with an accurate measurement of temperature, the ratios can include lines from different energy levels. From among, for example, the lines of multiplet 1 and 5 (see Fig. 12.7) as many links as possible in the array are measured, and then the measured *gf*-values are adjusted by small amounts consistent with the experimental uncertainty to form an internally consistent set of ratios based on no residual closure errors around any set of links.

A general and more powerful means of comparing linked ratios of measured *gf*-values to obtain a set of optimized values on a consistent relative scale requires that measurements in both emission and absorption be made of the *gf*-values of lines making up the bowtie-shaped array outlined in Fig. 12.7. The absorption and the emission data are formed separately into absorption and emission ratios with their experimental uncertainties. These ratios are then combined to form the bowtie ratio and its uncertainty.

If both the absorption measurement and the emission measurements are perfect then this ratio would be unity. However, all measurements have experimental uncertainties and these will cause the bowtie ratio to depart from unity within the limits set by the calculated experimental uncertainties

Fig. 12.8. Comparison of Fe I *gf*-values for the lines of multiplets 1 and 5 measured by the Oxford absorption method (Blackwell, Ibbsetson & Petford (1975)) and by the Harvard hook and absorption combined method (Huber & Tubbs (1972)). Both experiments used a King-type furnace. The absolute scale is set on log *gf* = −0.43 of the 3719.94 Å resonance line of multiplet 5.

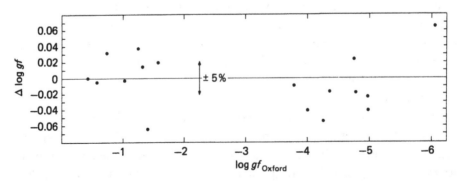

for the measurement. Because the sources of errors in an absorption measurement are different from an emission method measurement, the possibility of compensating errors is unlikely, and therefore the bowtie ratio can reveal errors in the data. In addition, the ratio is independent of the physical characteristics of the absorption or of the emission source.

Once the optimum and consistent sets of gfs are found, the relative gf-scale can be made absolute by basing it on an absolute gf-value, if it is known from Hanle effect or lifetime measurements of one of the upper levels.

A vivid example of gf-value measurements which use the dynamic range of the hook method is shown in Fig. 12.9, which is a hook spectrogram of the principal series ($5s^2\,^1S_0$–$5snp\,^1P^0$, $n = 10 \rightarrow > 26$) of Sr I. In this particular series of lines, the method is applied to gf-values spanning a range of about four orders of magnitude.

The alkaline-earth metal is heated in a furnace which maintains the absorbing atomic vapor column in the test arm of a Mach–Zehnder interferometer. The background continuum light source and spectrograph, as described earlier, are used to record the spectra. These complex spectra require a computer-assisted iteration calculation using the multi-line hook equation (12.49).

The absolute gf-value scale is set by using (12.50) and the density independent measurement of the gf-value of the resonance line, the first member of the series. Typically, the uncertainty in the measurement of the gf-values of the lines of these series is ± 12 percent.

12.3.3.4 *Phase-matching method*

The relative gf-values for members of the principal series of Ca I ($4s^2\,^1S_0$–$4snp\,^1P$, $n = 10$–14) have been measured, with a possible 6-fold improvement in the accuracy over the hook method, by a new approach based on measurements of the frequencies at which a nonlinear optical sum mixing process is phase matched. As we stated earlier, the high-intensity light beam from a laser makes possible the study and use of non-linear optical properties of a medium.

For an electric dipole, the polarizability can be expressed as a power series of the electric field, with the susceptibility $\chi^n(\omega)$ as coefficients. The non-linearities allow the light waves and the medium to interact with the resulting generation of new frequencies, v_s. For efficient generation, the phase velocities or refractive indices should match between the interacting waves and the medium. This phase-matching condition can be satisfied in a medium by mixing gases that have the correct refractivity or by working near a region of anomalous dispersion. With co-linear laser beams v_1, v_2, and

Fig. 12.9. A hook spectrogram of the principal series of Sr I recorded on Kodak SWR plates with a 3-m, Czerny–Turner mounted spectrograph, equipped with a 2400 lines/mm grating. The Mach–Zehnder interferometer contained a heat-pipe type furnace in the test arm. Quartz compensation plates which differ in thickness by 0.5 mm were used to change the hook constant for each exposure (Parkinson, Reeves & Tomkins (1976)).

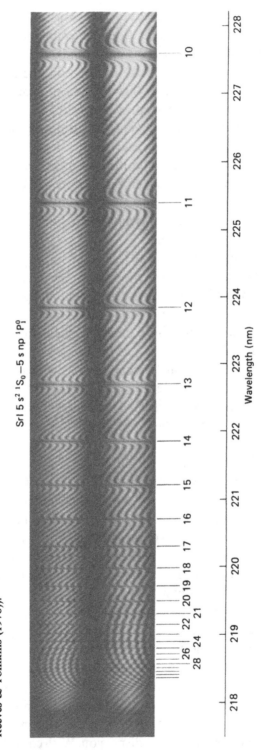

Sr I $5s^2\ ^1S_0 - 5s\,np\ ^1P_1^0$

Wavelength (nm)

v_3, the frequency v_s is generated for the phase-matching condition $k_s = k_1 + k_2 + k_3$ or $\Delta k = k_1 + k_2 + k_3 - k_s$ where k_i is the wave number and $k = n(v/c)$. The intensity of the generated beam when absorption is neglected is

$$I_s \alpha I_1 I_2 I_3 |\chi^3(v_s = v_1 + v_2 + v_3)|^2 \frac{\sin(1/2\,\Delta kl)^2}{1/2\,\Delta kl} Nl, \tag{12.65}$$

where $\chi^2(v_s = v_1 + v_2 + v_3)$ is the third-order, nonlinear susceptibility and Nl is the column density of atoms in the common lowest level.

The Sellmeier formula (12.26),

$$n - 1 = \frac{e^2}{4\pi m v_{ij}} \frac{N_i f_{ij}}{(v_{ij} - v)}$$

is used to relate the refractive indices to the f-values, when the laser is tuned in the wing of the line so that linewidth factors are negligible.

The first measurement of a gf-value with the phase-matching method was of the Rb-resonance line. In this case Xe was used to phase-match in Rb vapor and the f-value determined relative to the refractive index of Xe.

In the measurements on Ca I, the ratios of the f-values of series members, f_n, to the f-value of the resonance line, f_R, have been determined. The accuracy of these ratios depends on the accuracy of the f-values of all of the other series members and on the spectral density of oscillator strength $(\partial f/\partial v_c)$ and its integral (see (12.22a)). Nonlinear least squares fits of the experimental data with the parameter f_n/f_R and column density, according to (12.65), led to f-values for lines of the series. The estimated uncertainty in the measured f-values was ± 2 percent. From Table 12.1, we see that the gf-values measured by the hook method were systematically 10 percent higher than those also measured by the phase-matching condition. However, because the hook data have not been analyzed with the hook vernier approach, the narrow hooks associated with the weaker lines of the series may have been over-estimated. The absolute gf-value scale has been set by reference to the density independent measurement of the lifetime of $4s4p\ {}^1P_1$ by the Hanle effect.

12.3.4 *Measurement of electron impact excitation cross sections*

In general there are two types of experimental approaches to the quantitative investigation of electron–ion collisions: methods using well-defined plasma devices, and methods using well-defined beams of colliding electrons and positive ions. The method in which a plasma device is used depends on the plasma being sufficiently diffuse that electron impact excitation is the dominant process for populating the state of interest and

Table 12.1. *Ca I, log gf-values*

Transition	n	λ (nm) air	T_n (cm^{-1})	Hook[a] log gf	Hook[a] Uncertainty[a]	Phase-matched[f] log gf	Phase-matched[f] Uncertainty[a]
$4s^2\ ^1S_0-4snp\ ^1P_1^0$	4	422.6727	23 653.09[e]	+0.253[d]	±0.04	+0.253[d]	±0.009
	5	272.1645	36 731.61[e]	−3.03	+0.08 / −0.10		±0.009
	6	239.8559	41 679.01[e]	−1.38	±0.06		
	7	220.0728	45 425.33[e]	−1.49	±0.06		
	8	215.0796	46 479.79[e]	−1.90	±0.06		
	9	211.8672	47 184.45[e]	−2.25	±0.05		
	10	209.735	47 662.10[c]	−2.48	±0.05	−2.51	±0.009
	11	208.278	47 997.4[b]	−2.68	±0.05	−2.75	±0.009
	12	207.228	48 240.53[b]	−2.85	±0.05	−2.91	±0.009
	13	206.452	48 422.09[b]	−3.05	±0.06	−3.08	±0.009
	14	205.860	48 561.10[b]	−3.15	+0.08 / −0.10	−3.18	±0.02
	15	205.400	48 669.83[b]	−3.25	+0.08 / −0.10		
	16	205.035	48 756.45[b]	−3.35	+0.08 / −0.10		
−3d4p $^1P_1^0$		227.5462	43 933.477[e]	−1.17	±0.06		
−3d5p $^1P_1^0$		188.32 (λ vac)	53 100[b]				

[a] Uncertainty in DEX; designates intervals in powers of 10, e.g. 0.06 DEX is $10^{0.06}$, corresponding to ±15 percent. [b] Brown et al. (1973). [c] Garton & Codling (1965). [d] Kelly & Mathur (1980). [e] Risberg (1968). [f] Wynne & Beigang (1981). [g] Parkinson et al. (1976).

spontaneous decay is the dominant depopulating mechanism. Plasma modeling techniques are applied to determine the rate coefficient for electron impact excitation. It is necessary, however, to know the radiative transfer properties through the plasma, the ion densities, electron density, and electron temperature. The emergent intensity must be measured on an absolute scale. Usually the result of such a plasma-based measurement is the electron excitation rate coefficient. On the other hand, the experimental method in which well-defined beams of colliding electrons and ions are used yields the basic electron impact excitation cross section as a function of energy. This method is independent of the plasma parameters.

In this case the experimental arrangement uses a modulated beam of mass analyzed and selected ions, which intersects a pulsed electron beam at an angle θ. The ions are excited by inelastic collisions with the electrons within a collision volume defined by the overlapping beams. Photons are emitted by the radiative decay of the excited ions. The collision volume is centered in a vacuum chamber which contains the means to measure beam currents, the spatial distribution of interacting beam particles, and emitted photons.

Consider a beam of positive ions colliding at an angle θ with a beam of electrons. The number of collision events R per second in the overlapping volume element dW is

$$R\,dW = \sigma(E)N(V)n(v)U\,dW \tag{12.66}$$

where V and v are the particle velocities of the ions and electrons, respectively, U is the relative collision velocity, $N(V)$ and $n(v)$ are the ion and electron beam particle number densities, and $\sigma(E)$ is the electron impact excitation cross section for the collision energy E. The ion and electron particle velocities are related to the relative collision velocity by $U^2 = V^2 + v^2 - 2Vv\cos\theta$.

The particle number density can be expressed in terms of the measurable ion and electron currents I and J, which flow through the volume defined by the cross-sectional areas and lengths of the intersecting beams. By integrating over the volume defined by intersection of the beams,

$$R = [\sigma(E)IJU/Ze^2Vv]F \tag{12.67}$$

or, expressing the excitation cross section in the terms of all the measurable quantities,

$$\sigma(E) = Ze^2RVv/UIJ\,F \tag{12.68}$$

where Ze is the charge of the ion and F is a measurable factor which depends on the details of the shape of the overlapping volume and on the particular experimental approach.

The electron excitation process for a two-level ion yields two products: an electron which has lost the energy (ε) of the excited transition and a photon at the energy ($h\nu$) of the transition. For example, the excitation of the transition $2s^2 2p$–$2s2p^2$ in C^+ proceeds:

$$C^+(2s^2 2p\ ^2P^0) + e(E_1) \rightarrow C^+(2s2p^2\ ^2D) + e(E_1 - \varepsilon)$$

$$\rightarrow C^+(2s^2 2p\ ^2P^0) + e(E_1 - \varepsilon) + h\nu(\lambda = 133.5\ \text{nm}).$$

To determine the electron impact excitation cross section, it is necessary to measure the collision event rate R at which either one of these products is formed in the collision of the ion and electron beams.

If the number of collision events per second is determined by a measurement of the emitted photon rate, then the absolute number of photons per second per steradian emitted from the intersection of the particle beams must be measured and related to the number of photons emitted in 4π steradian. An absolute photometric calibration of the detection and optical system must be made. The problem of the anisotropy of the radiation can be avoided by collecting photons at a position angle of $54°\ 46'$ (the magic angle) with respect to the electron beam axis. At this angle the total photon flux is equal to the average photon rate over the total solid angle. Alternatively the anisotropy can be accounted for by collecting photons at two position angles with respect to the electron beam axis. If, however, efficient linear polarizers are available at the wavelength to be studied, the anisotropy can be measured by rotating the polarizer and observing at one position. In most cases the polarization of the detection and optical system must be measured or limited to one known component of polarization.

There are competing background processes which produce photons from collisions of ions and electrons with residual gas and with parts of the apparatus. These background photon rates must be minimized and measured. Clean surfaces, the reduction of absorbed gas by baking the apparatus, and the application of ultra-high vacuum technology are mandatory. Pressures in the 10^{-9}–10^{-11} Torr range should be used in the beam's interaction region. The photon background rate can be reduced further by narrowing the wavelength band pass of the detection system to the region containing the radiation of interest and by imaging only the beams intersection region onto the photon detector.

The signal can be sorted from the background and noise rates by modulating the beams in cycles to measure rates from each beam separately, both beams together, and both beams off. The period of the modulation must be short compared with the times required for transport of gas about the vacuum chamber.

It is necessary to determine particle beam densities in the volume defined by the intersection of the beams and the region viewed by the photon optical system. The particle densities can be measured by mapping the current density distribution of each particle beam by scanning a Faraday cup with a small entrance aperture of known size over the cross-sectional area of each beam. Total beam currents measured in larger-apertured Faraday cups, located in beam dumps which trap the beams, can be used to monitor and to scale the current densities during the course of the full measurement.

The distribution of the ions in the beam among the various states of the ion and, in particular, in the metastable states must be known in order to determine the number populating the lower level of the transition of interest. It is possible to determine the metastable particle densities of the beams by measuring the absolute rate of photon production from the metastable level and by applying the radiative lifetime of the level (see Section 12.3.2) to yield a measure of the number of metastable ions per unit length of the ion beam.

If the number of collision events per second is to be determined by a measurement of the number of inelastically scattered electrons that have lost the energy required to excite the energy stage of interest, some of the problems associated with the photon detection method are removed, but new problems arise. Because the electron energy loss specifies both the initial and final energy states of the ion excitation process, the ambiguities of the photon detection methods are eliminated. Excitations from the various energy states that may be populated in the beam can be distinguished from each other, and excitation to higher states is also distinguished by the electron energy loss. The electron method is not affected by excitation to higher energy levels followed by cascades or excitation from any lower levels which may be populated in the incident ion beam. The electron detection method is also independent of measured polarization and absolute spectral irradiance. It remains necessary to measure the total absolute number of inelastically scattered electrons per steradian. However, the high background rate from inelastic electron scattering from residual gas in the apparatus is a serious source of noise with this method. Near threshold energies for excitation, electrons are scattered in the center of mass rest frame into a broad angular distribution, but in the laboratory frame they are scattered into a narrow cone centered about the ion beam. For example, with a 4.5-keV beam of C^+ ions intersecting an electron beam at 45°, near a threshold energy of 10 eV in the center of mass frame, inelastically scattered electrons of 1 eV would be peaked in the forward direction at an angle of 35° with respect to the ion beam axis in the laboratory frame. Various schemes to deflect the low-energy electrons further from the ion axis and to focus them and detect them are under development and in use.

Fig. 12.10 shows the results of the measurements of electron impact excitation cross sections of C^+ by detecting and measuring the absolute number of photons at 133.5 nm produced by the excited ions. In this experiment a 4.2-keV ion beam intersected an electron beam at an angle of 45°. The experimental data are the filled circles and triangles. The dashed curve is the result of the Gaunt factor predictor formula, the small crosses and dots are from Coulomb–Born calculations, and the open squares are

Fig. 12.10. The electron impact excitation cross sections of C^+ for $2s2p^2$ ^2D. The filled points are the measurements and have been taken from Lafyatis (1982); Lafyatis & Kohl (1987). The dashed curve is the result of the Gaunt factor predictor formula, the small crosses and dots from Coulomb–Born calculations and the open squares from close-coupling calculations.

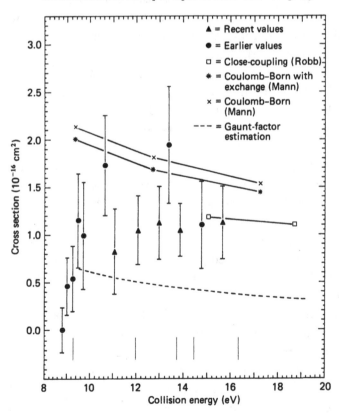

from close coupling calculations. The failure of the Gaunt factor estimator formula should be noted.

12.4 Measurement of bound–free transition parameters
12.4.1 *Measurement of photoionization cross sections*
When the absolute *f*-values are available for the lines of a series, it is possible to plot them, and with the aid of (12.38) and (12.39), to obtain the value at the limit for the adjoining photoionization continuum.

Fig. 12.11 is a quantum defect plot of the differential oscillator strength for the transition $4s^2\,{}^1S_0$–$4snp\,{}^1P_1^0$ in Ca I, using the *f*-values from Table 12.1 obtained by the hook method. The additional points marked 3d4p and 3d5p are lines of the $3dnp\,{}^1P_1^0$ series. The extrapolation of the bound state *gf-*

Fig. 12.11. The differential oscillator strength distribution in the discrete and continuous spectra of Ca I. The filled circles, at negative energies, are from the *f*-value measurements (see Table 12.1) of Parkinson, Reeves & Tomkins (1976). The points marked, 3d4p and 3d5p, are for lines of the $3dnp\,{}^1P_1^0$ series. The absorption cross section for the point at $+0.46\,\text{eV}$ has been taken from the measurements of McIlrath & Sandeman (1972). The broken curve has been drawn to follow the wavelength dependence of the photoionization cross section as published in the literature. The smooth extrapolation of the bound state data to the threshold energy leads to a value of the photoionization cross section in agreement with the revised scale of McIlrath & Sandeman.

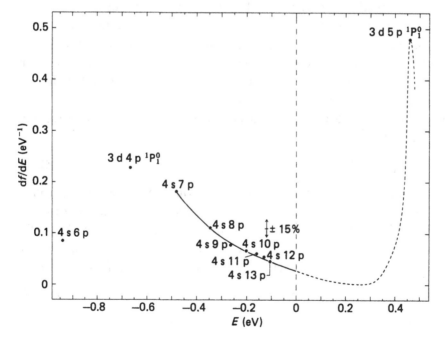

values yields 2.75×10^{-18} cm^2 at threshold. Photoionization cross sections can also be derived directly by detection of photoionized particles in a beam or, less directly, by measuring the continuous photoabsorption cross section. As Fig. 12.11 demonstrates, the Ca I result from *gf*-value extrapolation is in good agreement with direct photoionization cross section measurements of Ca I. These are based on photoelectric absorption measurements combined with hook method measurements on the 422.6 nm resonance line which established the column densities for the absorption measurement.

Fig. 12.11 shows the principal series 4snp strongly perturbed by a level which is mainly of 3d4p $^1P_1^0$ character and the broad resonance in the continuum due to 3d5p $^1P_1^0$. This double excited state is broadened because in LS coupling it autoionizes.

12.4.2 *Measurement of autoionization parameter*

In the Sr I photograph and in the differential oscillator strengths plot of the Ca I series, the series converge to the limit corresponding to the ion in the ground state plus an electron with zero kinetic energy. However, when there are bound states beyond the limit and in the photoionization continuum, corresponding to excitation of one of the inner shell electrons or to two electron excitation, e.g., 3d5p, of the same angular momentum as the continuum, the properties of the states and continuum are mixed and there is a possibility of a radiationless transition and spontaneous ionization. This process is called autoionization. The resulting lines can be very wide, corresponding to upper level lifetimes of 10^{-15} s for the 3d5p $^3D^0$ state of Ca I, and can have asymmetrically shaped line profiles. These profiles are now often called Beutler–Fano profiles after H. Beutler, who reported many lines with the characteristic shape, and U. Fano, who provided the physical explanation for the shape.

The Fano theory for these profiles describes them in terms of a reduced energy variable

$$\varepsilon = [2(E - E_r)/\Gamma] \tag{12.69}$$

where ε represents the energy displacement from the resonance position E_r and Γ is one half-width of the profile. The mean lifetime of the autoionized state is $\hbar/\Gamma$. The ratio of the photoionization cross section $\sigma(E)$ in the profile of the autoionized broadened line to the background unperturbed continuum, σ plotted as a function of photon energy, is a family of curves

$$\sigma(E)/\sigma = g(\varepsilon) = (q + \varepsilon)^2/1 + \varepsilon^2 \tag{12.70}$$

where the parameter q is the 'line-profile index'. We should note that when

$\varepsilon = -q$ the profile goes to zero, and when q becomes very large the profile shape becomes a dispersion (Lorentzian) shape.

12.4.3 *Example of measurement of both photoionization and autoionization*

An excellent example of both photoionization and autoionization features in an astronomical spectrum is found in the Sun at wavelengths just shortward of 210.0 nm where, in atomic aluminum (Al I), these structures are strong and well developed. The ground state configuration of Al I is $3s^2\, 3p\, {}^2P^0_{1/2,\,3/2}$. Discrete transitions from the ground term to the ^2S and ^2D levels give rise to series terminating at the $3s^2\, {}^1S_0$ first ionization limit ($48279.19\ \mathrm{cm}^{-1}$). Beyond this limit there are two continua of importance here: the $3s^2\varepsilon s\, {}^2S$ and the $3s^2\varepsilon d\, {}^2D$ continua. Also in these continua is the $3s3p^2\, {}^2S_{1/2}$ level which, for transitions from the ground term, gives rise to two strong autoionization lines at 193.2 nm and 193.6 nm. We can describe this process as

$$\mathrm{Al}(3s^2\, 3p\, {}^2P^0_{1/2,\,3/2}) + h\nu \rightarrow \mathrm{Al}^*(3s\, 3p^2\, {}^2S_{1/2})$$

$$\rightarrow \mathrm{Al}^+(3s^2\, {}^1S_0) + e^-(\mathrm{ns}).$$

Such photoionization continua and autoionization features are important sources of opacity in the solar atmosphere.

Although the Al I spectrum can be produced in a conventional laboratory source such as a resistively heated furnace, there is also produced a number of unidentified, strong and diffuse features which blend with and seriously mask the Al series and beyond. For this reason the measurements of the Al I photoionization cross section and the parameters of the autoionization doublet were carried out by producing atomic vapor at high temperature with an aerodynamic shock tube. The column population density Nl of Al I was measured with the hook method (12.50) using the absolute gf-value of the lines 394.4 nm and 396.15 nm ($3p\ {}^2P^0$–$4s\ {}^2S_{1/2}$), while also photographing nearly simultaneously the clear and unblended Al I absorption spectrum. The measured value of the photoionization cross section at the first ionization limit is $65 \pm 7 \times 10^{-18}\ \mathrm{cm}^2$ and decreases to about $30 \times 10^{-18}\ \mathrm{cm}^2$ by 1750 Å. The measured cross section is the average from both the $3p\ {}^2P^0_{1/2}$ and the $3p\ {}^2P^0_{3/2}$ levels of the ground term and is the sum of the average cross sections to both the ^2S and ^2D continua. However, a quantum defect plot of the differential oscillator strength using (12.38) and (12.39) indicates that the photoionization cross section to the ^2S continuum is weaker than that to the ^2D by a factor of 10^2. Detailed theoretical calculations, using the R-matrix method, produced Al I photoionization cross sections in good agreement with the measurements.

When the measured photoionization cross sections, which are considerably greater than earlier values, are used in the solar models the agreement between observed and predicted solar emergent intensity is greatly improved.

Fig. 12.12 shows plots of the absolute emergent intensity from the central portion of the disk of the Sun over the wavelength ration 130.0 nm–300.0 nm. Measured data are shown as dots. Also plotted are the solar intensities that are predicted by the Vernazza, Avrett and Loeser model solar atmosphere. The upper broken curves are the predictions assuming continuum opacities only and showing the marked improvement with the observed intensities when the much larger measured Al photoionization cross sections are used. The central intensities predicted by the solar model

Fig. 12.12. A plot of the absolute, emergent solar intensity from the center of the disk over the wavelength range 130 nm–300 nm. The axes are logarithm specific intensity ($W\,cm^{-2}\,sr^{-1}\,nm^{-1}$) and wavelength (nm). The data, marked as dots, are from measurements by Kohl, Parkinson & Reeves and have been tabulated by Vernazza, Avrett & Loeser (1976). The upper dashed curve is an earlier solar model calculation (VAL, 1973) whereas the other broken curve is a recent version of the model (VALC, 1980) which includes the much larger, measured A1 photoionization cross sections of Kohl & Parkinson (1973). Both of the models use continuous opacities only. The full curve is a plot of the prediction of VALC when line blanketing is included with the continuum opacities (Avrett & Kurucz (1984)).

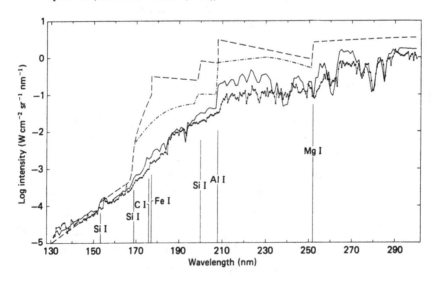

are again further improved in comparison with observations when line blanketing is included with the continuum opacity in the calculations.

In addition to the direct measurement of the parameters and cross sections of the Al I autoionization lines, $3s^2 \, 3p \, ^2P^0_{1/2, \, 3/2} - 3s \, 3p^2 \, ^2S_{1/2}$, the experiment led to the analysis of quantum defect plots of the differential oscillator strengths of the $3s^2 \, 3p \, ^2P \rightarrow 3s^2nd \, ^2D$ transitions. The plot produced good agreement with the measured threshold photoionization cross section but also revealed a broad Beutler–Fano profile indicating that the strong $3s^2\varepsilon d \, ^2D$ continuum must be represented by including the autoionizing perturber, $3s3p^2 \, ^2D$.

Other possible autoionizing levels of Al I that result from the $3s3p^2$ configuration are of the $3s3p^2 \, ^2P$ term. In combination with the ground terms, there are four Al I lines identified: 176.29, 176.56, 176.64, and 176.91 nm. The oscillator strengths and line widths of these lines were measured by applying the hook method in conjunction with absorption equivalent width measurements for lines of large optical depths. The $3s3p^2 \, ^2P_{1/2, \, 3/2}$ levels do not strongly autoionize in comparison with the other levels of the $3s3p^2$ configuration, but they do interact weakly with the ^2S and the ^2D continua. In particular the $3s3p^2 \, ^2P_{3/2}$ level autoionizes to the $^2D_{3/2}$ continuum and the $3s3p^2 \, ^2S_{1/2}$ level to the ^2S continuum. These autoionizing lines can be identified in the solar spectrum, but higher spectral resolution than is now available is required before they could be used to study the solar atmosphere.

12.4.4 *Measurement of dielectronic recombination cross sections*

Autoionizing states play their most important role in astrophysics in the process known as dielectronic recombination. This is a two-part process which we can describe as involving, first, the combination of a free electron of energy E with a positive ion X(i) of charge $Z+$ in configuration (i). This creates, by inverse autoionization, an ion of charge $(Z - 1)$ in an intermediate autoionizing state (j, nl) where nl represents the quantum numbers of the captured electron;

$$X^{Z+}(i) + e^-(E) \leftrightarrows X^{(Z-1)+}(j, nl).$$

It is possible for the (nl) electron in the (j, nl) state to return to the continuum by autoionization and, thus, to decrease the recombination cross sections. This process is observable as a resonance in the elastic scattering cross section. Secondly, an allowed radiative decay occurs to stabilize and form the $X^{(Z-1)+}$ ion in a bound state (k, nl)

$$X^{(Z-1)+}(j, nl) \rightarrow X^{(Z-1)+}(k, nl) + h\nu.$$

The $X^{(Z-1)+}$ ions proceed to the ground state through a cascade of radiative decays. All members of a Rydberg series of autoionizing states can contribute to the dielectronic recombination rate. External electric fields can enhance dielectronic recombination by causing a decrease in the autoionization probability for high Rydberg states. Thus the branching ratio involving autoionization and radiative decay favors stabilization.

It has been found that, for example, at the high temperature of the solar corona the calculated total recombination rate can increase by more than an order of magnitude when the importance of including dielectronic recombination is recognized. This process removes the factor of two discrepancy between the electron temperature deduced from Doppler widths and the temperature based on balancing rates of electron impact ionization and rates of recombination. The dielectronic recombination process is important in many high-temperature astrophysical sources and in laboratory plasmas such as potential controlled fusion devices.

There are as yet few successful experimental measurements of dielectronic recombination cross sections and none of field-free cross sections. There are difficulties in separating the much larger competing processes of charge transfer that tend to dominate low-temperature laboratory conditions. It is also very difficult to cause dielectronic recombination in the laboratory without directly exciting the $X^{Z+}(j)$ state whose radiative decay photons nearly match in wavelength the stabilizing photons. Experiments are also affected by re-ionization of the ions in the highly excited and fragile Rydberg states and by electric field enhancement. The idea of Lafyatis and Kohl that led to the successful measurement approach is to record the two products, the photon and the $X^{(Z-1)}$ ion, in a coincidence measurement. Intersecting beams of electrons and ions (X^{Z+}) are used (see Section 12.3.4) where the number of $X^{(Z-1)+}$ ions due to charge transfer is reduced as much as practicable prior to the intersection. The primary beam ions of X^{Z+} and the newly formed $X^{(Z-1)+}$ ions are separated and detected after intersection. The photons are detected with an optical system whose axis is at right angles to the beams and passes through the intersection point. Only $X^{(Z-1)+}$ ions that arrive at the detector in precise synchronization with the detection of photons are counted. The much larger numbers of $X^{(Z-1)+}$ ions due to charge transfer are rejected. The cross section for electric field-enhanced dielectronic recombination of Mg^+ was the first to be measured by Dunn and his colleagues using this technique.

Acknowledgments

I thank my colleagues who have helped me with this chapter and with the laboratory astrophysics that went before it. I especially acknowledge and appreciate the suggestions of Bartley Cardon, Alex Dalgarno, John Kohl, Harriet Griesinger, and Peter Smith who read the chapter at some stage of its preparation. I also sincerely thank Nancy Galluccio for typing and preparing all stages of it. My work was supported in part by Harvard College and the Smithsonian Institution.

Compilations of atomic data

Wavelength and multiplet tables

Kaufman, V. & Edlén, B. (1974). Reference wavelengths from atomic spectra in the range 15 Å to 25 000 Å. *J. Phys. Chem. Ref. Data*, **3**, 825.

Kelly, R. L. (1982). *Atomic and Ionic Spectrum Lines Below 2000 Å Hydrogen through Argon*, ORNL 5922. Controlled Fusion Atomic Data Center, Physics Division, Oak Ridge National Laboratory, Oak Ridge, Tennessee 37830.

Moore, C. E. (1950–1962). *An Ultraviolet Multiplet Table*, NBS Circular 488. Sect. 1 (1950): hydrogen through vanadium; Sect. 2 (1952): chromium through niobium; Sect. 3 (1962): molybdenum through lanthanum, hafnium through radium; Sect. 4 (1962): finding list-hydrogen through niobium; Sect. 5 (1962): finding list-molybdenum through lanthanum, hafnium through radium.

Moore, C. E. (1965–1983). *Selected Tables of Atomic Spectra*, NSRDS-NBS 3. Sect. 1 (1965): (Si II–IV); Sect. 2 (1967): (Si I); Sect. 3 (1970): (C I–VI); Sect. 4 (1971): (N IV–VII); Sect. 5 (1975): (N I–III); Sect. 6 (1972): (H, D, T); Sect. 7 (1976): (O I); Sect. 8 (1979): (O VI–VIII); Sect. 9 (1980): (O V); Sect. 10 (1983): (O IV).

Moore, C. E. (1972). *A Multiplet Table of Astrophysical Interest*, Revised Edition, NSRDS-NBS 40. (Magnetic-tape version has been prepared by L. Gratton and F. Querci and is obtainable from National Space Science Data Center.)

Reader, J. & Corliss, C. H. (1980). *Wavelengths and Transition Probabilities for Atoms and Atomic Ions, Part I. Wavelengths*. NSRDS-NBS 68.

Grotrian diagrams and energy levels

Bashkin, S. & Stoner, J. O., Jr. (1975–1981). *Atomic Energy-Level and Grotrian Diagrams*. Vol. I. Hydrogen I – Phosphorus XV (1975); Vol. I. Addenda (1978); Vol. II. Sulfur I – Titanium XXII (1978); Vol. III. Vanadium I – Chromium XXIV (1981). Amsterdam: North-Holland.

Corliss, C. & Sugar, J. (1977). Energy levels of manganese, Mn I through Mn XXV. *J. Phys. Chem. Ref. Data*, **6**, 1253.

Corliss, C. & Sugar, J. (1979). Energy levels of potassium, K I through K XIX, *J. Phys. Chem. Ref. Data*, **8**, 1109.

Corliss, C. & Sugar, J. (1979). Energy levels of titanium, Ti I through Ti XXII. *J. Phys. Chem. Ref. Data,* **8**, 1.

Corliss, C. & Sugar, J. (1981). Energy levels of nickel, Ni I through Ni XXVIII. *J. Phys. Chem. Ref. Data,* **10**, 197.

Corliss, C. & Sugar, J. (1982). Energy levels of iron, Fe I through Fe XXVI. *J. Phys. Chem. Ref. Data,* **11**, 135.

Erickson, G. W. (1977). Energy levels of one-electron atoms. *J. Phys. Chem. Ref. Data,* **6**, 831.

Martin, W. C. (1973). Energy levels of neutral helium (^{4}He I). *J. Phys. Chem. Ref. Data,* **2**, 257.

Martin, W. C. & Zalubas, R. (1979). Energy levels of aluminum, Al I through Al XIII. *J. Phys. Chem. Ref. Data,* **8**, 817.

Martin, W. C. & Zalubas, R. (1980). Energy levels of magnesium, Mg I through Mg XII. *J. Phys. Chem. Ref. Data,* **9**, 1.

Martin, W. C. & Zalubas, R. (1981). Energy levels of sodium, Na I through Na XI. *J. Phys. Chem. Ref. Data,* **10**, 153.

Martin, W. C. & Zalubas, R. (1983). Energy levels of silicon, Si I through Si XIV. *J. Phys. Chem. Ref. Data,* **12**, 323.

Martin, W. C., Zalubas, R. & Hagan, L. (1978). *Atomic Energy Levels – The Rare-Earth Elements.* NSRDS-NBS 60.

Moore, C. E. (1971). *Atomic Energy Levels (As Derived from the Analyses of Optical Spectra).* NSRDS-NBS 35, reprinted from NBS Circular 467: Vol. I (Hydrogen through Vanadium); Vol. II (Chromium through Niobium); Vol. III (Molybdenum through Lanthanum; Hafnium through Actinium).

Sugar, J. & Corliss, C. (1977). Energy levels of chromium, Cr I through Cr XXIV. *J. Phys. Chem. Ref. Data,* **6**, 317.

Sugar, J. & Corliss, C. (1978). Energy levels of vanadium, V I through V XXIII. *J. Phys. Chem. Ref. Data,* **7**, 1191.

Sugar, J. & Corliss, C. (1979). Energy levels of calcium, Ca I through Ca XX. *J. Phys. Chem. Ref. Data,* **8**, 865.

Sugar, J. & Corliss, C. (1980). Energy levels of scandium, Sc I through Sc XXI. *J. Phys. Chem. Ref. Data,* **9**, 473.

Sugar, J. & Corliss, C. (1981). Energy levels of cobalt, Co I through Co XXVII. *J. Phys. Chem. Ref. Data,* **10**, 1097.

Transition probabilities (oscillator strengths, line strengths)

Eidelsberg, M., Crifo-Magnant, F. & Zeippen, C. J. (1981). Forbidden lines in hot astronomical sources. *Astron. Astrophys. Suppl.* **43**, 455.

Fuhr, J. R., Martin, G. A., Wiese, W. L. & Younger, S. M. (1981). Atomic transition probabilities for iron, cobalt, and nickel (A critical data compilation of allowed lines). *J. Phys. Chem. Ref. Data,* **10**, 305.

Wiese, W. L. & Fuhr, J. R. (1975). Atomic transition probabilities for scandium and titanium (A critical data compilation of allowed lines). *J. Phys. Chem. Ref. Data,* **4**, 263.

Wiese, W. L. & Martin, G. A. (1980). *Wavelengths and Transition Probabilities for Atoms and Atomic Ions, Part II. Transition Probabilities.* NSRDS-NBS 68.

Wiese, W. L., Smith, M. W. & Glennon, B. M. (1966). *Atomic Transition Probabilities, Vol. I. Hydrogen through Neon.* NSRDS-NBS 4.

Wiese, W. L., Smith, M. W. & Miles, B. M. (1969). *Atomic Transition Probabilities, Vol. II. Sodium through Calcium.* NSRDS-NBS 22.

Younger, S. M., Fuhr, J. R., Martin, G. A. & Wiese, W. L. (1978). Atomic transition probabilities for vanadium, chromium, and manganese (A critical data compilation of allowed lines). *J. Phys. Chem. Ref. Data,* **7**, 495.

References

The following books and papers are a selection of references which are suggested as sources of additional and often more complete information on the subjects of this chapter.

12.1 Introduction

Goldberg, L. (1938). *The Intensities of Helium Lines.* Thesis:Harvard University.

Goldberg, L., Muller, E. A. & Aller, L. H. (1960). The abundances of the elements in the solar atmosphere. *Astrophys. J. Suppl.* **5**, 1–138.

Kohl, J. L., Parkinson, W. H. & Kurucz, R. L. (1978). *Center and Limb Solar Spectra in High Spectral Resolution 225.2 nm to 319.6 nm.* Cambridge: Harvard University Printing Office.

Ross, J. E. & Aller, L. H. (1976). The chemical composition of the Sun. *Science,* **191**, 1223–9.

12.2 Basic formulae and definitions

Breit, G. (1933). Quantum theory of dispersion continued. *Rev. Mod. Phys.* **5**, 91–140.

Foster, E. W. (1964). The measurement of oscillator strengths in atomic spectra. *Rep. Prog. Phys.* XXVII, 469–551.

Huber, M. C. E. & Sandeman, R. J. (1986). The measurement of oscillator strengths. *Rep. Prog. Phys.* **49**, 397–490.

Imhof, R. E. & Reed, F. H. (1977). Measurement of lifetimes of atoms, molecules, and ions. *Rep. Prog. Phys.* **40**, 1–104.

Lu, K. T. & Fano, U. (1970). Graphic analysis of perturbed Rydberg series. *Phys. Rev. A,* **2**, 81–6.

Menzel, D. H. (1953). *Mathematical Physics.* New York: Prentice-Hall, Inc.

Mitchell, A. C. G. & Zemansky, M. W. (1961). *Resonance Radiation and Excited Atoms.* Cambridge: The University Press.

Moiseiwitsch, B. L. & Smith, S. J. (1968). Electron impact excitation of atoms. *Rev. Mod. Phys.* **40**, 238–322.

Seaton, M. J. (1958). The quantum defect method. *Mon. Not. Roy. Astron. Soc.* **118**, 504–18.

Seaton, M. J. (1966). Quantum defect theory I. General formulation. *Proc. Phys. Soc.* **88**, 801–14.

Seaton, M. J. (1966). Quantum defect theory II. Illustrative one-channel and two-channel problems. *Proc. Phys. Soc.* **88**, 815–32.

Thorne, A. P. (1974). *Spectrophysics.* London: Chapman & Hall Ltd.

Wiese, W. L. (1979). Atomic Transition Probabilities and Lifetimes. In *Progress in Atomic Spectroscopy, Part B*, ed. W. Hanle & H. Kleinpoppen, pp. 1101–49. New York: Plenum Press.

12.3 Measurement of bound–bound radiative transition parameters
12.3.1 Measurement of radiative lifetimes

Berry, H. G. (1977). Beam-foil spectroscopy. *Rep. Prog. Phys.* **40**, 155–217.

deZafra, R. L. & Kerk, W. (1967). Measurement of atomic lifetimes by the Hanle effect. *Amer. J. Phys.* **35**, 573–82.

Hanle, W. (1924). The influence of magnetic fields on the polarisation of resonance radiation. *Zeits. f. Phys.* **30**, 93–105.

Kelly, F. M. & Mathur, M. S. (1980). Density dependence of the Hanle effect of the 4s4p $^1P_1$ level of neutral calcium. *Canad. J. Phys.* **58**, 1004–9.

Wood, R. W. & Ellett, A. (1923). On the influence of magnetic fields on the polarisation of resonance radiation. *Proc. Roy. Soc. London*, **103**, 396–403.

12.3.2 Measurement of radiative lifetimes of metastable levels

Dehmelt, H. G. (1967). Radiofrequency spectroscopy of stored ions I: storage. In *Advances in Atomic and Molecular Physics*, Vol. 3, ed. D. R. Bates & Immanuel Estermann, pp. 53–72. New York: Academic Press.

Dehmelt, H. G. (1969). Radiofrequency spectroscopy of stored ions II: spectroscopy. In *Advances in Atomic and Molecular Physics*, Vol. 5, ed. D. R. Bates & Immanuel Estermann, pp. 109–54. New York: Academic Press.

Kwong, H. S., Johnson, B. C., Smith, P. L. & Parkinson, W. H. (1983). Transition probability of the Si III 189.2 nm intersystem line. *Phys. Rev. A*, **27**, 3040–3.

Smith, P. L., Johnson, B. C., Kwong, H. S., Parkinson, W. H. & Knight, R. D. (1984). Measurements of transition probabilities for spin-changing lines of atomic ions used in diagnostics of astrophysical plasmas. *Phys. Scrip.* **T8**, 88–94.

12.3.3 Measurement of *f*-values
12.3.3.1 *Anomalous dispersion and the hook method*

Huber, M. C. E. (1971). Interferometric gas diagnostics by the hook method. In *Modern Optical Methods in Gas Dynamics Research*, ed. D. C. Dosanjh, pp. 85–112. New York: Plenum Press.

Marlow, W. C. (1967). Hakenmethode. *Appl. Opt.* **6**, 1715–24.

Sandeman, R. J. (1979). Hook vernier. *Appl. Opt.* **18**, 3873–4.

12.3.3.2 *Absorption coefficient and equivalent width method*

Armstrong, B. H. (1967). Spectrum line profile: The Voigt function. *J. Quant. Rad. Trans.* **7**, 61–88.

Hummer, D. G. (1965). The Voigt function. An eight-significant-figure table and generating procedure. *Mem. Roy. Astron. Soc.* **70**, 1–32.

12.3.3.3 *Example of f-value measurement by both absorption and hook methods*

Blackwell, D. E. & Collins, B. S. (1972). Precision measurement of relative oscillator strengths I. Fundamental technique: a first application to Mn I. *Mon. Not. Roy. Astron. Soc.* **157**, 255–71.

Blackwell, D. E., Ibbetson, P. A. & Petford, A. D. (1975). Precision measurement of relative oscillator strengths II. Fe I transitions from levels a $^5D_4$ (0.00 eV) and a $^5D_3$ (0.05 eV). *Mon. Not. Roy. Astron. Soc.* **171**, 195–208.

Brown, C. M., Tilford, S. G. & Ginter, M. L. (1973). Absorption spectrum of Ca I in the 1580–2090 Å region. *J. Opt. Soc. Amer.* **63**, 1454–62.

Cardon, B. L., Smith, P. L., Scalo, J. M., Testerman, L. & Whaling, W. (1982). Absolute oscillator strengths for lines of neutral cobalt between 2276 Å and 9357 Å and a redetermination of the solar cobalt abundance. *Astrophys. J.* **260**, 395–412.

Cardon, B. L., Smith, P. L. & Whaling, W. (1979). New methods for determining relative oscillator strengths of atoms through combined absorption and emission measurements: application to titanium (Ti I). *Phys. Rev. A*, **20**, 2411–19.

Garton, W. R. S. & Codling, K. (1965). Ultraviolet extensions of the arc spectrum of the alkaline earths: the absorption spectrum of calcium vapour. *Proc. Roy. Soc. London*, **86**, 1067–75.

Geiger, J. (1979). The oscillator strength density of the $4s^2$–4snp, 3dnp, $^1P_1$, $^3P_1$ series of calcium as a two-channel case in quantum defect theory. *J. Phys. B*, **12**, 2277–90.

Huber, M. C. E. & Tubbs, E. F. (1972). Oscillator strengths of weak Fe I resonance lines measured by combined hook and absorption techniques. *Astrophys. J.* **177**, 847–54.

King, R. B. & King, A. S. (1935). Relative *f*-values for lines of Fe I from electric-furnace absorption spectra. *Astrophys. J.* **82**, 377–95.

Parkinson, W. H., Reeves, E. M. & Tomkins, F. S. (1976). Neutral calcium, strontium and barium: determination of the *f*-values of the principal series by the hook method. *J. Phys. B.* **9**, 157–65.

Risberg, G. (1968). The spectrum of atomic calcium, Ca I, and extensions to the analysis of Ca II. *Ark. f. Fys.* **37**, 231–49.

12.3.3.4 *Phase-matching method*

Puell, H. & Vidal, C. R. (1976). Determination of oscillator strengths with the phase-matching condition. *Opt. Comm.* **19**, 279–83.

Vidal, C. R. (1980). Coherent VUV sources for high resolution spectroscopy. *Appl. Opt.* **19**, 3897–903.

Wynne, J. J. & Beigang, R. (1981). Accurate measurement of relative oscillator strengths by phase-matched nonlinear optics. *Phys. Rev. A.* **23**, 2736–9.

12.3.4 Measurement of electron impact excitation cross sections

Chutjian, A., Msezane, A. Z. & Henry, R. J. W. (1983). Angular distribution for electron excitation of the $4^2S \rightarrow 4^2P$ transition in Zn II: comparison of experiment and theory. *Phys. Rev. Lett.* **50**, 1357–60.

Cochrane, D. M. & McWhirter, R. W. P. (1983). Collisional excitation rate coefficients for lithium-like ions. *Phys. Scrip.* **28**, 25–44.

Dolder, K. T. & Peart, B. (1976). Collisions between electrons and ions. *Rep. Prog. Phys.* **39**, 697–749.

Dunn, G. H. (1980). Electron–ion collisions. In *Physics of Ionized Gases*, ed.
M. Matíc, pp. 49–96. Belgrade: Boris Kidric Institute of Nuclear Sciences.

Lafyatis, G. P. (1982). *An Experimental Study of the Electron Impact Excitation
of Positive Ions*. Thesis: Harvard University.

Lafyatis, G. P. & Kohl, J. L. (1987). Measurement of electron impact excitation
in boron-like carbon. *Phys. Rev. A* (in press).

Lafyatis, G. P., Kohl, J. L. & Gardner, L. D. (1987). An experimental
apparatus for measurements of electron impact excitation. *Rev. Sci.
Instrum.* **58** (in press).

12.4 Measurement of bound–free transition parameters

12.4.1 Measurement of photoionization cross sections

Berkowitz, J. (1979). *Photoabsorption, Photoionization, and Photoelectron
Spectroscopy*. New York: Academic Press.

Goldberg, L. G. (1966). Astrophysical implication of autoionization. In
Autoionization, ed. Aaron Temkin, pp. 1–23. Baltimore: Mono Book Corp.

McIlrath, T. J. & Sandeman, R. J. (1972). Revised absolute absorption cross
sections of Ca I at 1886.5 and 1765.1 Å. *J. Phys. B*, **5**, L217–9.

12.4.3 Example of measurement of both photoionization and autoionization

Avrett, E. H. & Kurucz, R. L. (1984). VAL with line blanketing, private
communication.

Kohl, J. L. & Parkinson, W. H. (1973). Measurement of the neutral-aluminum
photoionization cross section and parameters of the 3p ^{2}P^0–3s3p^2 ^{2}S$_{1/2}$
autoionization doublet. *Astrophys. J.* **184**, 641–52.

Le Dourneuf, M., Lan, Vo Ky, Burke, P. G. & Taylor, K. T. (1975). The
photoionization of neutral aluminium. *J. Phys. B*, **8**, 2640–53.

Lin, C. D. (1974). Theoretical analysis of the Al I absorption spectrum.
Astrophys. J. **187**, 385–7.

Lombardi, G. G., Cardon, B. L. & Kurucz, R. L. (1981). Measurement of the
oscillator strengths and autoionization widths of the neutral-aluminum
multiplet 3s^{2}3p ^{2}P^0–3s3p^2 ^{2}P. *Astrophys. J.* **248**, 1202–8.

Vernazza, J. E., Avrett, E. H. & Loeser, R. (1976). Structure of the solar
chromosphere II: the underlying photosphere and temperature-minimum
region. *Astrophys. J. Suppl.* **30**, 1–60.

12.4.4 Measurement of dielectronic recombination cross sections

Belic, D. S., Dunn, G. H., Morgan, T. J., Mueller, D. W. & Timmer, C. (1983).
Dielectronic recombination: a crossed-beams observation and measurement
of cross section. *Phys. Rev. Lett.* **50**, 339–42.

Burgess, A. (1964). Dielectronic recombination and the temperature of the solar
corona. *Astrophys. J.* **139**, 776–80.

Dittner, P. F., Datz, S., Miller, P. D., Moak, C. D., Stelson, P. H., Bottcher, C.,
Dress, W. B., Alton, G. D. & Nesković, N. (1983). Cross sections for
dielectronic recombination of B^{2+} and C^{3+} via 2s $\rightarrow$ 2p excitation. *Phys.
Rev. Lett.* **51**, 31–4.

Gardner, L. D., Kohl, J. L., Lafyatis, G. P., Young, A. R. & Chutjian, A. (1986). An experimental apparatus for photon–ion coincidence measurements of dielectronic recombination. *Rev. Sci. Instrum.* **57**, 2254–65.

Jacobs, V. L., Davis, J. & Kepple, P. C. (1976). Enhancement of dielectronic recombination by plasma electric microfields. *Phys. Rev. Lett.* **37**, 1390–3.

LaGattuta, K. & Hahn, Y. (1982). Dielectronic recombination cross section for Mg^+. *J. Phys. B*, **15**, 2101–7.

LaGattuta, K. & Hahn, Y. (1983). Effect of extrinsic electric fields upon dielectronic recombination: Mg^{1+}. *Phys. Rev. Lett.* **51**, 558–61.

Mitchell, J. B. A., Ng, C. T., Forand, J. L., Levac, D. P., Mitchell, R. E., Sen, A., Miko, D. B. & McGowan, J. Wm. (1983). Dielectronic-recombination cross-section measurements for C^+ ions. *Phys. Rev. Lett.* **50**, 335–8.

Index